T0205675

Information Security and Cryptography

More information about this series at http://www.springer.com/series/4752

Joan Daemen • Vincent Rijmen

The Design of Rijndael

The Advanced Encryption Standard (AES)

Second Edition

Joan Daemen
Digital Security Group
Radboud University Nijmegen
Nijmegen, The Netherlands

Vincent Rijmen
COSIC Group
KU Leuven
Heverlee, Belgium

ISSN 1619-7100 ISSN 2197-845X (electronic)
Information Security and Cryptography
ISBN 978-3-662-60771-8 ISBN 978-3-662-60769-5 (eBook)
https://doi.org/10.1007/978-3-662-60769-5

This Springer imprint is published by the registered company Springer-Verlag GmbH, DE part of Springer Nature.
The registered company address is: Heidelberger Platz 3, 14197 Berlin, Germany

Foreword

Rijndael was the surprise winner of the contest for the new Advanced Encryption Standard (AES) for the United States. This contest was organized and run by the National Institute of Standards and Technology (NIST) beginning in January 1997; Rijndael was announced as the winner in October 2000. It was the "surprise winner" because many observers (and even some participants) expressed scepticism that the US government would adopt as an encryption standard any algorithm that was not designed by US citizens.

Yet NIST ran an open, international, selection process that should serve as a model for other standards organizations. For example, NIST held their 1999 AES meeting in Rome, Italy. The five finalist algorithms were designed by teams from all over the world.

In the end, the elegance, efficiency, security, and principled design of Rijndael won the day for its two Belgian designers, Joan Daemen and Vincent Rijmen, over the competing finalist designs from RSA, IBM, Counterpane Systems, and an English/Israeli/Danish team.

This book is the story of the design of Rijndael, as told by the designers themselves. It outlines the foundations of Rijndael in relation to the previous ciphers the authors have designed. It explains the mathematics needed to understand the operation of Rijndael, and it provides reference C code and test vectors for the cipher.

Most importantly, this book provides justification for the belief that Rijndael is secure against all known attacks. The world has changed greatly since the DES was adopted as the national standard in 1976. Then, arguments about security focused primarily on the length of the key (56 bits). Differential and linear cryptanalysis (our most powerful tools for breaking ciphers) were then unknown to the public. Today, there is a large public literature on block ciphers, and a new algorithm is unlikely to be considered seriously unless it is accompanied by a detailed analysis of the strength of the cipher against at least differential and linear cryptanalysis.

This book introduces the "wide trail" strategy for cipher design, and explains how Rijndael derives strength by applying this strategy. Excellent resistance to differential and linear cryptanalysis follows as a result. High efficiency is also a result, as relatively few rounds are needed to achieve strong security.

The adoption of Rijndael as the AES is a major milestone in the history of cryptography. It is likely that Rijndael will soon become the most widely used cryptosystem in the world. This wonderfully written book by the designers themselves is a "must read" for anyone interested in understanding this development in depth.

Ronald L. Rivest
Viterbi Professor of Computer Science
MIT

Preface

This book is about the design of Rijndael, the block cipher that became the Advanced Encryption Standard (AES). According to the 'Handbook of Applied Cryptography' [110], a block cipher can be described as follows:

> A block cipher is a function which maps n-bit plaintext blocks to n-bit ciphertext blocks; n is called the block length. [...] The function is parameterized by a key.

Although block ciphers are used in many interesting applications, such as e-commerce and e-security, this book is *not* about applications. Instead, this book gives a detailed description of Rijndael and explains the design strategy that was used to develop it.

Structure of this book

When we wrote this book, we had basically two kinds of readers in mind. Perhaps the largest group of readers will consist of people who want to read a full and unambiguous description of Rijndael. For those readers, the most important chapter of this book is Chap. 3, which gives its comprehensive description. In order to follow our description, it might be helpful to read the preliminaries given in Chap. 2. Advanced implementation aspects are discussed in Chap. 4. A short overview of the AES selection process is given in Chap. 1.

A large part of this book is aimed at readers who want to know *why* we designed Rijndael in the way we did. For them, we explain the ideas and principles underlying the design of Rijndael, culminating in our wide trail design strategy. In Chap. 5 we explain our approach to block cipher design and the criteria that played an important role in the design of Rijndael. Our design strategy has grown out of our experiences with linear and differential cryptanalysis, two cryptanalytical attacks that have been applied with some success to the previous standard, the Data Encryption Standard (DES). In Chap. 6 we give a short overview of the DES and of the differential and the linear attacks that are applied to it. Our framework to describe linear cryptanalysis is explained in Chap. 7; differential cryptanalysis is described in Chap. 8. Finally, in Chap. 9, we explain how the wide trail design strategy follows from these considerations.

Chap. 10 gives an overview of the published attacks on reduced-round variants of Rijndael. Chap. 11 gives an overview of ciphers related to Rijndael. We describe its predecessors and discuss their similarities and differences. This is followed by a short description of a number of block ciphers that have been strongly influenced by Rijndael and its predecessors.

In Chap. 12 we show how linear and differential analysis can be applied to ciphers that are defined in terms of finite field operations rather than Boolean functions. In Chap. 13 we discuss extensions of differential and linear cryptanalysis. In Chap. 14 we study the probability of differentials and trails over two rounds of Rijndael, and in Chap. 15 we define plateau trails. To assist programmers, Appendix A lists some tables that are used in various descriptions of Rijndael, Appendix B gives a set of test vectors, and Appendix C consists of an example implementation of Rijndael in the C programming language.

Acknowledgments

This book would not have been written without the support and help of many people. It is impossible for us to list all people who contributed along the way. Although we probably will make oversights, we would like to name some of our supporters here.

First of all, we would like to thank the many cryptographers who contributed to developing the theory on the design of symmetric ciphers, and from who we learned much of what we know today. We would like to mention explicitly the people who gave us feedback in the early stages of the design process: Johan Borst, Antoon Bosselaers, Paulo Barreto, Craig Clapp, Erik De Win, Lars R. Knudsen and Bart Preneel.

Elaine Barker, James Foti and Miles Smid, and all the other people at NIST who worked very hard to make the AES process possible and visible.

The moral support of our family and friends, without whom we would never have persevered.

Brian Gladman, who provided test vectors.

Othmar Staffelbach, Elisabeth Oswald, Lee McCulloch and other proofreaders, who provided very valuable feedback and corrected numerous errors and oversights.

The financial support of KU Leuven, the Fund for Scientific Research – Flanders (Belgium), Banksys, Proton World and Cryptomathic is also greatly appreciated.

November 2001 *Joan Daemen and Vincent Rijmen*

Preface to the second edition

This edition contains updates of several chapters as well as four new chapters (Chaps. 12 to 15). We adapted our text to new terminology that has come into use since the first edition and removed some material that is now obsolete, including the original Appendix A (Propagation Analysis in Galois Fields) and the original Appendix B (Trail Clustering).

We thank Ronan Nugent of Springer for his persistent encouragements to finalize this second edition. We thank Dave, Eric Bach, Nicolas T. Courtois, Praveen Gauravaram, Jorge Nakahara Jr., Ralph Wernsdorf, Shengbo Xu and Uyama Yasumasa for pointing out errors in the first edition of this book. We thank Bart Mennink for proofreading some of the updates. All remaining errors and those newly introduced are of course our own.

July 2019 *Joan Daemen and Vincent Rijmen*

Contents

1. The Advanced Encryption Standard Process

The main subject of this book would probably have remained an esoteric topic of cryptographic research — with a name unpronounceable to most of the world — without the Advanced Encryption Standard (AES) process. Therefore, we thought it proper to include a short overview of the AES process.

1.1 In the Beginning ...

In January 1997, the US National Institute of Standards and Technology (NIST) announced the start of an initiative to develop a new encryption standard: the AES. The new encryption standard was to become a Federal Information Processing Standard (FIPS), replacing the old Data Encryption Standard (DES) and Triple DES.

Unlike the selection process for the DES, the Secure Hash Algorithm (SHA-1) and the Digital Signature Algorithm (DSA), NIST had announced that the AES selection process would be open. Anyone could submit a candidate cipher. Each submission, provided it met the requirements, would be considered on its merits. NIST would not perform any security or efficiency evaluation itself, but instead invited the cryptology community to mount attacks and try to cryptanalyze the different candidates, and anyone who was interested to evaluate implementation cost. All results could be sent to NIST as public comments for publication on the NIST AES web site or be submitted for presentation at AES conferences. NIST would merely collect contributions, using them as the basis for their selection. NIST would motivate their choices in evaluation reports.

1.2 AES: Scope and Significance

The official scope of a FIPS standard is quite limited: the FIPS only applies to the US Federal Administration. Furthermore, the new AES would only be used for documents that contain *sensitive but not classified* information.

J. Daemen, V. Rijmen, *The Design of Rijndael*, Information Security and Cryptography,
https://doi.org/10.1007/978-3-662-60769-5_1

However, it was anticipated that the impact of the AES would be much larger than this: for the AES is the successor of the DES, the cipher that ever since its adoption has been used as a worldwide de facto cryptographic standard by banks, administrations and industry.

The major factors for the quick acceptance of Rijndael are that it is available royalty-free, and that it can be implemented easily on a wide range of platforms without reducing bandwidth in a significant way.

1.3 Start of the AES Process

In September 1997, the final request for candidate nominations for the AES was published. The minimum functional requirements asked for symmetric block ciphers capable of supporting block lengths of 128 bits and key lengths of 128, 192 and 256 bits. An early draft of the AES functional requirements had asked for block ciphers also supporting block sizes of 192 and 256 bits, but this requirement was dropped later on. Nevertheless, since the request for proposals mentioned that extra functionality in the submissions would be received favorably, some submitters decided to keep the variable block length in the designs. (Examples include RC6 and Rijndael.)

NIST declared that it was looking for a block cipher *as secure as Triple DES, but much more efficient.* Another mandatory requirement was that the submitters agreed to make their cipher available on a worldwide royalty-free basis if it were to be selected as the AES. In order to qualify as an official AES candidate, the designers had to provide:

1. A complete written specification of the block cipher in the form of an algorithm.
2. A reference implementation in ANSI C, and mathematically optimized implementations in ANSI C and Java.
3. Implementations of a series of known-answer and Monte Carlo tests, as well as the expected outputs of these tests for a correct implementation of their block cipher.
4. Statements concerning the estimated computational efficiency in both hardware and software, the expected strength against cryptanalytic attacks, and the advantages and limitations of the cipher in various applications.
5. An analysis of the cipher's strength against known cryptanalytic attacks.

It turned out that the required effort to produce a 'complete and proper' submission package would already filter out several of the proposals. Early in the submission stage, the Cryptix team announced that they would provide Java implementations for all submitted ciphers, as well as Java implementations of the known-answer and Monte Carlo tests. This generous offer took some weight off the designers' shoulders, but still the effort required to compile a submission package was too heavy for some designers. The fact that the AES Application Programming Interface (API), which all submissions were required to follow, was updated twice during the submission stage increased the workload. Table 1.1 lists (in alphabetical order) the 15 submissions that were completed in time and accepted.

Table 1.1. The 15 AES candidates accepted for the first evaluation round

Submissions	Submitter(s)	Submitter type
CAST-256	Entrust (CA)	Company
Crypton	Future Systems (KR)	Company
DEAL	Outerbridge and Knudsen (USA–DK)	Researchers
DFC	ENS-CNRS (FR)	Researchers
E2	NTT (JP)	Company
FROG	TecApro (CR)	Company
HPC	Schroeppel (USA)	Researcher
LOKI97	Brown et al. (AU)	Researchers
Magenta	Deutsche Telekom (DE)	Company
Mars	IBM (USA)	Company
RC6	RSA (USA)	Company
Rijndael	Daemen and Rijmen (BE)	Researchers
SAFER+	Cylink (USA)	Company
Serpent	Anderson, Biham and Knudsen (UK–IL–DK)	Researchers
Twofish	Counterpane (USA)	Company

1.4 The First Round

The selection process was divided into several stages, with a public workshop to be held near the end of each stage. The process started with a *submission stage*, which ended on 15 May 1998. All accepted candidates were presented at *The First Advanced Encryption Standard Candidate Conference*, held in Ventura, California, on 20-22 August 1998. This was the official start of the first evaluation round, during which the international cryptographic community was asked for comments on the candidates.

1.5 Evaluation Criteria

The evaluation criteria for the first round were divided into three major categories: security, cost and *algorithm and implementation characteristics*. NIST invited the cryptology community to mount attacks and try to cryptanalyze the different candidates, and anyone interested to evaluate implementation cost. The result could be sent to NIST as public comments or be submitted for presentation at the second AES conference. NIST collected all contributions and would use these to select five finalists. In the following sections we discuss the evaluation criteria.

1.5.1 Security

Security was the most important category, but perhaps the most difficult to assess. Only a small number of candidates showed some theoretical design flaws. The large majority of the candidates fell into the category 'no weakness demonstrated'.

1.5.2 Costs

The 'costs' of the candidates were divided into different subcategories. A first category was formed by costs associated with intellectual property (IP) issues. First of all, each submitter was required to make his cipher available for free if it were to be selected as the AES. Secondly, each submitter was also asked to make a signed statement that he would not claim ownership or exercise patents on ideas used in *another submitter's proposal* that would eventually be selected as the AES. A second category of 'costs' was formed by costs associated with the implementation and execution of the candidates. This covers aspects such as computational efficiency, program size and working memory requirements in software implementations, and chip area in dedicated hardware implementations.

1.5.3 Algorithm and Implementation Characteristics

The category *algorithm and implementation characteristics* grouped a number of features that are harder to quantify. The first one is *versatility*, meaning the ability to be implemented efficiently on different platforms. At one end of the spectrum the AES should fit 8-bit micro-controllers and smart cards, which have limited storage for the program and a very restricted amount of RAM for working memory. At the other end of the spectrum the AES should be implementable efficiently in dedicated hardware, e.g. to provide on-the-fly encryption/decryption of communication links at gigabit-per-second rates. In between there is the whole range of processors that are used in servers, workstations, PCs, palmtops etc., which are all devices in need of cryptographic support. A prominent place in this range is taken by the Pentium family of processors due to its presence in most personal computers.

A second feature is *key agility*. In most block ciphers, key setup takes some processing. In applications where the same key is used to encrypt large amounts of data, this processing is relatively unimportant. In applications where the key often changes, such as the encryption of Internet Protocol (IP) packets in Internet Protocol Security (IPSEC), the overhead due to key setup may become quite relevant. Obviously, in those applications it is an advantage to have a fast key setup.

Finally, there is the criterion of *simplicity*, which may even be harder to evaluate than cryptographic security. Simplicity is related to the size of the description, the number of different operations used in the specification, the symmetry or lack of symmetry in the cipher and the ease with which the algorithm can be understood. All other things being equal, NIST considered it to be an advantage for an AES candidate to be more simple for reasons of ease of implementation and confidence in security.

1.6 Selection of Five Finalists

In March 1999, the second AES conference was held in Rome, Italy. The remarkable fact that a US Government department organized a conference on a future US Standard in Europe is easily explained. NIST chose to combine the conference with the yearly Fast Software Encryption Workshop, which had for the most part the same target audience and was scheduled to be in Rome.

1.6.1 The Second AES Conference

The papers presented at the conference ranged from crypto-attacks, cipher cross-analysis, smart-card-related papers, and so-called *algorithm observations*. In the session on cryptographic attacks, it was shown that FROG, Magenta and LOKI97 did not satisfy the security requirements imposed by NIST. For DEAL it was already known in advance that the security requirements were not satisfied. For HPC weaknesses had been demonstrated in a paper previously sent to NIST. This eliminated five candidates.

Some cipher cross-analysis papers focused on performance evaluation. The paper of B. Gladman [65], a researcher who had no link with any submission, considered performance on the Pentium processor. From this paper it became clear that RC6, Rijndael, Twofish, MARS and Crypton were the five fastest ciphers on this processor. On the other hand, the candidates DEAL, Frog,

Magenta, SAFER+ and Serpent appeared to be problematically slow. Other papers by the Twofish team (Bruce Schneier et al.) [138] and a French team of 12 cryptographers [10] essentially confirmed this.

A paper by E. Biham warned that the security margins of the AES candidates differed greatly and that this should be taken into account in the performance evaluation [21]. The lack of speed of Serpent (with E. Biham in the design team) was seen to be compensated for by a very high margin of security. Discussions on how to measure and take into account security margins lasted until after the third AES conference.

In the session on smart cards there were two papers comparing the performance of AES candidates on typical 8-bit processors and a 32-bit processor: one by G. Keating [78] and one by G. Hachez et al. [69]. From these papers and results from other papers, it became clear that some candidates simply did not fit onto a smart card and that Rijndael was by far the best suited for this platform. In the same session there were some papers that discussed power analysis attacks and the suitability of the different candidates for implementations that can resist against these attacks [27, 39, 49].

Finally, in the *algorithm observations* session, there were a number of papers in which AES submitters re-confirmed their confidence in their submission by means of a considerable amount of formulas, graphs and tables and some *loyal cryptanalysis* (the demonstration of having found no weaknesses after attacks on their own cipher).

1.6.2 The Five Finalists

After the workshop there was a relatively calm period that ended with the announcement of the five candidates by NIST in August 1999. The finalists were (in alphabetical order): MARS, RC6, Rijndael, Serpent and Twofish.

Along with the announcement of the finalists, NIST published a status report [118] in which the selection was motivated. The choice coincided with the top five that resulted from the response to a questionnaire handed out at the end of the second AES workshop. Despite its moderate performance, Serpent made it thanks to its high security margin. The candidates that had not been eliminated because of security problems were not selected mainly for the following reasons:

1. CAST-256: comparable to Serpent but with a higher implementation cost.
2. Crypton: comparable to Rijndael and Twofish but with a lower security margin.
3. DFC: low security margin and bad performance on anything other than 64-bit processors.

4. E2: comparable to Rijndael and Twofish in structure, but with a lower security margin and higher implementation cost.
5. SAFER+: high security margin similar to Serpent but even slower.

1.7 The Second Round

After the announcement of the five candidates NIST made another open call for contributions focused on the finalists. Intellectual property issues and performance and chip area in dedicated hardware implementations entered the picture. A remarkable contribution originated from NSA, presenting the results of hardware-performance simulations performed for the finalists. This third AES conference was held in New York City in April 2000. As in the year before, it was combined with the Fast Software Encryption Workshop.

In the sessions on cryptographic attacks there were some interesting results but no breakthroughs, since none of the finalists showed any weaknesses that could jeopardize their security. Most of the results were attacks on reduced-round versions of the ciphers. All attacks presented are only of academic relevance in that they are only slightly faster than an exhaustive key search. In the sessions on software implementations, the conclusions of the second workshop were confirmed.

In the sessions on dedicated hardware implementations attention was paid to Field Programmable Gate Arrays (FPGAs) and Application-Specific Integrated Circuits (ASICs). In the papers Serpent came out as a consistently excellent performer. Rijndael and Twofish also proved to be quite suited for hardware implementation while RC6 turned out to be expensive due to its use of 32-bit multiplication. Dedicated hardware implementations of MARS seemed in general to be quite costly. The Rijndael-related results presented at this conference are discussed in more detail in Chap. 4 (which is on efficient implementations) and Chap. 10 (which is on cryptanalytic results).

At the end of the conference a questionnaire was handed out asking about the preferences of the attendants. Rijndael was resoundingly voted to be the public's favorite.

1.8 The Selection

On 2 October 2000, NIST officially announced that Rijndael, without modifications, would become the AES. NIST published an excellent 116-page report in which they summarize all contributions and motivate the choice [117]. In the conclusion of this report, NIST motivates the choice of Rijndael with the following words:

Rijndael appears to be consistently a very good performer in both hardware and software across a wide range of computing environments regardless of its use in feedback or non-feedback modes. Its key setup time is excellent, and its key agility is good. Rijndael's very low memory requirements make it very well suited for restricted-space environments, in which it also demonstrates excellent performance. Rijndael's operations are among the easiest to defend against power and timing attacks. Additionally, it appears that some defense can be provided against such attacks without significantly impacting Rijndael's performance.

Finally, Rijndael's internal round structure appears to have good potential to benefit from instruction-level parallelism.

2. Preliminaries

In this chapter we introduce a number of mathematical concepts and explain the terminology that we need in the specification of Rijndael (in Chap. 3), in the treatment of some implementation aspects (in Chap. 4) and when we discuss our design choices (Chaps. 5–9).

The first part of this chapter starts with a discussion of finite fields, the representation of their elements and the impact of this on their operations of addition and multiplication. Subsequently, there is a short introduction to linear codes. Understanding the mathematics is not necessary for a full and correct implementation of the cipher. However, the mathematics are necessary for a good understanding of our design motivations. Knowledge of the underlying mathematical constructions also helps for doing optimized implementations. Not all aspects will be covered in detail; where possible, we refer to books dedicated to the topics we introduce.

In the second part of this chapter we introduce the terminology that we use to indicate different common types of Boolean functions and block ciphers.

When the discussion moves from a general level to an example specific to Rijndael, the text is put in a grey box.

Notation. We use in this book two types of indexing:

subscripts: Parts of a larger, named structure are denoted with subscripts. For instance, the bytes of a state $\mathbf{a}$ are denoted by $a_{i,j}$ (see Chap. 3).

superscripts: In an enumeration of more or less independent objects, where the objects are denoted by their own symbols, we use superscripts. For instance the elements of a nameless set are denoted by $\{\mathbf{a}^{(1)}, \mathbf{a}^{(2)}, \ldots\}$, and consecutive rounds of an iterative transformation are denoted by $\rho^{(1)}, \rho^{(2)}, \ldots$ (see Sect. 2.3.4).

J. Daemen, V. Rijmen, *The Design of Rijndael*, Information Security and Cryptography,
https://doi.org/10.1007/978-3-662-60769-5_2

2.1 Finite Fields

In this section we present a basic introduction to the theory of finite fields. For a more formal and elaborate introduction, we refer to the work of Lidl and Niederreiter [95].

2.1.1 Groups, Rings and Fields

We start with the formal definition of a group.

Definition 2.1.1. *An* Abelian group $(G, +)$ *consists of a set* G *and an operation defined on its elements, here denoted by '+':*

$$+ : G \times G \to G : (a, b) \mapsto a + b. \tag{2.1}$$

For the group to qualify as an Abelian group, *the operation has to fulfill the following conditions:*

$$\begin{aligned} \textit{closed:} &\ \forall\ a, b \in G : a + b \in G & (2.2)\\ \textit{associative:} &\ \forall\ a, b, c \in G : (a + b) + c = a + (b + c) & (2.3)\\ \textit{commutative:} &\ \forall\ a, b \in G : a + b = b + a & (2.4)\\ \textit{neutral element:} &\ \exists \mathbf{0} \in G, \forall\ a \in G : a + \mathbf{0} = a & (2.5)\\ \textit{inverse elements:} &\ \forall\ a \in G, \exists\ b \in G : a + b = \mathbf{0}. & (2.6) \end{aligned}$$

Example 2.1.1. The best-known example of an Abelian group is $(\mathbb{Z}, +)$, the set of integers, with the operation 'addition'. The structure $(\mathbb{Z}_n, +)$ is a second example. The set contains the integer numbers 0 to $n - 1$ and the operation is addition modulo n.

Since the addition of integers is the best-known example of a group, usually the symbol '+' is used to denote an arbitrary group operation. In this book, both an arbitrary group operation and integer addition will be denoted by the symbol '+'.

Both rings and fields are formally defined as structures that consist of a set of elements with *two* operations defined on these elements.

Definition 2.1.2. *A* ring $(R, +, \cdot)$ *consists of a set* R *with two operations defined on its elements, here denoted by '+' and '·'. For* R *to qualify as a ring, the operations have to fulfill the following conditions:*

1. *The structure* $(R, +)$ *is an Abelian group.*
2. *The operation '·' is closed, and associative over* R*. There is a neutral element for '·' in* R*.*

3. *The two operations '+' and '·' are related by the law of distributivity:*

$$\forall\, a, b, c \in R:\ (a + b) \cdot c = (a \cdot c) + (b \cdot c). \tag{2.7}$$

The neutral element for '·' is usually denoted by **1**. A ring $(R, +, \cdot)$ is called a *commutative ring* if the operation '·' is commutative.

Example 2.1.2. The best-known example of a ring is $(\mathbb{Z}, +, \cdot)$: the set of integers, with the operations 'addition' and 'multiplication'. This ring is a commutative ring. The set of matrices with n rows and n columns, with the operations 'matrix addition' and 'matrix multiplication' is a ring, but not a commutative ring (if $n > 1$).

Definition 2.1.3. *A structure $(F, +, \cdot)$ is a* field *if the following two conditions are satisfied:*

1. *$(F, +, \cdot)$ is a commutative ring.*
2. *For all elements of F, there is an inverse element in F with respect to the operation '·', except for the element* **0**, *the neutral element of $(F, +)$.*

A structure $(F, +, \cdot)$ is a field iff both $(F, +)$ and $(F \backslash \{\mathbf{0}\}, \cdot)$ are Abelian groups and the law of distributivity applies. The neutral element of $(F \backslash \{\mathbf{0}\}, \cdot)$ is called the *unit element* of the field.

Example 2.1.3. The best-known example of a field is the set of real numbers, with the operations 'addition' and 'multiplication.' Other examples are the set of complex numbers and the set of rational numbers, with the same operations. Note that for these examples the number of elements is infinite.

2.1.2 Vector Spaces

Let $(F, +, \cdot)$ be a field, with unit element **1**, and let $(V, +)$ be an Abelian group. Let '$\odot$' be an operation on elements of F and V:

$$\odot:\ F \times V \rightarrow V. \tag{2.8}$$

Definition 2.1.4. *The structure $(F, V, +, +, \cdot, \odot)$ is a* vector space *over F if the following conditions are satisfied:*

1. *distributivity:*

$$\forall\, a \in F, \forall\, \mathbf{v}, \mathbf{w} \in V:\ a \odot (\mathbf{v} + \mathbf{w}) = (a \odot \mathbf{v}) + (a \odot \mathbf{w}) \tag{2.9}$$

$$\forall\, a, b \in F, \forall\, \mathbf{v} \in V:\ (a + b) \odot \mathbf{v} = (a \odot \mathbf{v}) + (b \odot \mathbf{v}). \tag{2.10}$$

2. *associativity:*

$$\forall\, a, b \in F, \forall\, \mathbf{v} \in V: (a \cdot b) \odot \mathbf{v} = a \odot (b \odot \mathbf{v}). \tag{2.11}$$

3. *neutral element:*

$$\forall\, \mathbf{v} \in V : \mathbf{1} \odot \mathbf{v} = \mathbf{v}. \tag{2.12}$$

The elements of V are called *vectors*, and the elements of F are the *scalars*. The operation '+' is called the *vector addition*, and '$\odot$' is the *scalar multiplication.*

Example 2.1.4. For any field F, the set of n-tuples $(a_0, a_1, \ldots, a_{n-1})$ forms a vector space, where '+' and '$\odot$' are defined in terms of the field operations:

$$(a_1, \ldots, a_n) + (b_1, \ldots, b_n) = (a_1 + b_1, \ldots, a_n + b_n) \tag{2.13}$$

$$a \odot (b_1, \ldots, b_n) = (a \cdot b_1, \ldots, a \cdot b_n). \tag{2.14}$$

A vector $\mathbf{v}$ is a *linear combination* of the vectors $\mathbf{w}^{(1)}, \mathbf{w}^{(2)}, \ldots, \mathbf{w}^{(s)}$ if there exist scalars $a^{(i)}$ such that:

$$\mathbf{v} = a^{(1)} \odot \mathbf{w}^{(1)} + a^{(2)} \odot \mathbf{w}^{(2)} + \cdots + a^{(s)} \odot \mathbf{w}^{(s)}. \tag{2.15}$$

In a vector space we can always find a set of vectors such that all elements of the vector space can be written in exactly one way as a linear combination of the vectors of the set. Such a set is called a *basis* of the vector space. We will consider only vector spaces where the bases have a finite number of elements. We denote a basis by a column vector e:

$$\mathsf{e} = \left[\mathbf{e}^{(1)}, \mathbf{e}^{(2)}, \ldots \mathbf{e}^{(n)}\right]^{\mathrm{T}}. \tag{2.16}$$

In this expression the T superscript denotes taking the transpose of the row vector in the right-hand side of the equation. The scalars used in this linear combination are called the *coordinates* of $\mathbf{x}$ with respect to the basis e:

$$\mathrm{co}(\mathbf{x}) = \boldsymbol{x} = (c_1, c_2, \ldots, c_n) \Leftrightarrow \mathbf{x} = \textstyle\sum_{i=1}^{n} c_i \odot \mathbf{e}^{(i)}. \tag{2.17}$$

In order to simplify the notation, from now on we will denote vector addition by the same symbol as the field addition ('+'), and the scalar multiplication by the same symbol as the field multiplication ('$\cdot$'). It should always be clear from the context what operation the symbols are referring to.

A function f is called *a linear function* of a vector space V over a field F, if it has the following properties:

$$\forall\, \mathbf{x}, \mathbf{y} \in V : \; f(\mathbf{x} + \mathbf{y}) = f(\mathbf{x}) + f(\mathbf{y}) \tag{2.18}$$

$$\forall\, a \in F, \forall\, \mathbf{x} \in V : \; f(a\mathbf{x}) = af(\mathbf{x}). \tag{2.19}$$

The linear functions of a vector space can be represented by a matrix multiplication on the coordinates of the vectors. A function f is a linear function of the vector space $\mathrm{GF}(p)^n$ iff there exists a matrix M such that

$$\mathrm{co}(f(\mathbf{x})) = \mathsf{M} \cdot \boldsymbol{x}, \forall\, \mathbf{x} \in \mathrm{GF}(p)^n. \tag{2.20}$$

2.1.3 Fields with a Finite Number of Elements

A *finite field* is a field with a finite number of elements. The number of elements in the set is called the *order* of the field. A field with order m exists iff m is a *prime power*, i.e. $m = p^n$ for some integer n and with p a prime integer. p is called the *characteristic* of the finite field.

All finite fields used in the description of Rijndael have characteristic 2.

Fields of the same order are *isomorphic*: they display exactly the same algebraic structure, differing only in the *representation* of the elements. In other words, for each prime power there is exactly one finite field, denoted by $\mathrm{GF}(p^n)$. From now on, we will only consider fields with a finite number of elements.

Perhaps the most intuitive examples of finite fields are the fields of prime order p. The elements of a finite field $\mathrm{GF}(p)$ can be represented by the integers $0, 1, \ldots, p-1$. The two operations of the field are then 'integer addition modulo p' and 'integer multiplication modulo p'.

For finite fields with an order that is not prime, the operations *addition* and *multiplication* cannot be represented by addition and multiplication of integers modulo a number. Instead, slightly more complex representations must be introduced. Finite fields $\mathrm{GF}(p^n)$ with $n > 1$ can be represented in several ways. The representation of $\mathrm{GF}(p^n)$ by means of polynomials over $\mathrm{GF}(p)$ is quite popular and is the one we have adopted in Rijndael and its predecessors. In the next sections, we explain this representation.

2.1.4 Polynomials over a Field

A polynomial over a field F is an expression of the form

$$b(x) = b_{n-1}x^{n-1} + b_{n-2}x^{n-2} + \cdots + b_2x^2 + b_1x + b_0, \tag{2.21}$$

x being called the *indeterminate* of the polynomial, and the $b_i \in F$ the *coefficients*. The *degree* of a polynomial equals ℓ if $b_j = 0, \forall j > \ell$, and ℓ is the smallest number with this property. The set of polynomials over a field F is denoted by $F[x]$. The set of polynomials over a field F that have a degree below ℓ is denoted by $F[x]|_\ell$.

In computer memory, the polynomials in $F[x]|_\ell$ with F a finite field can be stored efficiently by storing the ℓ coefficients as a string.

Example 2.1.5. Let the field F be GF(2), and let $\ell = 8$. The polynomials can conveniently be stored as 8-bit values, or bytes:

$$b(x) \mapsto b_7b_6b_5b_4b_3b_2b_1b_0. \tag{2.22}$$

Strings of bits are often abbreviated using the hexadecimal notation.

Example 2.1.6. The polynomial in $\mathrm{GF}(2)|_8$

$$x^6 + x^4 + x^2 + x + 1$$

corresponds to the bit string 01010111, or `57` in hexadecimal notation.

2.1.5 Operations on Polynomials

We define the following operations on polynomials.

Addition. Summing of polynomials consists of summing the coefficients with equal powers of x, where the summing of the coefficients occurs in the underlying field F:

$$c(x) = a(x) + b(x) \Leftrightarrow c_i = a_i + b_i, \quad 0 \leq i < n. \tag{2.23}$$

The neutral element for the addition **0** is the polynomial with all coefficients equal to 0. The inverse element of a polynomial can be found by replacing each coefficient by its inverse element in F. The degree of $c(x)$ is at most the maximum of the degrees of $a(x)$ and $b(x)$, hence the addition is closed. The structure $(F[x]|_\ell, +)$ is an Abelian group.

Example 2.1.7. Let F be the field GF(2). The sum of the polynomials denoted by `57` and `83` is the polynomial denoted by `D4`, since:

$$\begin{aligned}
&(x^6 + x^4 + x^2 + x + 1) + (x^7 + x + 1) \\
&\quad = x^7 + x^6 + x^4 + x^2 + (1 + 1)x + (1 + 1) \\
&\quad = x^7 + x^6 + x^4 + x^2.
\end{aligned}$$

In binary notation we have: 01010111 + 10000011 = 11010100. Clearly, the addition can be implemented with the bitwise XOR instruction.

Multiplication. Multiplication of polynomials is associative (2.3), commutative (2.4) and distributive (2.7) with respect to addition of polynomials. There is a neutral element: the polynomial of degree 0 and with coefficient of x^0 equal to 1. In order to make the multiplication closed (2.2) over $F[x]|_\ell$, we select a polynomial $m(x)$ of degree ℓ, called the *reduction polynomial.* The multiplication of two polynomials $a(x)$ and $b(x)$ is then defined as the algebraic product of the polynomials modulo the polynomial $m(x)$:

$$c(x) = a(x) \cdot b(x) \Leftrightarrow c(x) \equiv a(x) \times b(x) \pmod{m(x)}. \tag{2.24}$$

Hence, the structure $(F[x]|_\ell, +, \cdot)$ is a commutative ring. For special choices of the reduction polynomial $m(x)$, the structure becomes a field.

Definition 2.1.5. *A polynomial $d(x)$ is* irreducible *over the field* GF(p) *iff there exist no two polynomials $a(x)$ and $b(x)$ with coefficients in* GF(p) *such that $d(x) = a(x) \times b(x)$, where $a(x)$ and $b(x)$ are of degree* > 0.

The inverse element for the multiplication can be found by means of the extended Euclidean algorithm (see, e.g. [110, p. 81]). Let $a(x)$ be the polynomial we want to find the inverse for. The extended Euclidean algorithm can then be used to find two polynomials $b(x)$ and $c(x)$ such that:

$$a(x) \times b(x) + m(x) \times c(x) = \gcd(a(x), m(x)). \tag{2.25}$$

Here $\gcd(a(x), m(x))$ denotes the greatest common divisor of the polynomials $a(x)$ and $m(x)$, which is always equal to $\mathbf{1}$ iff $m(x)$ is irreducible. Applying modular reduction to (2.25), we get:

$$a(x) \times b(x) \equiv \mathbf{1} \pmod{m(x)}, \tag{2.26}$$

which means that $b(x)$ is the inverse element of $a(x)$ for the definition of the multiplication '$\cdot$' given in (2.24). A polynomial of degree n is *primitive* if its roots generate the multiplicative group of $\mathrm{GF}(p^n)$, or equivalently, the multiplicative polynomial x has order $p^n - 1$. It can be shown that a primitive polynomial is irreducible.

Conclusion. Let F be the field $\mathrm{GF}(p)$. With a suitable choice for the reduction polynomial, the structure $(F[x]|_n, +, \cdot)$ is a field with p^n elements, usually denoted by $\mathrm{GF}(p^n)$.

Example 2.1.8. Consider the field $\mathrm{GF}(2^3)$. Let α be a root of $x^3 + x + 1 = 0$. Then the elements of $\mathrm{GF}(2^3)$ can be denoted by $0, 1, \alpha, \alpha+1, \alpha^2, \alpha^2+1, \alpha^2+\alpha$ and $\alpha^2 + \alpha + 1$.

2.1.6 Polynomials and Bytes

According to (2.22) a byte can be considered as a polynomial with coefficients in $\mathrm{GF}(2)$:

$$b_7 b_6 b_5 b_4 b_3 b_2 b_1 b_0 \mapsto b(x) \tag{2.27}$$

$$b(x) = b_7 x^7 + b_6 x^6 + b_5 x^5 + b_4 x^4 + b_3 x^3 + b_2 x^2 + b_1 x + b_0. \tag{2.28}$$

The set of all possible byte values corresponds to the set of all polynomials with degree less than eight. Addition of bytes can be defined as addition of the corresponding polynomials. In order to define the multiplication, we need to select a reduction polynomial $m(x)$.

In the specification of Rijndael, we consider the bytes as polynomials. Byte addition is defined as addition of the corresponding polynomials. In order to define byte multiplication, we use the following irreducible polynomial as reduction polynomial:

$$m(x) = x^8 + x^4 + x^3 + x + 1. \tag{2.29}$$

Since this reduction polynomial is irreducible, we have constructed a representation for the field GF(2^8). Hence we can state the following: In the Rijndael specification, bytes are considered as elements of GF(2^8). Operations on bytes are defined as operations in GF(2^8).

Example 2.1.9. In our representation for GF(2^8), the product of the elements denoted by `57` and `83` is the element denoted by `C1`, since:

$$\begin{aligned}(x^6 + x^4 + x^2 + x + 1) \times (x^7 + x + 1) \\ &= (x^{13} + x^{11} + x^9 + x^8 + x^7) + (x^7 + x^5 + x^3 + x^2 + x) \\ &\quad + (x^6 + x^4 + x^2 + x + 1) \\ &= x^{13} + x^{11} + x^9 + x^8 + x^6 + x^5 + x^4 + x^3 + 1\end{aligned}$$

and

$$\begin{aligned}(x^{13} + x^{11} + x^9 + x^8 + x^6 + x^5 + x^4 + x^3 + 1) \\ &\equiv x^7 + x^6 + 1 \pmod{x^8 + x^4 + x^3 + x + 1}.\end{aligned}$$

2.1.7 Polynomials and Columns

In the Rijndael specification, 4-byte columns are considered as polynomials over GF(2^8), having a degree smaller than four. In order to define the multiplication operation, the following reduction polynomial is used:

$$l(x) = x^4 + 1. \tag{2.30}$$

This polynomial is reducible, since in GF(2^8)

$$x^4 + 1 = (x + 1)^4. \tag{2.31}$$

In the definition of Rijndael, one of the inputs of the multiplication is a constant polynomial.

Since $l(x)$ is reducible over GF(2^8), not all polynomials have an inverse element for the multiplication modulo $l(x)$. A polynomial $b(x)$ has an inverse if the polynomial $x + 1$ does not divide it.

Multiplication with a fixed polynomial. We work out in more detail the multiplication with the fixed polynomial used in Rijndael.

Let $b(x)$ be the fixed polynomial with degree three:

$$b(x) = b_0 + b_1 x + b_2 x^2 + b_3 x^3 \tag{2.32}$$

and let $c(x)$ and $d(x)$ be two variable polynomials with coefficients c_i and d_i, respectively ($0 \leq i < 4$). We derive the matrix representation of the transformation that takes as input the coefficients of polynomial c, and produces as output the coefficients of the polynomial $d = b \times c$. We have:

$$d = b \cdot c \tag{2.33}$$

$$\Updownarrow$$

$$(b_0 + b_1 x + b_2 x^2 + b_3 x^3) \times (c_0 + c_1 x + c_2 x^2 + c_3 x^3)$$
$$\equiv (d_0 + d_1 x + d_2 x^2 + d_3 x^3) \pmod{x^4 + 1}. \tag{2.34}$$

Working out the product and separating the conditions for different powers of x, we get:

$$\begin{bmatrix} d_0 \\ d_1 \\ d_2 \\ d_3 \end{bmatrix} = \begin{bmatrix} b_0 & b_3 & b_2 & b_1 \\ b_1 & b_0 & b_3 & b_2 \\ b_2 & b_1 & b_0 & b_3 \\ b_3 & b_2 & b_1 & b_0 \end{bmatrix} \times \begin{bmatrix} c_0 \\ c_1 \\ c_2 \\ c_3 \end{bmatrix}. \tag{2.35}$$

2.1.8 Functions over Fields

All functions over $\mathrm{GF}(p^n)$ can be expressed as a polynomial over $\mathrm{GF}(p^n)$ of degree at most $p^n - 1$:

$$f(a) = \sum_{i=0}^{p^n - 1} c_i a^i , \tag{2.36}$$

with $c_i \in \mathrm{GF}(p^n)$. For simplicity of notation we follow the convention that $0^0 = 1$. Given a table specification where the output value $f(a)$ is given for the p^n different input values a, the p^n coefficients of this polynomial can be found by applying Lagrange interpolation [95, p. 28]. On the other hand, given a polynomial expression, the table specification can be found by evaluating the polynomial for all values of a.

The trace function is a function from $\mathrm{GF}(p^n)$ to $\mathrm{GF}(p)$ that will turn out to be useful later.

Definition 2.1.6. *Let x be an element of $\mathrm{GF}(p^n)$. The* trace *of x over $\mathrm{GF}(p)$ is defined by*

$$\mathrm{Tr}(x) = x + x^p + x^{p^2} + x^{p^3} + \cdots + x^{p^{n-1}}.$$

The trace is linear over GF(p):

$$\forall\, x, y \in \mathrm{GF}(p^n): \ \mathrm{Tr}(x+y) = \mathrm{Tr}(x) + \mathrm{Tr}(y)$$
$$\forall\, a \in \mathrm{GF}(p), \forall\, x \in \mathrm{GF}(p^n): \ \mathrm{Tr}(ax) = a\mathrm{Tr}(x).$$

and it can be shown that $\mathrm{Tr}(x)$ is an element of GF(p).

2.1.9 Representations of GF(p^n)

Finite fields can be represented in different ways.

Cyclic representation of GF(p^n). It can be proven that the multiplicative group of GF(p^n) is cyclic. The elements of this group (different from **0**) can be represented as the $p^n - 1$ powers of a generator $\alpha \in \mathrm{GF}(p^n)$:

$$\forall x \in \mathrm{GF}(p^n)\backslash\{\mathbf{0}\}, \exists\ a_x \in \mathbb{Z}_{p^n-1} : x = \alpha^{a_x}\ . \tag{2.37}$$

In this representation, multiplication of two non-zero elements corresponds to addition of their powers modulo $p^n - 1$:

$$x \cdot y = \alpha^{a_x} \cdot \alpha^{a_y} = \alpha^{a_x + a_y \bmod p^n - 1}. \tag{2.38}$$

Operations such as taking the multiplicative inverse and exponentiation are trivial in this representation. For addition, however, the vector representation (discussed next) is more appropriate. In computations involving both addition and multiplication, one may switch between the two different representations by means of conversion tables. The table used for conversion from the vector representation to the cyclic representation is called a *log table*, and the table for the inverse conversion is called an *alog table*. We have used this principle in our reference implementation (see Appendix C).

Vector space representation of GF(p^n). The additive group of the finite field GF(p^n) and the n-dimensional vector space over GF(p) are isomorphic. The addition of vectors in this vector space corresponds to the addition in GF(p^n). We can choose a basis e consisting of n elements $e^{(i)} \in \mathrm{GF}(p^n)$. We depict the basis e by a column vector that has as elements the elements of the basis:

$$\mathsf{e} = \left[e^{(1)}\ e^{(2)}\ \cdots\ e^{(n)}\right]^{\mathrm{T}}$$

The elements of GF(p^n) can be represented by their coordinates with respect to this basis. We have

$$a = \sum_i a_i e^{(i)} = \boldsymbol{a}^{\mathrm{T}}\mathsf{e}. \tag{2.39}$$

where $a_i \in \mathrm{GF}(2)$ are the coordinates of a with respect to the basis e and where $\boldsymbol{a}$ is the column vector consisting of coordinates a_i. The map

$$\phi_{\mathsf{e}} : \mathrm{GF}(p^n) \to \mathrm{GF}(p)^n : a \mapsto \phi_{\mathsf{e}}(a) = \boldsymbol{a}$$

forms an isomorphism.

Dual bases. Coordinates of a field element with respect to a basis can be expressed in terms of the *dual basis* and the *trace* map.

Definition 2.1.7. *Two bases* e *and* d *are called* dual bases *if for all* i *and* j *with* $1 \leq i, j \leq n$, *it holds that*

$$\mathrm{Tr}(d^{(i)} e^{(j)}) = \delta(i - j) \tag{2.40}$$

with $\delta(x)$ the Kronecker delta that is 1 if $x = 0$ and 0 otherwise. Every basis has exactly one dual basis. Let e and d be dual bases. Then we have

$$\mathrm{Tr}(d^{(j)} a) = \mathrm{Tr}\left(d^{(j)} \sum_{i=1}^{n} a_i e^{(i)} \right) = \sum_{i=1}^{n} a_i \mathrm{Tr}(d^{(j)} e^{(i)}) = a_j.$$

Hence the coordinates with respect to basis e can be expressed in an elegant way by means of the trace map and the dual basis d [95]:

$$\phi_{\mathsf{e}}(a) = \boldsymbol{a} = \left[\mathrm{Tr}(d^{(1)} a) \; \mathrm{Tr}(d^{(2)} a) \; \ldots \; \mathrm{Tr}(d^{(n)} a) \right]. \tag{2.41}$$

Applying (2.39) gives:

$$a = \sum_{i=1}^{n} \mathrm{Tr}(d^{(i)} a) e^{(i)} = \sum_{i=1}^{n} \mathrm{Tr}(e^{(i)} a) d^{(i)}. \tag{2.42}$$

Example 2.1.10. By choosing a basis, we can represent the elements of $\mathrm{GF}(2^3)$ as vectors. We choose the basis e as follows:

$$\mathsf{e} = [\alpha^2 + \alpha + 1, \alpha + 1, 1]^{\mathrm{T}}.$$

The dual basis of e can be determined by solving (2.40). It is given by

$$\mathsf{d} = [\alpha, \alpha^2 + \alpha, \alpha^2 + 1]^{\mathrm{T}}.$$

Table 2.1 shows the coordinates of the elements of $\mathrm{GF}(2^3)$, with respect to both bases.

Boolean functions and functions in GF(p^n). Functions of $\mathrm{GF}(p^n)$ can be mapped to functions of $\mathrm{GF}(p)^n$ by choosing a basis e in $\mathrm{GF}(p^n)$. Given

$$f : \mathrm{GF}(p^n) \to \mathrm{GF}(p^n) : a \mapsto b = f(a),$$

we can define a Boolean function $\boldsymbol{f}$:

$$\boldsymbol{f} : \mathrm{GF}(p)^n \to \mathrm{GF}(p)^n : \boldsymbol{a} \mapsto \boldsymbol{b} = \boldsymbol{f}(\boldsymbol{a})$$

where

Table 2.1. Coordinates of the field elements, with respect to the bases e and d

a	$\boldsymbol{a}$	$\boldsymbol{a}_{\mathrm{d}}$
0	000	000
1	001	111
$\alpha+1$	010	011
α	011	100
$\alpha^2+\alpha+1$	100	101
$\alpha^2+\alpha$	101	010
α^2	110	110
α^2+1	111	001

$$\boldsymbol{a} = [a_1 \; a_2 \; \dots \; a_n]$$
$$\boldsymbol{b} = [b_1 \; b_2 \; \dots \; b_n],$$

and

$$a_i = \mathrm{Tr}(ad^{(i)})$$
$$b_i = \mathrm{Tr}(bd^{(i)}).$$

On the other hand, given a Boolean function $\boldsymbol{g}$, a function over GF(p^n) can be defined as

$$a = \boldsymbol{a}^{\mathrm{T}}\mathsf{e}$$
$$b = \boldsymbol{b}^{\mathrm{T}}\mathsf{e}.$$

So in short, $\boldsymbol{f} = \phi_{\mathsf{e}} \circ f \circ \phi_{\mathsf{e}}^{-1}$ and $f = \phi_{\mathsf{e}}^{-1} \circ \boldsymbol{f} \circ \phi_{\mathsf{e}}$. This can be extended to functions operating on vectors of elements of GF(p^n).

2.2 Linear Codes

In this section we give a short introduction to the theory of linear codes. For a more detailed treatment, we refer the interested reader to the work of MacWilliams and Sloane [100]. In a code, message words are represented by codewords that have some redundancy. In code theory textbooks, it is customary to write codewords as $1 \times n$ matrices, or *row vectors.* We will follow that custom here. In subsequent chapters, one-dimensional arrays will as often be denoted as $n \times 1$ matrices, or *column vectors.*

2.2.1 Definitions

The *Hamming weight* of a codeword is defined as follows.

Definition 2.2.1. *The* Hamming weight $\mathrm{w_h}(\mathbf{x})$ *of a vector* $\mathbf{x}$ *is the number of non-zero components of the vector* $\mathbf{x}$.

Based on the definition of Hamming weight, we can define the *Hamming distance* between two vectors.

Definition 2.2.2. *The* Hamming distance *between two vectors* $\mathbf{x}$ *and* $\mathbf{y}$ *is* $\mathrm{w_h}(\mathbf{x}-\mathbf{y})$, *which is equal to the Hamming weight of the difference of the two vectors.*

Now we are ready to define linear codes.

Definition 2.2.3. *A* linear $[n, k, d]$ code *over* $\mathrm{GF}(2^p)$ *is a* k*-dimensional subspace of the vector space* $\mathrm{GF}(2^p)^n$, *where any two different vectors of the subspace have a Hamming distance of at least* d *(and* d *is the largest number with this property).*

The distance d of a linear code equals the minimum weight of a non-zero codeword in the code. A linear code can be described by each of the two following matrices:

1. A *generator* matrix G for an $[n, k, d]$ code $\mathcal{C}$ is a $k \times n$ matrix whose rows form a vector space basis for $\mathcal{C}$ (only generator matrices of full rank are considered). Since the choice of a basis in a vector space is not unique, a code has many different generator matrices, which can be reduced to one another by performing elementary row operations. The *echelon form* of the generator matrix is the following:
$$\mathsf{G}_e = \left[\mathsf{I}_{k\times k} \; \mathsf{A}_{k\times(n-k)}\right], \tag{2.43}$$
where $\mathsf{I}_{k\times k}$ is the $k \times k$ identity matrix.
2. A *parity-check* matrix H for an $[n, k, d]$ code $\mathcal{C}$ is an $(n-k) \times n$ matrix with the property that a vector $\mathbf{x}$ is a codeword of $\mathcal{C}$ iff
$$\mathsf{H}\mathbf{x}^{\mathrm{T}} = 0. \tag{2.44}$$

If G is a generator matrix and H a parity-check matrix of the same code, then

$$\mathsf{G}\mathsf{H}^{\mathrm{T}} = 0. \tag{2.45}$$

Moreover, if $\mathsf{G} = [\mathsf{I}\ \mathsf{C}]$ is a generator matrix of a code, then $\mathsf{H} = \left[-\mathsf{C}^{\mathrm{T}}\ \mathsf{I}\right]$ is a parity-check matrix of the same code.

The *dual code* $\mathcal{C}^{\perp}$ of a code $\mathcal{C}$ is defined as the set of vectors that are orthogonal to all the vectors of $\mathcal{C}$:

$$\mathcal{C}^{\perp} = \{\mathbf{x} \mid \mathbf{x}\mathbf{y}^{\mathrm{T}} = 0, \forall\ \mathbf{y} \in \mathcal{C}\}. \tag{2.46}$$

It follows that a parity-check matrix of $\mathcal{C}$ is a generator matrix of $\mathcal{C}^{\perp}$ and vice versa.

2.2.2 MDS Codes

The theory of linear codes addresses the problems of determining the distance of a linear code and the construction of linear codes with a given distance. We review a few well-known results.

The Singleton bound gives an upper bound for the distance of a code with given dimensions.

Theorem 2.2.1 (The Singleton bound). *If C is an $[n, k, d]$ code, then $d \leq n - k + 1$.*

A code that meets the Singleton bound is called a *maximal distance separable (MDS) code.* The following theorems relate the distance of a code to properties of the generator matrix G.

Theorem 2.2.2. *A linear code C has distance d iff every $d - 1$ columns of the parity-check matrix H are linearly independent and there exists some set of d columns that are linearly dependent.*

By definition, an MDS code has distance $n - k + 1$. Hence, every set of $n - k$ columns of the parity-check matrix are linearly independent. This property can be translated into a requirement for the matrix A:

Theorem 2.2.3 ([100]). *An $[n, k, d]$ code with generator matrix*

$$\mathsf{G} = \left[\mathsf{I}_{k \times k} \ \mathsf{A}_{k \times (n-k)}\right],$$

is an MDS code iff every square submatrix of A is nonsingular.

A well-known class of MDS codes is formed by the Reed-Solomon codes, for which efficient construction algorithms are known.

2.3 Boolean Functions

The smallest finite field has an order of 2: GF(2). Its two elements are denoted by **0** and **1**. Its addition is the integer addition modulo 2 and its multiplication is the integer multiplication modulo 2. Variables that range over GF(2) are called *Boolean variables*, or *bits* for short. The addition of 2 bits corresponds to the Boolean operation *exclusive or*, denoted by XOR. The multiplication of 2 bits corresponds to the Boolean operation AND. The operation of changing the value of a bit is called *complementation.*

A vector whose coordinates are bits is called a *Boolean vector.* The operation of changing the value of all bits of a Boolean vector is called complementation. If we have two Boolean vectors $\mathbf{a}$ and $\mathbf{b}$ of the same dimension, we can apply the following operations:

1. Bitwise XOR: results in a vector whose bits consist of the XOR of the corresponding bits of $\mathbf{a}$ and $\mathbf{b}$.
2. Bitwise AND: results in a vector whose bits consist of the AND of the corresponding bits of $\mathbf{a}$ and $\mathbf{b}$.

A function $\mathbf{b} = \phi(\mathbf{a})$ that maps a Boolean vector to another Boolean vector is called a *Boolean function*:

$$\phi : \mathrm{GF}(2)^n \to \mathrm{GF}(2)^m : \mathbf{a} \mapsto \mathbf{b} = \phi(\mathbf{a}), \tag{2.47}$$

where $\mathbf{b}$ is called the output Boolean vector and $\mathbf{a}$ the input Boolean vector. This Boolean function has n input bits and m output bits.

A *binary Boolean function* $b = f(\mathbf{a})$ is a Boolean function with a single output bit, in other words $m = 1$:

$$f : \mathrm{GF}(2)^n \to \mathrm{GF}(2) : \mathbf{a} \mapsto b = f(\mathbf{a}), \tag{2.48}$$

where b is called the output bit. Each bit of the output of a Boolean function is itself a binary Boolean function of the input vector. These functions are called the *component binary Boolean functions* of the Boolean function.

A Boolean function can be specified by providing the output value for the 2^n possible values of the input Boolean vector. A Boolean function with the same number of input bits as output bits can be considered as operating on an n-bit *state*. We call such a function a *Boolean transformation*. A Boolean transformation is called *invertible* if it maps all input states to different output states. An invertible Boolean transformation is called a *Boolean permutation*.

2.3.1 Tuple Partitions

In several instances it is useful to see the bits of a state as being partitioned into a number of subsets. Boolean transformations operating on a state can be expressed in terms of these subsets rather than in terms of the individual bits of the state. In the context of this book we restrict ourselves to partitions that divide the state bits into a number of equally sized subsets. We will use the term *tuples* to describe these subsets.[1] Because in the case of Rijndael, tuples contain eight bits, we will often write *bytes* instead. However, the reasoning applies as well to tuples of other sizes (larger than two).

Consider an n_b-bit state $\mathbf{a}$ consisting of bits a_i where $i \in \mathcal{I}$. $\mathcal{I}$ is called the index space. In its simplest form, the index space is just equal to $\{1, \ldots, n_\mathrm{b}\}$. However, for clarity the bits may be indexed in another way to ease specifications. A partitioning of the state bits may be reflected by having an index with two components: one component indicating the byte position within

[1] In the first edition we used the term *bundles*, but this term now gets another meaning in Chap. 14.

the state, and one component indicating the bit position within the byte. In this representation, $a_{(i,j)}$ would mean the state bit in byte i at bit position j within that byte. The value of the byte itself can be indicated by a_i. On some occasions, even the byte index can be decomposed. For example, in Rijndael the bytes are arranged in a two-dimensional array with the byte index composed of a column index and a row index.

Next to the obvious 8-bit bytes also the 32-bit *columns* in Rijndael define a partitioning. The nonlinear steps in the round transformations of the AES finalist Serpent [4] operate on 4-bit bytes. The nonlinear step in the round transformation of 3-Way [43] and BaseKing [45] operate on 3-bit tuples. The tuples can be considered as representations of elements in some group, ring or field. Examples are the integers modulo 2^m or elements of GF(2^m). In this way, steps of the round transformation, or even the full round transformation can be expressed in terms of operations in these mathematical structures.

2.3.2 Transpositions

A transposition is a Boolean permutation that only moves the positions of bits of the state without affecting their value. For a transposition $\mathbf{b} = \pi(\mathbf{a})$ we have:

$$b_i = a_{p(i)}, \tag{2.49}$$

where $p(i)$ is a permutation over the index space.

A byte transposition is a transposition that changes the positions of the bytes but leaves the positions of the bits within the bytes intact. This can be expressed as:

$$b_{(i,j)} = a_{(p(i),j)}. \tag{2.50}$$

An example is shown in Fig. 2.1. Figure 2.2 shows the pictogram that we will use to represent a byte transposition in this book.

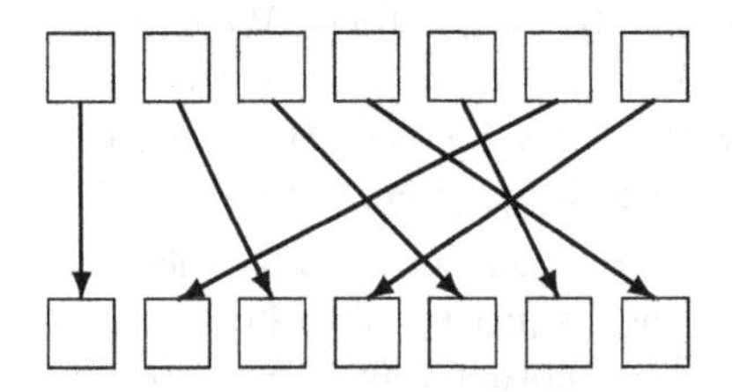

Fig. 2.1. Example of a byte transposition

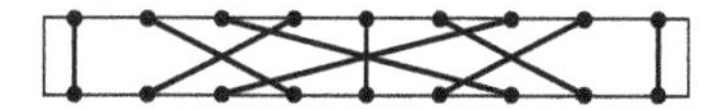

Fig. 2.2. Pictogram for a byte transposition

2.3.3 Bricklayer Functions

A *bricklayer function* is a function that can be decomposed into a number of Boolean functions operating independently on subsets of bits of the input vector. These subsets form a partition of the bits of the input vector. A bricklayer function can be considered as the parallel application of a number of Boolean functions operating on smaller inputs. If nonlinear, these Boolean functions are called *S-boxes*. If linear, we use the term *D-box*, where D stands for *diffusion*.

A bricklayer function operating on a state is called a *bricklayer transformation*. As a bricklayer transformation operates on a number of subsets of the state independently, it defines a byte partition. The component transformations of the bricklayer transformation operate independently on a number of bytes. A graphical illustration is given in Fig. 2.3. An invertible bricklayer transformation is called a bricklayer permutation. For a bricklayer transformation to be invertible, all of its S-boxes (or D-boxes) must be permutations. The pictogram that we will use is shown in Fig. 2.4.

For a bricklayer transformation $\mathbf{b} = \phi(\mathbf{a})$ we have:

$$(b_{(i,1)}, b_{(i,2)}, \ldots, b_{(i,m)}) = \phi_i(a_{(i,1)}, a_{(i,2)}, \ldots, a_{(i,m)}), \tag{2.51}$$

for all values of i. If the bytes within $\mathbf{a}$ and $\mathbf{b}$ are represented by a_i and b_i, respectively, this becomes:

$$b_i = \phi_i(a_i). \tag{2.52}$$

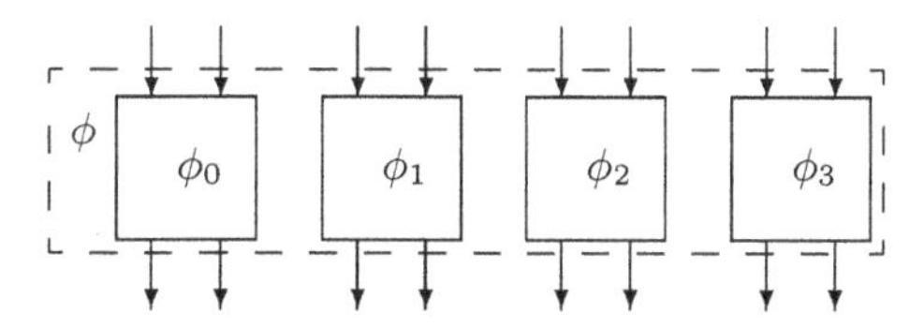

Fig. 2.3. Example of a bricklayer transformation

Fig. 2.4. Pictogram for a bricklayer transformation

2.3.4 Iterative Boolean Transformations

A Boolean vector can be transformed iteratively by applying a sequence of Boolean transformations, one after the other. Such a sequence is referred to as an *iterative Boolean transformation*. If the individual Boolean transformations are denoted by $\rho^{(i)}$, an iterative Boolean transformation is of the form

$$\beta = \rho^{(r)} \circ \ldots \circ \rho^{(2)} \circ \rho^{(1)}. \tag{2.53}$$

A schematic illustration is given in Fig. 2.5. We have $\mathbf{b} = \beta(\mathbf{d})$, where $\mathbf{d} = \mathbf{a}^{(0)}, \mathbf{b} = \mathbf{a}^{(m)}$ and $\mathbf{a}^{(i)} = \rho^{(i)}(\mathbf{a}^{(i-1)})$. The value of $\mathbf{a}^{(i)}$ is called an *intermediate state*. An iterative Boolean transformation that is a sequence of Boolean permutations is an iterative Boolean permutation.

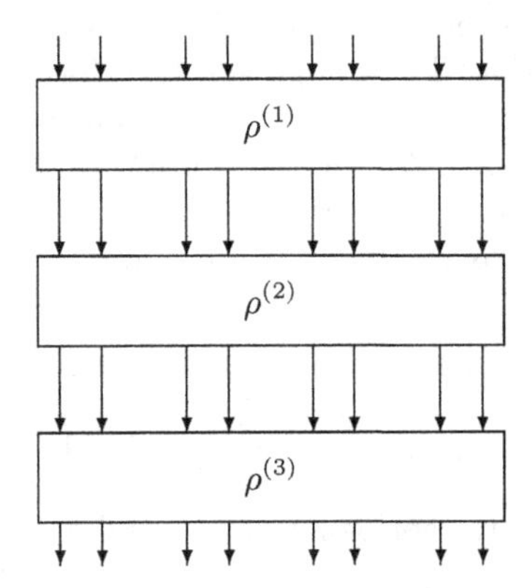

Fig. 2.5. Iterative Boolean transformation

2.4 Block Ciphers

A *block cipher* transforms *plaintext blocks* of a fixed length n_b into *ciphertext blocks* of the same length under the influence of a cipher key k. More precisely, a block cipher is a set of Boolean permutations operating on n_b-bit vectors. This set contains a Boolean permutation for each value of the *cipher key* k. In this book we only consider block ciphers in which the cipher key is a Boolean vector. If the number of bits in the cipher key is denoted by n_k, a block cipher consists of 2^{n_k} Boolean permutations.

The operation of transforming a plaintext block into a ciphertext block is called *encryption*, and the operation of transforming a ciphertext block into a plaintext block is called *decryption*.

Usually, block ciphers are specified by an *encryption algorithm*, being the sequence of transformations to be applied to the plaintext to obtain the ciphertext. These transformations are operations with a relatively simple description. The resulting Boolean permutation depends on the cipher key due to the fact that key material, computed from the cipher key, is used in the transformations.

For a block cipher to be up to its task, it has to fulfill two requirements:

1. **Efficiency.** Given the value of the cipher key, applying the corresponding Boolean permutation, or its inverse, is efficient, preferably on a wide range of platforms.
2. **Security.** It must be impossible to exploit knowledge of the internal structure of the cipher in cryptographic attacks.

All block ciphers of any significance satisfy these requirements by iteratively applying Boolean permutations that are relatively simple to describe.

2.4.1 Iterative Block Ciphers

In an *iterative block cipher*, the Boolean permutations are iterative. The block cipher is defined as the application of a number of key-dependent Boolean permutations. The Boolean permutations are called the *round transformations* of the block cipher. Every application of a round transformation is called a *round*.

Example 2.4.1. The DES has 16 rounds. Since every round uses the same round transformation, we say the DES has only one round transformation.

We denote the number of rounds by r. We have:

$$B[\mathbf{k}] = \rho^{(r)}[\mathbf{k}^{(r)}] \circ \cdots \circ \rho^{(2)}[\mathbf{k}^{(2)}] \circ \rho^{(1)}[\mathbf{k}^{(1)}]. \tag{2.54}$$

In this expression, $\rho^{(i)}$ is called the ith *round* of the block cipher and $\mathbf{k}^{(i)}$ is called the ith round key.

The round keys are computed from the cipher key. Usually, this is specified with an algorithm. The algorithm that describes how to derive the round keys from the cipher key is called the *key schedule*. The concatenation of all round keys is called the *expanded key*, denoted by K:

$$\mathsf{K} = \mathbf{k}^{(0)}|\mathbf{k}^{(1)}|\mathbf{k}^{(2)}|\ldots|\mathbf{k}^{(r)} \tag{2.55}$$

The length of the expanded key is denoted by n_{K}. The iterative block cipher model is illustrated in Fig. 2.6. Almost all block ciphers known can be modeled this way. There is however a large variety in round transformations and key schedules. An iterative block cipher in which all rounds (with the exception of the initial or final round) use the same round transformation is called an *iterated block cipher*.

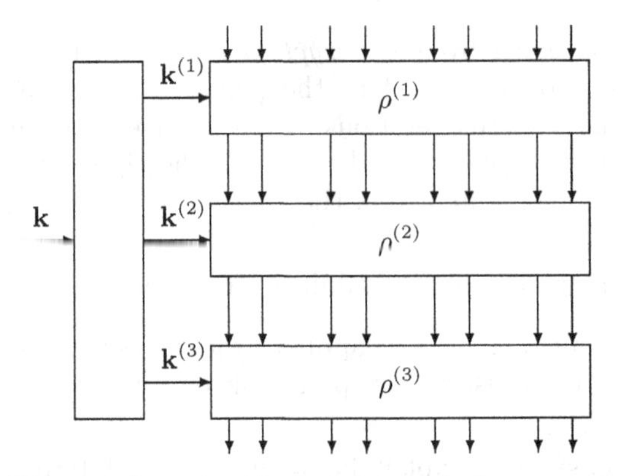

Fig. 2.6. Iterative block cipher with three rounds

2.4.2 Key-Alternating Block Ciphers

Rijndael belongs to a class of block ciphers in which the round key is applied in a particularly simple way: the key-alternating block ciphers. A key-alternating block cipher is an iterative block cipher with the following properties:

1. **Alternation.** The cipher is defined as the alternated application of key-independent round transformations and key additions. The first round key is added before the first round and the last round key is added after the last round.
2. **Simple key addition.** The round keys are added to the state by means of a simple XOR. A key addition is denoted by $\sigma[k]$.

We have:

$$B[\mathbf{k}] = \sigma[\mathbf{k}^{(r)}] \circ \rho^{(r)} \circ \sigma[\mathbf{k}^{(r-1)}] \circ \cdots \circ \sigma[\mathbf{k}^{(1)}] \circ \rho^{(1)} \circ \sigma[\mathbf{k}^{(0)}]. \quad (2.56)$$

A graphical illustration is given in Fig. 2.7.

Key-alternating block ciphers are a class of block ciphers that lend themselves to analysis with respect to their resistance against cryptanalysis. This will become clear in Chaps. 7–9 and in Chaps. 13–15. A special class of key-alternating block ciphers are the *key-iterated block ciphers*. In this class, all rounds (except maybe the first or the last) of the cipher use the same round transformation. We have:

$$B[\mathbf{k}] = \sigma[\mathbf{k}^{(r)}] \circ \rho \circ \sigma[\mathbf{k}^{(r-1)}] \circ \cdots \circ \sigma[\mathbf{k}^{(1)}] \circ \rho \circ \sigma[\mathbf{k}^{(0)}]. \quad (2.57)$$

In this case, ρ is called *the* round transformation of the block cipher. The relations between the different classes of block ciphers that we define here are shown in Fig. 2.8.

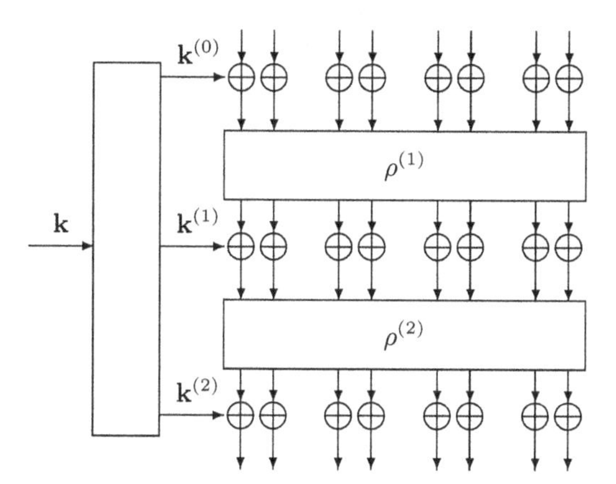

Fig. 2.7. Key-alternating block cipher with two rounds

Key-iterated block ciphers lend themselves to efficient implementations. In dedicated hardware implementations, one can hardwire the round transformation and the key addition. The block cipher can be executed by simply iterating the round transformation alternated with the right round keys. In software implementations, the program needs to code only the one round transformation in a loop and the cipher can be executed by executing this loop the required number of times. In practice, for performance reasons, block ciphers in software will often have code for all rounds: so-called *loop unrolling.* In these implementations, it is less important to have identical rounds. Nevertheless, the most-used block ciphers all consist of a number of identical rounds. Some other advantages of the key-iterated structure are discussed in Chap. 5.

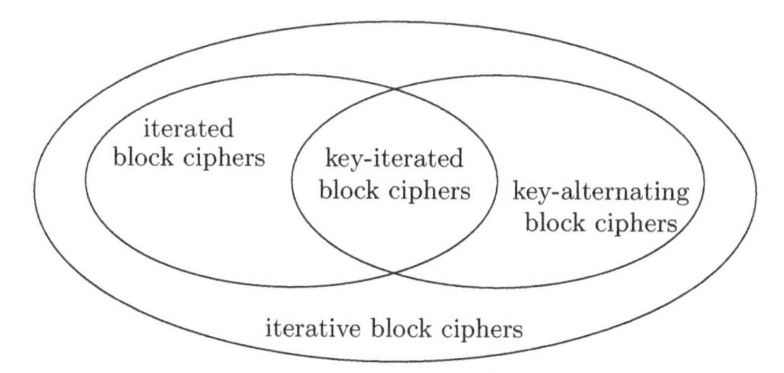

Fig. 2.8. Block cipher taxonomy

A block cipher is a cryptographic primitive that can convert a fixed-length plaintext block to a fixed-length ciphertext block and vice versa under a given cipher key. In order to use a cipher to protect the confidentiality or integrity of messages of arbitrary length, it must be specified how the cipher is used. These specifications are the so-called *modes of operation* of a block cipher. Modes of operation are out of scope of this book. We refer the reader to [102].

3. Specification of Rijndael

In this chapter we specify the cipher structure and the building blocks of Rijndael. After explaining the difference between the Rijndael specifications and the AES standard, we specify the external interface to the ciphers. This is followed by the description of the Rijndael structure and the steps of its round transformation. Subsequently, we specify the number of rounds as a function of the block and key length, and describe the key schedule. We conclude this chapter with a treatment of algorithms for implementing decryption with Rijndael. This chapter is not intended as an implementation guideline. For implementation aspects, we refer to Chap. 4.

3.1 Differences Between Rijndael and the AES

The *only* difference between Rijndael and the AES is the range of supported values for the block length and cipher key length.

Rijndael is a block cipher with both a variable block length and a variable key length. The block length and the key length can be independently specified to any multiple of 32 bits, with a minimum of 128 bits and a maximum of 256 bits. It would be possible to define versions of Rijndael with a higher block length or key length, but currently there seems no need for it.

The AES fixes the block length to 128 bits, and supports key lengths of 128, 192 or 256 bits only. The extra block and key lengths in Rijndael were not evaluated in the AES selection process, and consequently they are not adopted in the current FIPS standard.

3.2 Input and Output for Encryption and Decryption

The input and output of Rijndael are considered to be one-dimensional arrays of 8-bit bytes. For encryption the input is a *plaintext block* and a *key*, and the output is a *ciphertext block*. For decryption, the input is a ciphertext block and a key, and the output is a plaintext block. The round transformation of Rijndael, and its steps, operate on an intermediate result, called the *state*.

J. Daemen, V. Rijmen, *The Design of Rijndael*, Information Security and Cryptography,
https://doi.org/10.1007/978-3-662-60769-5_3

The state can be pictured as a rectangular array of bytes, with four rows. The number of columns in the state is denoted by $\mathrm{N_b}$ and is equal to the block length divided by 32. Let the plaintext block be denoted by

$$p_0 p_1 p_2 p_3 \ldots p_{4 \cdot \mathrm{N_b}-1},$$

where p_0 denotes the first byte,and $p_{4 \cdot \mathrm{N_b}-1}$ denotes the last byte of the plain text block. Similarly, a ciphertext block can be denoted by

$$c_0 c_1 c_2 c_3 \ldots c_{4 \cdot \mathrm{N_b}-1}.$$

Let the state be denoted by

$$a_{i,j}, \quad 0 \leq i < 4, 0 \leq j < \mathrm{N_b},$$

where $a_{i,j}$ denotes the byte in row i and column j. The input bytes are mapped onto the state bytes in the order $a_{0,0}$, $a_{1,0}$, $a_{2,0}$, $a_{3,0}$, $a_{0,1}$, $a_{1,1}$, $a_{2,1}$, $a_{3,1}$, ... For encryption, the input is a plaintext block and the mapping is

$$a_{i,j} = p_{i+4j}, \quad 0 \leq i < 4, 0 \leq j < \mathrm{N_b}. \tag{3.1}$$

For decryption, the input is a ciphertext block and the mapping is

$$a_{i,j} = c_{i+4j}, \quad 0 \leq i < 4, 0 \leq j < \mathrm{N_b}. \tag{3.2}$$

At the end of the encryption, the ciphertext is extracted from the state by taking the state bytes in the same order:

$$c_i = a_{i \bmod 4, i/4}, \quad 0 \leq i < 4\mathrm{N_b}. \tag{3.3}$$

At the end of decryption, the plaintext block is extracted from the state according to

$$p_i = a_{i \bmod 4, i/4}, \quad 0 \leq i < 4\mathrm{N_b}. \tag{3.4}$$

Similarly, the key is mapped onto a two-dimensional cipher key. The cipher key is pictured as a rectangular array with four rows similar to the state. The number of columns of the cipher key is denoted by $\mathrm{N_k}$ and is equal to the key length divided by 32. The bytes of the key are mapped onto the bytes of the cipher key in the order: $k_{0,0}$, $k_{1,0}$, $k_{2,0}$, $k_{3,0}$, $k_{0,1}$, $k_{1,1}$, $k_{2,1}$, $k_{3,1}$, $k_{0,2}$ .... If we denote the key by

$$z_0 z_1 z_2 z_3 \ldots z_{4 \cdot \mathrm{N_k}-1},$$

then

$$k_{i,j} = z_{i+4j}, \quad 0 \leq i < 4, 0 \leq j < \mathrm{N_k}. \tag{3.5}$$

The representation of the state and cipher key and the mappings plaintext–state and key–cipher key are illustrated in Fig. 3.1.

p_0	p_4	p_8	p_{12}
p_1	p_5	p_9	p_{13}
p_2	p_6	p_{10}	p_{14}
p_3	p_7	p_{11}	p_{15}

k_0	k_4	k_8	k_{12}	k_{16}	k_{20}
k_1	k_5	k_9	k_{13}	k_{17}	k_{21}
k_2	k_6	k_{10}	k_{14}	k_{18}	k_{22}
k_3	k_7	k_{11}	k_{15}	k_{19}	k_{23}

Fig. 3.1. State and cipher key layout for the case $N_b = 4$ and $N_k = 6$

3.3 Structure of Rijndael

Rijndael is a key-iterated block cipher: it consists of the repeated application of a round transformation on the state. The number of rounds is denoted by N_r and depends on the block length and the key length.

Note that in this chapter, contrary to the definitions (2.54)–(2.57), the key addition is included in the *round* transformation. This is done in order to make the description in this chapter consistent with the description in the FIPS standard.

Following a suggestion of B. Gladman, we changed the names of some steps with respect to the description given in our original AES submission. The new names are more consistent, and are also adopted in the FIPS standard. We made some further changes, all in order to make the description more clear and complete. No changes have been made to the block cipher itself.

An encryption with Rijndael consists of an initial key addition, denoted by `AddRoundKey`, followed by $N_r - 1$ applications of the transformation `Round`, and finally one application of `FinalRound`. The initial key addition and every round take as input the `State` and a round key. The round key for round i is denoted by `ExpandedKey`$[i]$, and `ExpandedKey`$[0]$ denotes the input of the initial key addition. The derivation of `ExpandedKey` from the `CipherKey` is denoted by `KeyExpansion`. A high-level description of Rijndael in pseudocode notation is shown in List. 3.1.

3.4 The Round Transformation

The round transformation is denoted `Round`, and is a sequence of four transformations, called *steps*. This is shown in List. 3.2. The final round of the cipher is slightly different. It is denoted `FinalRound` and is also shown in List. 3.2. In the listings, the *transformations* (`Round`, `SubBytes`, `ShiftRows`,

```
procedure RIJNDAEL(State,Cipherkey)
   KeyExpansion(CipherKey,ExpandedKey)
   AddRoundKey(State,ExpandedKey[0])
   for i = 1 to N_r - 1 do
      Round(State,ExpandedKey[i])
   end for
   FinalRound(State,ExpandedKey[N_r])
end procedure
```

List. 3.1. High-level algorithm for encryption with Rijndael

...) operate on arrays to which pointers (State, ExpandedKey[i]) are provided. It is easy to verify that the transformation FinalRound is equal to the transformation Round, but with the MixColumns step removed. The steps are specified in the following subsections, together with the design criteria we used for each step. Besides the step-specific criteria, we also applied the following two general design criteria:

1. **Invertibility.** The structure of the Rijndael round transformation requires that all steps be invertible.
2. **Simplicity.** As explained in Chap. 5, we prefer simple components over complex ones.

```
procedure ROUND(State,ExpandedKey[i])
   SubBytes(State);
   ShiftRows(State);
   MixColumns(State);
   AddRoundKey(State,ExpandedKey[i]);
end procedure
procedure FINALROUND(State,ExpandedKey[N_r])
   SubBytes(State);
   ShiftRows(State);
   AddRoundKey(State,ExpandedKey[N_r]);
end procedure
```

List. 3.2. The Rijndael round transformation

3.4.1 The SubBytes Step

The SubBytes step is the only nonlinear transformation of the cipher. SubBytes is a bricklayer permutation consisting of an S-box applied to the bytes of the state. We denote the particular S-box being used in Rijndael by S_{RD}. Figure 3.2 illustrates the effect of the SubBytes step on the state. Figure 3.3 shows the pictograms that we will use to represent SubBytes and its inverse.

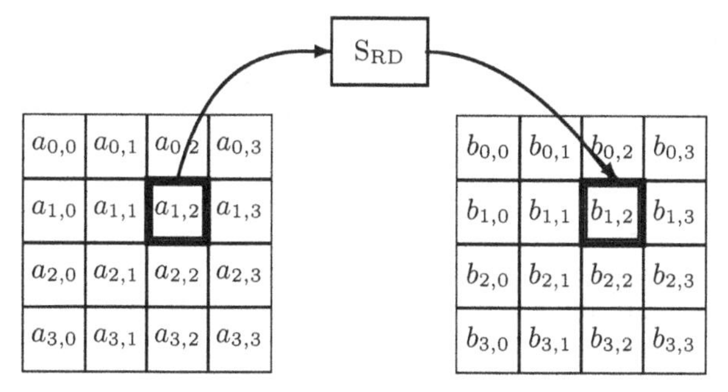

Fig. 3.2. `SubBytes` acts on the individual bytes of the state

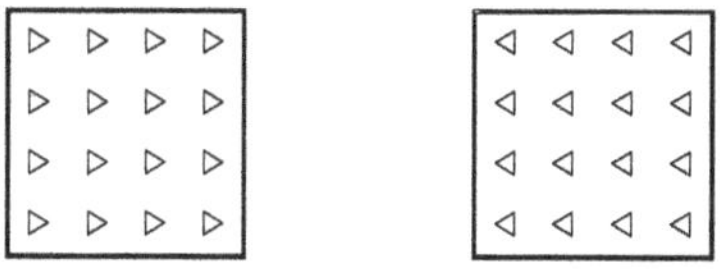

Fig. 3.3. The pictograms for `SubBytes` *(left)* and `InvSubBytes` *(right)*

Design criteria for S_{RD}. We have applied the following design criteria for S_{RD}, appearing in order of importance:

1. **Nonlinearity.**
   a) **Correlation.** The maximum input-output correlation amplitude must be as small as possible.
   b) **Difference propagation probability.** The maximum difference propagation probability must be as small as possible.
2. **Algebraic complexity.** The algebraic expression of S_{RD} in $\mathrm{GF}(2^8)$ has to be complex.

Only one S-box is used for all byte positions. This is certainly not a necessity: `SubBytes` could as easily be defined with different S-boxes for every byte. This issue is discussed in Chap. 5. The nonlinearity criteria are inspired by linear and differential cryptanalysis. Chap. 9 discusses this in depth.

Selection of S_{RD}. In [120], K. Nyberg gives several construction methods for S-boxes with good nonlinearity. For invertible S-boxes operating on bytes, the maximum correlation amplitude can be made as low as 2^{-3}, and the maximum difference propagation probability can be as low as 2^{-6}. We decided to choose — from the alternatives described in [120] — the S-box that is defined by the following function in $\mathrm{GF}(2^8)$:

$$\mathrm{Inv}_8 : a \rightarrow b = a^{254}. \tag{3.6}$$

This function is usually described as the mapping $a \rightarrow a^{-1}$ extended with 0 being mapped to 0. We use the polynomial representation of $\mathrm{GF}(2^8)$ defined in Sect. 2.1.6: the elements of GF(2^8) are considered as polynomials having a degree smaller than eight, with coefficients in the finite field GF(2). Multiplication is done modulo the irreducible polynomial $m(x) = x^8+x^4+x^3+x+1$; the functions a^{254} and a^{-1} are defined accordingly. By definition, g has a very simple algebraic expression. This could allow algebraic manipulations that can be used to mount attacks such as interpolation attacks. Therefore, we built the S-box as the sequence of Inv_8 and an invertible affine transformation Aff_8: $\text{S}_{\text{RD}} = \text{Aff}_8 \circ \text{Inv}_8$. This affine transformation has no impact on the nonlinearity properties, but if properly chosen, allows S_{RD} to have a complex algebraic expression. We have chosen an affine transformation that has a very simple description per se, but a complicated algebraic expression if combined with the transformation Inv_8. Because this still leaves many possibilities for the choice of Aff_8, we additionally imposed the restriction that S_{RD} should have no fixed points and no opposite fixed points:

$$\text{S}_{\text{RD}}[a] + a \neq \texttt{0},\ \forall a \tag{3.7}$$

$$\text{S}_{\text{RD}}[a] + a \neq \texttt{FF},\ \forall a. \tag{3.8}$$

Note that we are not aware of any attacks that would exploit the existence of (opposite) fixed points.

The affine transformation Aff_8 is defined by

$$b = \text{Aff}_8(a)$$
$$\Updownarrow$$
$$\begin{bmatrix} b_7\\ b_6\\ b_5\\ b_4\\ b_3\\ b_2\\ b_1\\ b_0 \end{bmatrix} = \begin{bmatrix} 1&1&1&1&1&0&0&0\\ 0&1&1&1&1&1&0&0\\ 0&0&1&1&1&1&1&0\\ 0&0&0&1&1&1&1&1\\ 1&0&0&0&1&1&1&1\\ 1&1&0&0&0&1&1&1\\ 1&1&1&0&0&0&1&1\\ 1&1&1&1&0&0&0&1 \end{bmatrix} \times \begin{bmatrix} a_7\\ a_6\\ a_5\\ a_4\\ a_3\\ a_2\\ a_1\\ a_0 \end{bmatrix} + \begin{bmatrix} 0\\ 1\\ 1\\ 0\\ 0\\ 0\\ 1\\ 1 \end{bmatrix}. \tag{3.9}$$

The affine transformation Aff_8 can also be described as a linearized polynomial over GF(2^8), followed by the addition (in GF(2^8)) with a constant:

$$\text{Aff}_8(a) = L(a) + q. \tag{3.10}$$

This is explained in Appendix A, where also a tabular description of S_{RD} is given.

Inverse operation. `InvSubBytes` is the inverse operation of `SubBytes`. It is a bricklayer permutation consisting of the inverse S-box $\text{S}_{\text{RD}}{}^{-1}$ applied to the bytes of the state. The inverse S-box $\text{S}_{\text{RD}}{}^{-1}$ is obtained by applying the inverse of the affine transformation (3.9) followed by taking the multiplicative inverse in GF(2^8). The inverse of Aff_8 is specified by:

$$\begin{bmatrix} x_7 \\ x_6 \\ x_5 \\ x_4 \\ x_3 \\ x_2 \\ x_1 \\ x_0 \end{bmatrix} = \begin{bmatrix} 0\,1\,0\,1\,0\,0\,1\,0 \\ 0\,0\,1\,0\,1\,0\,0\,1 \\ 1\,0\,0\,1\,0\,1\,0\,0 \\ 0\,1\,0\,0\,1\,0\,1\,0 \\ 0\,0\,1\,0\,0\,1\,0\,1 \\ 1\,0\,0\,1\,0\,0\,1\,0 \\ 0\,1\,0\,0\,1\,0\,0\,1 \\ 1\,0\,1\,0\,0\,1\,0\,0 \end{bmatrix} \times \begin{bmatrix} y_7 \\ y_6 \\ y_5 \\ y_4 \\ y_3 \\ y_2 \\ y_1 \\ y_0 \end{bmatrix} + \begin{bmatrix} 0 \\ 0 \\ 0 \\ 0 \\ 0 \\ 1 \\ 0 \\ 1 \end{bmatrix} . \qquad (3.11)$$

Tabular descriptions of $\mathrm{S_{RD}}^{-1}$ and Aff_8^{-1} are given in Appendix A.

3.4.2 The `ShiftRows` Step

The `ShiftRows` step is a byte transposition that cyclically shifts the rows of the state over different offsets. Row 0 is shifted over C_0 bytes, row 1 over C_1 bytes, row 2 over C_2 bytes and row 3 over C_3 bytes, so that the byte at position j in row i moves to position $(j - C_i) \bmod \mathrm{N_b}$. The shift offsets C_0, C_1, C_2 and C_3 depend on the value of $\mathrm{N_b}$.

Design criteria for the offsets. The design criteria for the offsets are the following:

1. **Diffusion optimal.** The four offsets have to be different (see Definition 9.4.1).
2. **Other diffusion effects.** The resistance against truncated differential attacks (see Chap. 10) and saturation attacks has to be maximized.

Diffusion optimality is important in providing resistance against differential and linear cryptanalysis. The other diffusion effects are only relevant when the block length is larger than 128 bits.

Selection of the offsets. The simplicity criterion dictates that one offset is taken equal to 0. In fact, for a block length of 128 bits, the offsets have to be 0, 1, 2 and 3. The assignment of offsets to rows is arbitrary. For block lengths larger than 128 bits, there are more possibilities. Detailed studies of truncated differential attacks and saturation attacks on reduced versions of Rijndael show that not all choices are equivalent. For certain choices, the attacks can be extended by one round. Among the choices that are best with respect to saturation and truncated differential attacks, we picked the simplest ones. The different values are specified in Table 3.1. Figure 3.4 illustrates the effect of the `ShiftRows` step on the state. Figure 3.5 shows the pictograms for `ShiftRows` and its inverse.

Inverse operation. The inverse operation of `ShiftRows` is called `InvShiftRows`. It is a cyclic shift of the three bottom rows over $\mathrm{N_b} - C_1$, $\mathrm{N_b} - C_2$ and $\mathrm{N_b} - C_3$ bytes respectively so that the byte at position j in row i moves to position $(j + C_i) \bmod \mathrm{N_b}$.

Table 3.1. `ShiftRows`: shift offsets for different block lengths

N_b	C_0	C_1	C_2	C_3
4	0	1	2	3
5	0	1	2	3
6	0	1	2	3
7	0	1	2	4
8	0	1	3	4

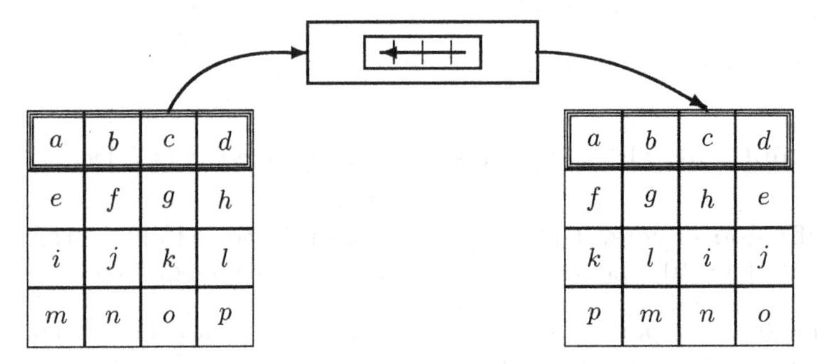

Fig. 3.4. `ShiftRows` operates on the rows of the state

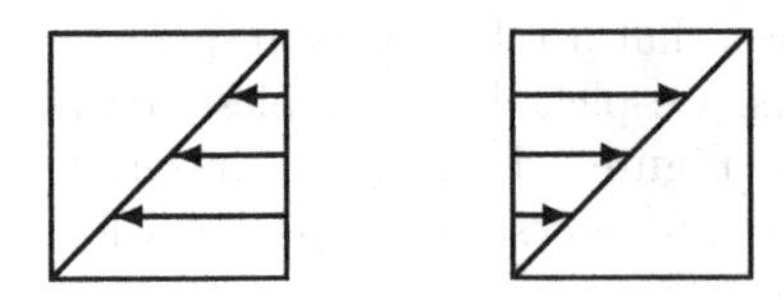

Fig. 3.5. Pictograms for `ShiftRows` *(left)* and `InvShiftRows` *(right)*

3.4.3 The MixColumns Step

The MixColumns step is a bricklayer permutation operating on the state column by column.

Design criteria. The design criteria for the MixColumns step are the following:

1. **Dimensions.** The transformation is a bricklayer transformation operating on 4-byte columns.
2. **Linearity.** The transformation is preferably linear over GF(2).
3. **Diffusion.** The transformation has to have *relevant* diffusion power.
4. **Performance on 8-bit processors.** The performance of the transformation on 8-bit processors has to be high.

The criteria about linearity and diffusion are requirements imposed by the wide trail strategy (see Chap. 9). The dimensions criterion of having columns consisting of 4 bytes is to make optimal use of 32-bit architectures in look-up table implementations (see Sect. 4.2.1). The performance on 8-bit processors is mentioned because MixColumns is the only step for which good performance on 8-bit processors is not trivial to obtain.

Selection. The diffusion and performance criteria have lead us to the following choice for the definition of the D-box in MixColumns. The columns of the state are considered as polynomials over $GF(2^8)$ and multiplied modulo $x^4 + 1$ with a fixed polynomial $c(x)$. The criteria about invertibility, diffusion and performance impose conditions on the coefficients of $c(x)$. The performance criterion can be satisfied if the coefficients have simple values, such as 0, 1, 2, 3, ... Multiplication with the value 0 or 1 implies no processing at all, multiplication with 2 can be implemented efficiently with a dedicated routine (see Sect. 4.1.1) and multiplication with 3 can be implemented as a multiplication with 2 plus an additional XOR operation with the operand. The diffusion criterion induces a more complicated condition on the coefficients of $c(x)$. We determined the coefficients in such a way that the branch number of MixColumns is five, i.e. the maximum possible for a transformation with these dimensions. Further explanation of the branch number of a function and the relation to the diffusion power can be found in Sect. 9.3.

The polynomial $c(x)$ is given by

$$c(x) = \mathtt{3} \cdot x^3 + \mathtt{1} \cdot x^2 + \mathtt{1} \cdot x + \mathtt{2}. \tag{3.12}$$

This polynomial is coprime to $x^4 + 1$ and therefore invertible. Observe that the algebraic order of $c(x)$ equals 4:

$$c(x) \cdot c(x) \cdot c(x) \cdot c(x) = \mathtt{1} \pmod{x^4 + 1}. \tag{3.13}$$

As described in Sect. 2.1.7, the modular multiplication with a fixed polynomial can be written as a matrix multiplication. Let $b(x) = c(x) \cdot a(x) \pmod{x^4 + 1}$. Then

$$\begin{bmatrix} b_0 \\ b_1 \\ b_2 \\ b_3 \end{bmatrix} = \begin{bmatrix} 2 & 3 & 1 & 1 \\ 1 & 2 & 3 & 1 \\ 1 & 1 & 2 & 3 \\ 3 & 1 & 1 & 2 \end{bmatrix} \times \begin{bmatrix} a_0 \\ a_1 \\ a_2 \\ a_3 \end{bmatrix} . \tag{3.14}$$

We denote the matrix in the previous expression by M_c. Figure 3.6 illustrates the effect of the MixColumns step on the state. Figure 3.7 shows the pictograms for MixColumns and its inverse.

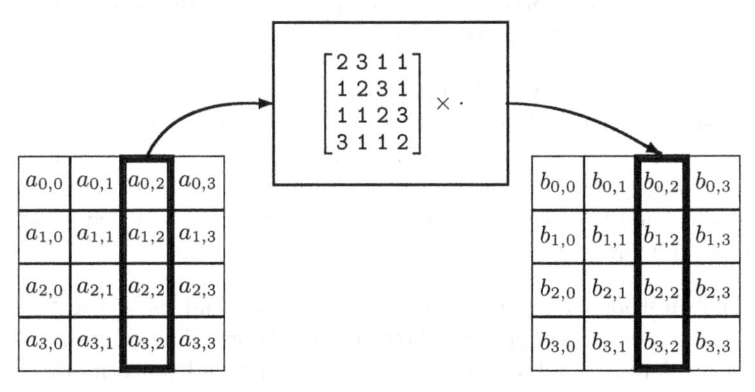

Fig. 3.6. MixColumns operates on the columns of the state

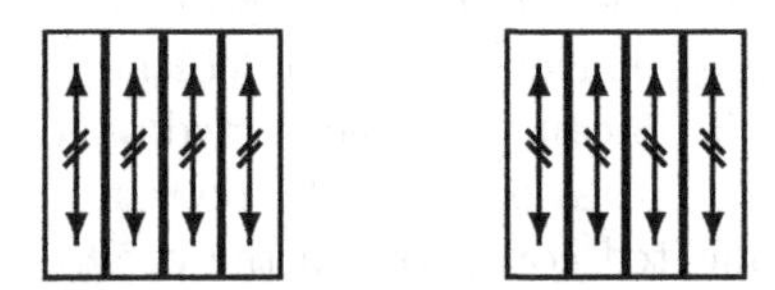

Fig. 3.7. Pictograms for MixColumns *(left)* and InvMixColumns *(right)*

Inverse operation. The inverse operation of MixColumns is called InvMixColumns. It is similar to MixColumns. Every column is transformed by multiplying it with a fixed multiplication polynomial $d(x)$, defined by

$$(3 \cdot x^3 + 1 \cdot x^2 + 1 \cdot x + 2) \cdot d(x) \equiv 1 \pmod{x^4 + 1}. \tag{3.15}$$

It is given by

$$d(x) = \mathtt{B} \cdot x^3 + \mathtt{D} \cdot x^2 + \mathtt{9} \cdot x + \mathtt{E}. \tag{3.16}$$

Written as a matrix multiplication, `InvMixColumns` transforms the columns in the following way:

$$\begin{bmatrix} b_0 \\ b_1 \\ b_2 \\ b_3 \end{bmatrix} = \begin{bmatrix} \mathtt{E} & \mathtt{B} & \mathtt{D} & \mathtt{9} \\ \mathtt{9} & \mathtt{E} & \mathtt{B} & \mathtt{D} \\ \mathtt{D} & \mathtt{9} & \mathtt{E} & \mathtt{B} \\ \mathtt{B} & \mathtt{D} & \mathtt{9} & \mathtt{E} \end{bmatrix} \times \begin{bmatrix} a_0 \\ a_1 \\ a_2 \\ a_3 \end{bmatrix} . \tag{3.17}$$

3.4.4 The Key Addition

The key addition is denoted `AddRoundKey`. In this transformation, the state is modified by combining it with a round key with the bitwise XOR operation. A round key is denoted by `ExpandedKey`$[i]$, $0 \leq i \leq \mathrm{N_r}$. The array of round keys `ExpandedKey` is derived from the cipher key by means of the key schedule (see Sect. 3.6). The round key length is equal to the block length. The `AddRoundKey` transformation is illustrated in Fig. 3.8. `AddRoundKey` is its own inverse. Figure 3.9 shows the pictogram for `AddRoundKey`.

$$\begin{array}{|c|c|c|c|} \hline a_{0,0} & a_{0,1} & a_{0,2} & a_{0,3} \\ \hline a_{1,0} & a_{1,1} & a_{1,2} & a_{1,3} \\ \hline a_{2,0} & a_{2,1} & a_{2,2} & a_{2,3} \\ \hline a_{3,0} & a_{3,1} & a_{3,2} & a_{3,3} \\ \hline \end{array} + \begin{array}{|c|c|c|c|} \hline k_{0,0} & k_{0,1} & k_{0,2} & k_{0,3} \\ \hline k_{1,0} & k_{1,1} & k_{1,2} & k_{1,3} \\ \hline k_{2,0} & k_{2,1} & k_{2,2} & k_{2,3} \\ \hline k_{3,0} & k_{3,1} & k_{3,2} & k_{3,3} \\ \hline \end{array} = \begin{array}{|c|c|c|c|} \hline b_{0,0} & b_{0,1} & b_{0,2} & b_{0,3} \\ \hline b_{1,0} & b_{1,1} & b_{1,2} & b_{1,3} \\ \hline b_{2,0} & b_{2,1} & b_{2,2} & b_{2,3} \\ \hline b_{3,0} & b_{3,1} & b_{3,2} & b_{3,3} \\ \hline \end{array}$$

Fig. 3.8. In `AddRoundKey`, the round key is added to the state with a bitwise XOR

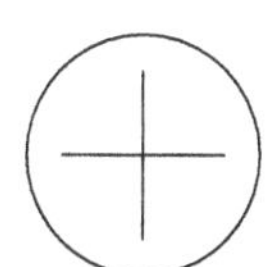

Fig. 3.9. Pictogram for `AddRoundKey`

3.4.5 The Rijndael Super Box

It is often convenient to describe Rijndael by means of a *super box*.

Definition 3.4.1. *A* super box *maps an array a of n_t elements a_i to an array e of n_t elements e_i. Each of the elements has size* n_s. *A super box takes a key k of size $n_t \times n_s = n_b$. It consists of the sequence of four transformations (or* steps*):*

- $b_i = S[a_i]$: *n_t parallel applications of an n_s-bit S-box*
- $c = M(b)$: *a linear map*
- $d = c + k$: *key addition*
- $e_i = S[d_i]$: *n_t parallel applications of an n_s-bit S-box.*

The S-boxes in the two S-box steps may also be all different.

For the Rijndael super box we have $n_t = 4$, $M = M_c$ and S applies S_{RD} four times. If we consider two Rijndael rounds, swap the steps `ShiftRows` and `SubBytes` in the first round, and remove the linear transformations before the first `SubBytes` transformation and after the second `SubBytes` transformation, then we obtain a map that can also be described as four parallel instances of the Rijndael super box.

3.5 The Number of Rounds

With the exception of slide attacks [31], the body of published cryptanalysis at the time of the design of Rijndael indicated that the resistance of iterative block ciphers against cryptanalytic attacks increases with the number of rounds, and this is still the case at time of writing this second edition of the book.

We have determined the number of rounds by considering the maximum number of rounds for which shortcut attacks (see Sect. 5.5.1) have been found that are significantly more efficient than an exhaustive key search. Subsequently, we added a considerable security margin. For Rijndael with a block length and key length of 128 bits, no shortcut attacks had been found for reduced versions with more than six rounds. We added four rounds as a security margin. This is a conservative approach, because

1. Two rounds of Rijndael provide 'full diffusion' in the following sense: every state bit depends on all state bits two rounds ago, or a change in one state bit is likely to affect half of the state bits after two rounds. Adding four rounds can be seen as adding a 'full diffusion step' at the beginning and at the end of the cipher. The high diffusion of the Rijndael round transformation is thanks to its uniform structure that operates on all state bits. For so-called Feistel ciphers, a round only operates on half of the state bits and full diffusion can at best be obtained after three rounds and in practice it typically takes four rounds or more.

2. Generally, linear cryptanalysis, differential cryptanalysis and truncated differential attacks exploit a propagation trail through n rounds in order to attack $n+1$ or $n+2$ rounds. This is also the case for the saturation attack (see Sect. 10.2) and the impossible-differential attack (see Sect. 10.8) that use a four-round propagation structure to attack six, respectively seven, rounds. In this respect, adding four rounds actually doubles the number of rounds through which a propagation trail has to be found.

For Rijndael versions with a longer key, the number of rounds was raised by one for every additional 32 bits in the cipher key. This was done for the following reasons:

1. One of the main objectives is the absence of shortcut attacks, i.e. attacks that are more efficient than an exhaustive key search. Since the workload of an exhaustive key search grows with the key length, shortcut attacks can afford to be less efficient for longer keys.
2. (Partially) known-key and related-key attacks exploit the knowledge of cipher key bits or the ability to apply different cipher keys. If the cipher key grows, the range of possibilities available to the cryptanalyst increases.

So far, there have been no publications that reveal any weakness in Rijndael that could be exploited to attacks relevant in the real world. For Rijndael versions with a higher block length, the number of rounds is raised by one for every additional 32 bits in the block length, for the following reasons:

1. For a block length above 128 bits, it takes three rounds to realize full diffusion, i.e. the diffusion power of the round transformation, relative to the block length, diminishes with the block length.
2. The larger block length causes the range of possible patterns that can be applied at the input/output of a sequence of rounds to increase. This additional flexibility may allow the extension of attacks by one or more rounds.

We have found that extensions of attacks by a single round are even hard to realize for the maximum block length of 256 bits. Therefore, this is a conservative margin.

Table 3.2 lists the value of $\mathrm{N_r}$ as a function of $\mathrm{N_b}$ and $\mathrm{N_k}$. For the AES, $\mathrm{N_b}$ is fixed to the value 4; $\mathrm{N_r} = 10$ for 128-bit keys ($\mathrm{N_k} = 4$), $\mathrm{N_r} = 12$ for 192-bit keys ($\mathrm{N_k} = 6$) and $\mathrm{N_r} = 14$ for 256-bit keys ($\mathrm{N_k} = 8$).

3.6 Key Schedule

The key schedule consists of two components: the key expansion and the round key selection. The key expansion specifies how `ExpandedKey` is derived

Table 3.2. Number of rounds (N_r) as a function of N_b (N_b = block length/32) and N_k (key length/32)

	N_b				
N_k	4	5	6	7	8
4	10	11	12	13	14
5	11	11	12	13	14
6	12	12	12	13	14
7	13	13	13	13	14
8	14	14	14	14	14

from the cipher key. The total number of bits in `ExpandedKey` is equal to the block length multiplied by the number of rounds plus 1, since the cipher requires one round key for the initial key addition, and one for each of the rounds. Please note that the `ExpandedKey` is always derived from the cipher key; it should never be specified directly.

3.6.1 Design Criteria

The key expansion has been chosen according to the following criteria:

1. **Efficiency.**
   a) **Working memory.** It should be possible to execute the key schedule using a small amount of working memory.
   b) **Performance.** It should have a high performance on a wide range of processors.
2. **Symmetry elimination.** It should use round constants to eliminate symmetries.
3. **Diffusion.** It should have an efficient diffusion of cipher key differences into the expanded key,
4. **Nonlinearity.** It should exhibit enough nonlinearity to prohibit the full determination of differences in the expanded key from cipher key differences only.

For a more thorough treatment of the criteria underlying the design of the key schedule, we refer to Sect. 5.8.

3.6.2 Selection

In order to be efficient on 8-bit processors, a lightweight, byte-oriented expansion scheme has been adopted. The application of the nonlinear S_{RD} ensures the nonlinearity of the scheme, without adding much in the way of temporary storage requirements on an 8-bit processor.

During the key expansion the cipher key is expanded into an expanded key array, consisting of four rows and $N_b(N_r+1)$ columns. This array is here denoted by $\mathsf{W}[4][N_b(N_r+1)]$. The round key of the ith round, $\texttt{ExpandedKey}[i]$, is given by the columns $N_b \cdot i$ to $N_b \cdot (i+1) - 1$ of W:

$$
\begin{aligned}
&\texttt{ExpandedKey}[i] = \\
&\quad \mathsf{W}[\cdot][N_b \cdot i] \parallel \mathsf{W}[\cdot][N_b \cdot i + 1] \parallel \cdots \parallel \mathsf{W}[\cdot][N_b \cdot (i+1) - 1], \\
&\quad 0 \le i \le N_r.
\end{aligned} \tag{3.18}
$$

The key expansion function depends on the value of N_k: there is a version for N_k equal to or below 6, shown in List. 3.3, and a version for N_k above 6, shown in List. 3.4. In both versions of the key expansion, the first N_k columns of W are filled with the cipher key. The following columns are defined recursively in terms of previously defined columns. The recursion uses the bytes of the previous column, the bytes of the column N_k positions earlier, and round constants $\mathrm{RC}[j]$.

The recursion function depends on the position of the column. If i is not a multiple of N_k, column i is the bitwise XOR of column $i - N_k$ and column $i-1$. Otherwise, column i is the bitwise XOR of column $i - N_k$ and a nonlinear function of column $i-1$. For cipher key length values $N_k > 6$, this is also the case if $i \bmod N_k = 4$. The nonlinear function is realized by means of the application of S_{RD} to the four bytes of the column, an additional cyclic rotation of the bytes within the column and the addition of a round constant (for elimination of symmetry). The round constants are independent of N_k, and defined by a recursion rule in $\mathrm{GF}(2^8)$:

$$\mathrm{RC}[1] = x^0 \text{ (i.e. 1)} \tag{3.19}$$

$$\mathrm{RC}[2] = x \text{ (i.e. 2)} \tag{3.20}$$

$$\mathrm{RC}[j] = x \cdot \mathrm{RC}[j-1] = x^{j-1}, \quad j > 2. \tag{3.21}$$

The key expansion process and the round key selection are illustrated in Fig. 3.10.

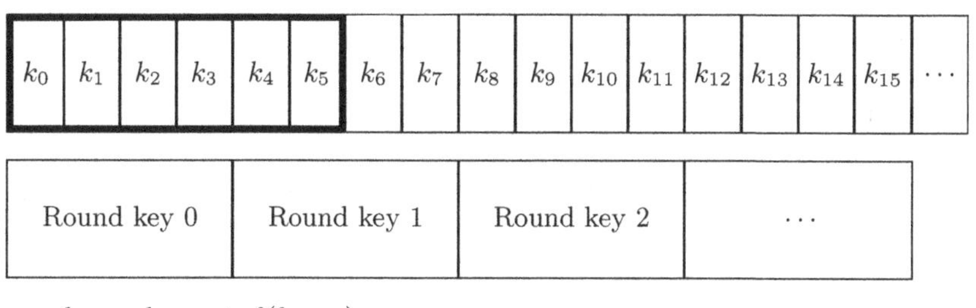

$$
\begin{aligned}
k_{6n} &= k_{6n-6} + f(k_{6n-1}) \\
k_i &= k_{i-6} + k_{i-1}, \quad i \neq 6n
\end{aligned}
$$

Fig. 3.10. Key expansion and round key selection for $N_b = 4$ and $N_k = 6$

```
procedure KeyExpansion(byte K[4][Nk], byte W[4][Nb(Nr + 1)])      ▷ Nk ≤ 6
    for j = 0 to Nk − 1 do
        for i = 0 to 3 do
            W[i][j] ← K[i][j]
        end for
    end for
    for j = Nk to Nb(Nr + 1) − 1 do
        if j mod Nk = 0 then
            W[0][j] ← W[0][j − Nk] + SRD[W[1][j − 1]] + RC[j/Nk]
            for i = 1 to 3 do
                W[i][j] ← W[i][j − Nk] + SRD[W[i + 1 mod 4][j − 1]]
            end for
        else
            for i = 0 to 3 do
                W[i][j] ← W[i][j − Nk] + W[i][j − 1]
            end for
        end if
    end for
end procedure
```

List. 3.3. The key expansion for $N_k \leq 6$

3.7 Decryption

The algorithm for decryption can be found in a straightforward way by using the inverses of the steps `InvSubBytes`, `InvShiftRows`, `InvMixColumns` and `AddRoundKey`, and reversing their order. We call the resulting algorithm the *straightforward decryption algorithm.* In this algorithm, not only the steps themselves differ from those used in encryption, but also the sequence in which the steps occur is different. For implementation reasons, it is often convenient that the only nonlinear step (`SubBytes`) is the first step of the round transformation (see Chap. 4). This aspect has been anticipated in the design. The structure of Rijndael is such that it is possible to define an *equivalent algorithm for decryption* in which the sequence of steps is equal to that for encryption, with the steps replaced by their inverses and a change in the key schedule. We illustrate this in Sect. 3.7.1–3.7.3 for a reduced version of Rijndael that consists of only one round followed by the final round. Note that this identity in *structure* differs from the identity of *components* and *structure* (cf. Sect. 5.3.5) that is found in most ciphers with the Feistel structure, but also in IDEA [93].

3.7.1 Decryption for a Two-Round Rijndael Variant

The straightforward decryption algorithm with a two-round Rijndael variant consists of the inverse of `FinalRound`, followed by the inverse of `Round`,

```
procedure KeyExpansion(byte K[4][Nk], byte W[4][Nb(Nr + 1)])      ▷ Nk > 6
    for j = 0 to Nk − 1 do
        for i = 0 to 3 do
            W[i][j] ← K[i][j]
        end for
    end for
    for j = Nk to Nb(Nr + 1) − 1 do
        if j mod Nk = 0 then
            W[0][j] ← W[0][j − Nk] + SRD[W[1][j − 1]] + RC[j/Nk]
            for i = 1 to 3 do
                W[i][j] ← W[i][j − Nk] + SRD[W[i + 1 mod 4][j − 1]]
            end for
        else if j mod Nk = 4 then
            for i = 0 to 3 do
                W[i][j] ← W[i][j − Nk] + SRD[W[i][j − 1]]
            end for
        else
            for i = 0 to 3 do
                W[i][j] ← W[i][j − Nk] + W[i][j − 1]
            end for
        end if
    end for
end procedure
```

List. 3.4. The key expansion for $N_k > 6$

followed by a key addition. The inverse transformation of `Round` is denoted `InvRound`. The inverse of `FinalRound` is denoted `InvFinalRound`. Both transformations are described in List. 3.5. Listing 3.6 gives the straightforward decryption algorithm for the two-round Rijndael variant.

```
procedure InvRound(State,ExpandedKey[i])
    AddRoundKey(State,ExpandedKey[i])
    InvMixColumns(State)
    InvShiftRows(State)
    InvSubBytes(State)
end procedure
procedure InvFinalRound(State,ExpandedKey[Nr])
    AddRoundKey(State,ExpandedKey[Nr])
    InvMixColumns(State)
    InvShiftRows(State)
    InvSubBytes(State)
end procedure
```

List. 3.5. Round transformations of the straightforward decryption algorithm

```
AddRoundKey(State,ExpandedKey[2])
InvShiftRows(State)
InvSubBytes(State)
AddRoundKey(State,ExpandedKey[1])
InvMixColumns(State)
InvShiftRows(State)
InvSubBytes(State)
AddRoundKey(State,ExpandedKey[0])
```

List. 3.6. Straightforward decryption algorithm for a two-round variant

3.7.2 Algebraic Properties

In order to derive the equivalent decryption algorithm, we use two properties of the steps:

1. The order of InvShiftRows and InvSubBytes is irrelevant.
2. The order of AddRoundKey and InvMixColumns can be inverted if the round key is adapted accordingly.

The first property can be explained as follows. InvShiftRows simply transposes the bytes and has no effect on the byte values. InvSubBytes operates on individual bytes, independent of their position. Therefore, the two steps commute.

The explanation of the second property is somewhat more sophisticated. For any linear transformation $A : \mathbf{x} \rightarrow \mathbf{y} = A(\mathbf{x})$, it holds by definition that

$$A(\mathbf{x} + \mathbf{k}) = A(\mathbf{x}) + A(\mathbf{k}). \tag{3.22}$$

Since AddRoundKey simply adds the constant ExpandedKey[i] to its input, and InvMixColumns is a linear operation, the sequence of steps

```
AddRoundKey(State,ExpandedKey[i])
InvMixColumns(State)
```

can be replaced by the following equivalent sequence of steps:

```
InvMixColumns(State)
AddRoundKey(State,EqExpandedKey[i])
```

where EqExpandedKey[i] is obtained by applying InvMixColumns to ExpandedKey[i]. This is illustrated graphically in Fig. 3.11.

3.7.3 The Equivalent Decryption Algorithm

Using the properties described above, we can transform the straightforward decryption algorithm given in List. 3.6 into the algorithm given in List. 3.7. Comparing List. 3.7 with the definition of the original round transformations

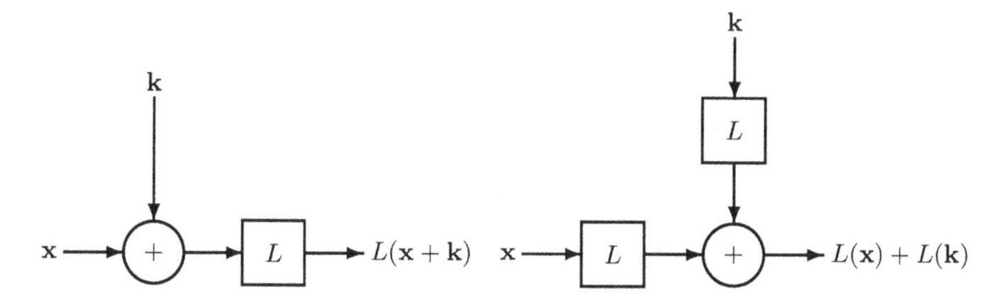

Fig. 3.11. A linear transformation L can be 'pushed through' an XOR

`Round` and `FinalRound` (List. 3.2), we see that we can regroup the operations of List. 3.7 into an initial key addition, a `Round`-like transformation and a `FinalRound`-like transformation. The `Round`-like transformation and the `FinalRound`-like transformation have the same structure as `Round` and `FinalRound`, but they use the inverse transformations. We can generalize this regrouping to any number of rounds.

```
AddRoundKey(State,ExpandedKey[2])
InvSubBytes(State)
InvShiftRows(State)
InvMixColumns(State)
AddRoundKey(State,ExpandedKey[1])
InvSubBytes(State)
InvShiftRows(State)
AddRoundKey(State,ExpandedKey[0])
```

List. 3.7. Equivalent decryption algorithm for a two-round variant

We define the equivalent round transformation `EqRound` and the equivalent final round transformation `EqFinalRound` to use in the equivalent decryption algorithm. The transformations are described in List. 3.8. Listing 3.9 gives the equivalent decryption algorithm. Figure 3.12 shows a graphical illustration of encryption with the two-round Rijndael variant, decryption according to the straightforward algorithm and decryption according to the equivalent algorithm. The dashed boxes enclose the steps that can be implemented together efficiently. In the straightforward decryption algorithm, the (inverse) steps appear in the wrong order and cannot be implemented as efficiently. By changing the order of `InvShiftRows` and `InvSubBytes`, and by pushing `MixColumns` through the XOR of `AddRoundKey`, the equivalent decryption algorithm is obtained. This structure has again the operations in a good order for efficient implementation.

```
procedure EqRound(State,EqExpandedKey[i])
    InvSubBytes(State)
    InvShiftRows(State)
    InvMixColumns(State)
    AddRoundKey(State,EqExpandedKey[i])
end procedure
procedure EqFinalRound(State,EqExpandedKey[0])
    InvSubBytes(State)
    InvShiftRows(State)
    AddRoundKey(State,EqExpandedKey[0])
end procedure
```

List. 3.8. Round transformations for the equivalent decryption algorithm

```
procedure InvRijndael(State,CipherKey)
    EqKeyExpansion(CipherKey,EqExpandedKey)
    AddRoundKey(State,EqExpandedKey[Nr])
    for i = Nr − 1 downto 1 do
        EqRound(State,EqExpandedKey[i])
    end for
    EqFinalRound(State,EqExpandedKey[0])
end procedure
```

List. 3.9. Equivalent decryption algorithm

`EqKeyExpansion`, the key expansion to be used in conjunction with the equivalent decryption algorithm, is defined as follows:

1. Apply the key expansion `KeyExpansion`.
2. Apply `InvMixColumns` to all round keys except the first one and the last one.

Listing 3.10 lists `EqKeyExpansion`.

```
procedure EqKeyExpansion(CipherKey,EqExpandedKey)
    KeyExpansion(CipherKey,EqExpandedKey)
    for i = 1 to Nr − 1 do
        InvMixColumns(EqExpandedKey[i])
    end for
end procedure
```

List. 3.10. Key expansion for the equivalent decryption algorithm

3.8 Conclusions

In this chapter we have given the specification of Rijndael encryption and decryption, and the motivation for some of the design choices.

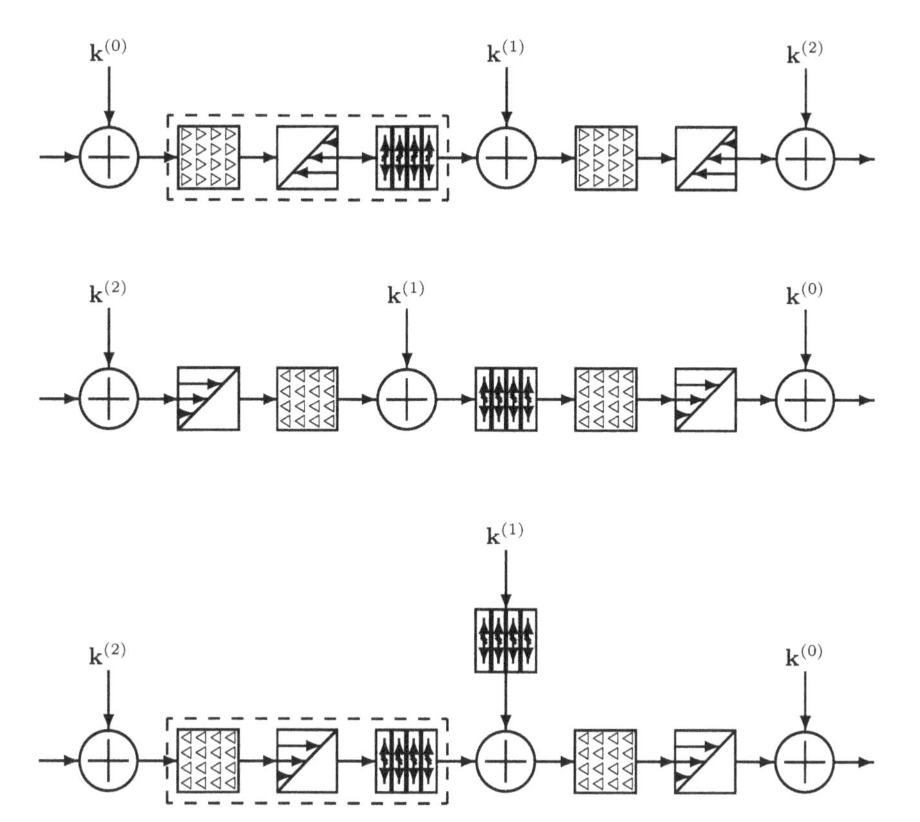

Fig. 3.12. Graphical representation of the algorithm for a two-round Rijndael variant: encryption *(top)*, decryption in the straightforward way *(middle)* and decryption in the equivalent way *(bottom)*. *Dashed boxes* enclose operations that can be implemented together efficiently

4. Implementation Aspects

In this chapter we discuss issues related to the implementation of Rijndael on different platforms. Most topics apply also to related ciphers such as Square, Anubis and Crypton, which are discussed in Chap. 11. We have grouped the material of this chapter into sections that deal with the most typical issues for one specific platform each. However, several of the discussed issues are relevant to more than one platform. If you want to squeeze out the best possible performance, we advise reading the whole chapter, with a critical mindset.

4.1 Eight-Bit Platforms

The performance on 8-bit processors is an important issue, since many cryptographic applications still run on such low-end processors.

4.1.1 Finite-Field Multiplication

In the algorithm of Rijndael there are no multiplications of two variables in $\mathrm{GF}(2^8)$, but only the multiplication of a variable with a constant. The latter is easier to implement than the former.

We describe here how multiplication by the value 2 can be implemented. The polynomial associated with 2 is x. Therefore, if we multiply an element b with 2, we get

$$\begin{aligned} b \cdot x &= b_7x^8 + b_6x^7 + b_5x^6 + b_4x^5 \\ &\quad + b_3x^4 + b_2x^3 + b_1x^2 + b_0x \pmod{m(x)} \qquad (4.1) \\ &= b_6x^7 + b_5x^6 + b_4x^5 + (b_3 + b_7)x^4 \\ &\quad + (b_2 + b_7)x^3 + b_1x^2 + (b_0 + b_7)x + b_7. \qquad (4.2) \end{aligned}$$

The multiplication by 2 is denoted `xtime`(x). `xtime` can be implemented with a shift operation and a conditional XOR operation. To prevent timing attacks, attention must be paid so that `xtime` is implemented in such a way that it takes a fixed number of cycles, independently of the value of

J. Daemen, V. Rijmen, *The Design of Rijndael*, Information Security and Cryptography,
https://doi.org/10.1007/978-3-662-60769-5_4

its argument. This can be achieved by inserting dummy instructions at the right places. However, this approach is likely to introduce weaknesses against power analysis attacks (see Sect. 10.9.2). A better approach seems to be to define a table M, where $M[a] = 2 \cdot a$. The routine `xtime` is then implemented as a table look-up using M.

Since all elements of $\mathrm{GF}(2^8)$ can be written as a sum of powers of 2, multiplication by any constant value can be implemented by a repeated use of `xtime`.

Example 4.1.1. The multiplication of an input b by the constant value `15` can be implemented as follows:

$$\begin{aligned} b \cdot \mathtt{15} &= b \cdot (\mathtt{1} + \mathtt{4} + \mathtt{10}) \\ &= b \cdot (\mathtt{1} + \mathtt{2}^2 + \mathtt{2}^4) \\ &= b + \mathtt{xtime}(\mathtt{xtime}(b)) + \mathtt{xtime}(\mathtt{xtime}(\mathtt{xtime}(\mathtt{xtime}(b)))) \\ &= b + \mathtt{xtime}(\mathtt{xtime}(b + \mathtt{xtime}(\mathtt{xtime}(b)))). \end{aligned}$$

4.1.2 Encryption

On an 8-bit processor, encryption with Rijndael can be programmed by simply implementing the different steps. The implementation of `ShiftRows` and `AddRoundKey` is straightforward from the description. The implementation of `SubBytes` requires a table of 256 bytes to store $\mathrm{S_{RD}}$.

`AddRoundKey`, `SubBytes` and `ShiftRows` can be efficiently combined and executed serially per state byte. Indexing overhead is minimized by explicitly coding the operation for every state byte.

`MixColumns`. In choosing the `MixColumns` polynomial, we took into account the efficiency on 8-bit processors. We illustrate in List. 4.1 how `MixColumns` can be realized in a small series of instructions. (The listing gives the algorithm to process one column.) The only finite-field multiplication used in this algorithm is multiplication with the element 2, denoted by '`xtime`'.

$t \leftarrow a[0] + a[1] + a[2] + a[3]$ ▷ a is a column
$u \leftarrow a[0]$
$v \leftarrow a[0] + a[1]$; $v \leftarrow \mathtt{xtime}(v)$; $a[0] \leftarrow a[0] + v + t$
$v \leftarrow a[1] + a[2]$; $v \leftarrow \mathtt{xtime}(v)$; $a[1] \leftarrow a[1] + v + t$
$v \leftarrow a[2] + a[3]$; $v \leftarrow \mathtt{xtime}(v)$; $a[2] \leftarrow a[2] + v + t$
$v \leftarrow a[3] + u$; $v \leftarrow \mathtt{xtime}(v)$; $a[3] \leftarrow a[3] + v + t$

List. 4.1. Efficient implementation of `MixColumns`

The key expansion. Implementing the key expansion in a single-shot operation is likely to occupy too much RAM in a low-end processor. Moreover, in most low-end applications the amount of data to be encrypted and decrypted or which is subject to a message authentication code (MAC) is typically only a few blocks per session. Hence, not much performance can be gained by storing the expanded key instead of regenerating it for every application of the block cipher.

In the design of the key schedule, we took into account the restrictions imposed by low-end processors. The key expansion can be implemented using a cyclic buffer of $4N_k$ bytes. When all bytes of the buffer have been used, the buffer content is updated. All operations in this key update can be implemented efficiently with byte-level operations.

4.1.3 Decryption

For implementations on 8-bit platforms, there is no benefit in following the equivalent decryption algorithm. Instead, the straightforward decryption algorithm is followed.

`InvMixColumns`. Decryption is similar in structure to encryption, but uses the `InvMixColumns` step instead of `MixColumns`. Where the `MixColumns` coefficients are limited to `1`, `2` and `3`, the coefficients of `InvMixColumns` are `9`, `E`, `B` and `D`. In our 8-bit implementation, these multiplications take significantly more time and this results in a performance degradation of the 8-bit implementation. A considerable speed-up can be obtained by using look-up tables at the cost of additional tables.

P. Barreto observes the following relation between the `MixColumns` polynomial $c(x)$ and the `InvMixColumns` polynomial $d(x)$:

$$d(x) = (4x^2 + 5)c(x) \pmod{x^4 + 1}. \tag{4.3}$$

In matrix notation, this relation becomes:

$$\begin{bmatrix} \mathtt{E} & \mathtt{B} & \mathtt{D} & \mathtt{9} \\ \mathtt{9} & \mathtt{E} & \mathtt{B} & \mathtt{D} \\ \mathtt{D} & \mathtt{9} & \mathtt{E} & \mathtt{B} \\ \mathtt{B} & \mathtt{D} & \mathtt{9} & \mathtt{E} \end{bmatrix} = \begin{bmatrix} \mathtt{2} & \mathtt{3} & \mathtt{1} & \mathtt{1} \\ \mathtt{1} & \mathtt{2} & \mathtt{3} & \mathtt{1} \\ \mathtt{1} & \mathtt{1} & \mathtt{2} & \mathtt{3} \\ \mathtt{3} & \mathtt{1} & \mathtt{1} & \mathtt{2} \end{bmatrix} \times \begin{bmatrix} \mathtt{5} & \mathtt{0} & \mathtt{4} & \mathtt{0} \\ \mathtt{0} & \mathtt{5} & \mathtt{0} & \mathtt{4} \\ \mathtt{4} & \mathtt{0} & \mathtt{5} & \mathtt{0} \\ \mathtt{0} & \mathtt{4} & \mathtt{0} & \mathtt{5} \end{bmatrix}. \tag{4.4}$$

The consequence is that `InvMixColumns` can be implemented as a simple preprocessing step, followed by a `MixColumns` step. An algorithm for the preprocessing step is given in List. 4.2. If the performance drop caused by this implementation of the preprocessing step is acceptable, no extra tables have to be defined.

```
u ← xtime(xtime(a[0] + a[2]))                    ▷ a is a column
v ← xtime(xtime(a[1] + a[3]))
a[0] ← a[0] + u
a[1] ← a[1] + v
a[2] ← a[2] + u
a[3] ← a[3] + v
```

List. 4.2. Preprocessing step for implementation of the decryption

The key expansion. The key expansion operation that generates W is defined in such a way that we can also start with the last N_k words of round key information and roll back to the original cipher key. When applications need to calculate frequently the decryption round keys 'on the fly', it is preferable to calculate the last N_k words of round key information once and store them for later reuse. The decryption round key calculation can then be implemented in such a way that it outputs the round keys in the order they are needed for the decryption process. Listings. 4.3 and 4.4 give a description of `InvKeyExpansion` in pseudo C notation. First note that K_i, the first input of the routine, is *not* the cipher key. Instead, K_i consists of the last N_k columns of the expanded key, generated from the cipher key by means of `KeyExpansion` (see Sect. 3.6). After running `InvKeyExpansion`, W_i contains the decryption round keys in the order they are used for decryption, i.e. columns with lower indices are used first. Secondly, note that this is the key expansion for use in conjunction with the straightforward decryption algorithm. If the equivalent decryption algorithm is implemented, all but two of the round keys have additionally to be transformed by `InvMixColumns` (see Sect. 3.7.3).

4.2 Thirty-Two-Bit Platforms

4.2.1 T-Table Implementation

The different steps of the round transformation can be combined in a single set of look-up tables, allowing for very fast implementations on processors with word lengths 32 or greater. In this section, we explain how this can be done.

Let the input of the round transformation be denoted by **a**, and the output of `SubBytes` by **b**:

$$b_{i,j} = \mathrm{S_{RD}}[a_{i,j}], \quad 0 \le i < 4; 0 \le j < \mathrm{N_b}. \tag{4.5}$$

Let the output of `ShiftRows` be denoted by **c** and the output of `MixColumns` by **d**:

```
procedure InvKeyExpansion(byte Ki[4][Nk], byte Wi[4][Nb(Nr + 1)])
                                                              ▷ Nk ≤ 6
    for j = 0 to Nk − 1 do
        for i = 0 to 3 do
            Wi[i][j] ← Ki[i][j]
        end for
    end for
    for j = Nk to Nb(Nr + 1) − 1 do
        if j mod Nk = 0 then
            Wi[0][j] ← Wi[0][j − Nk] + SRD[Wi[1][j − 1] + Wi[1][j − 2]]
                                                    + RC[Nr + 1 − j/Nk]
            for i = 1 to 3 do
                Wi[i][j] ← Wi[i][j − Nk] + SRD[Wi[i + 1 mod 4][j − 1]
                                                + Wi[i + 1 mod 4][j − 2]]
            end for
        else
            for i = 0 to 3 do
                Wi[i][j] ← Wi[i][j − Nk] + Wi[i][j − Nk − 1]
            end for
        end if
    end for
end procedure
```

List. 4.3. Algorithm for the inverse key expansion for $N_k \leq 6$

```
procedure InvKeyExpansion(byte Ki[4][Nk], byte Wi[4][Nb(Nr + 1)])
                                                              ▷ Nk > 6
    for j = 0 to Nk − 1 do
        for i = 0 to 3 do
            Wi[i][j] ← Ki[i][j]
        end for
    end for
    for j = Nk to Nb(Nr + 1) − 1 do
        if j mod Nk = 0 then
            Wi[0][j] ← Wi[0][j − Nk] + SRD[Wi[1][j − 1] + Wi[1][j − 2]]
                                                    + RC[Nr + 1 − j/Nk]
            for i = 1 to 3 do
                Wi[i][j] ← Wi[i][j − Nk] + SRD[Wi[i + 1 mod 4][j − 1]
                                                + Wi[i + 1 mod 4][j − 2]]
            end for
        else if j mod Nk = 4 then
            for i = 0 to 3 do
                Wi[i][j] ← Wi[i][j − Nk] + SRD[Wi[i][j − Nk − 1]]
            end for
        else
            for i = 0 to 3 do
                Wi[i][j] ← Wi[i][j − Nk] + Wi[i][j − Nk − 1]
            end for
        end if
    end for
end procedure
```

List. 4.4. Algorithm for the inverse key expansion for $N_k > 6$

$$\begin{bmatrix} c_{0,j} \\ c_{1,j} \\ c_{2,j} \\ c_{3,j} \end{bmatrix} = \begin{bmatrix} b_{0,j+C_0} \\ b_{1,j+C_1} \\ b_{2,j+C_2} \\ b_{3,j+C_3} \end{bmatrix}, \quad 0 \leq j < \mathrm{N_b} \tag{4.6}$$

$$\begin{bmatrix} d_{0,j} \\ d_{1,j} \\ d_{2,j} \\ d_{3,j} \end{bmatrix} = \begin{bmatrix} 2 & 3 & 1 & 1 \\ 1 & 2 & 3 & 1 \\ 1 & 1 & 2 & 3 \\ 3 & 1 & 1 & 2 \end{bmatrix} \cdot \begin{bmatrix} c_{0,j} \\ c_{1,j} \\ c_{2,j} \\ c_{3,j} \end{bmatrix}, \quad 0 \leq j < \mathrm{N_b}. \tag{4.7}$$

The addition in the indices of (4.6) must be done modulo $\mathrm{N_b}$. Equations (4.5)–(4.7) can be combined into

$$\begin{bmatrix} d_{0,j} \\ d_{1,j} \\ d_{2,j} \\ d_{3,j} \end{bmatrix} = \begin{bmatrix} 2 & 3 & 1 & 1 \\ 1 & 2 & 3 & 1 \\ 1 & 1 & 2 & 3 \\ 3 & 1 & 1 & 2 \end{bmatrix} \cdot \begin{bmatrix} \mathrm{S_{RD}}[a_{0,j+C_0}] \\ \mathrm{S_{RD}}[a_{1,j+C_1}] \\ \mathrm{S_{RD}}[a_{2,j+C_2}] \\ \mathrm{S_{RD}}[a_{3,j+C_3}] \end{bmatrix}, \quad 0 \leq j < \mathrm{N_b}. \tag{4.8}$$

The matrix multiplication can be interpreted as a linear combination of four column vectors:

$$\begin{bmatrix} d_{0,j} \\ d_{1,j} \\ d_{2,j} \\ d_{3,j} \end{bmatrix} = \begin{bmatrix} 2 \\ 1 \\ 1 \\ 3 \end{bmatrix} \mathrm{S_{RD}}[a_{0,j+C_0}] + \begin{bmatrix} 3 \\ 2 \\ 1 \\ 1 \end{bmatrix} \mathrm{S_{RD}}[a_{1,j+C_1}] + \begin{bmatrix} 1 \\ 3 \\ 2 \\ 1 \end{bmatrix} \mathrm{S_{RD}}[a_{2,j+C_2}] + \begin{bmatrix} 1 \\ 1 \\ 3 \\ 2 \end{bmatrix} \mathrm{S_{RD}}[a_{3,j+C_3}], \quad 0 \leq j < \mathrm{N_b}. \tag{4.9}$$

We define now the four T-tables: T_0, T_1, T_2 and T_3:

$$T_0[a] = \begin{bmatrix} 2 \cdot \mathrm{S_{RD}}[a] \\ 1 \cdot \mathrm{S_{RD}}[a] \\ 1 \cdot \mathrm{S_{RD}}[a] \\ 3 \cdot \mathrm{S_{RD}}[a] \end{bmatrix}, \quad T_1[a] = \begin{bmatrix} 3 \cdot \mathrm{S_{RD}}[a] \\ 2 \cdot \mathrm{S_{RD}}[a] \\ 1 \cdot \mathrm{S_{RD}}[a] \\ 1 \cdot \mathrm{S_{RD}}[a] \end{bmatrix}, \tag{4.10}$$

$$T_2[a] = \begin{bmatrix} 1 \cdot \mathrm{S_{RD}}[a] \\ 3 \cdot \mathrm{S_{RD}}[a] \\ 2 \cdot \mathrm{S_{RD}}[a] \\ 1 \cdot \mathrm{S_{RD}}[a] \end{bmatrix}, \quad T_3[a] = \begin{bmatrix} 1 \cdot \mathrm{S_{RD}}[a] \\ 1 \cdot \mathrm{S_{RD}}[a] \\ 3 \cdot \mathrm{S_{RD}}[a] \\ 2 \cdot \mathrm{S_{RD}}[a] \end{bmatrix}. \tag{4.11}$$

These tables each have 256 4-byte word entries and require 4 kB of storage space. Using these tables, (4.9) translates into

$$\begin{bmatrix} d_{0,j} \\ d_{1,j} \\ d_{2,j} \\ d_{3,j} \end{bmatrix} = T_0[a_{0,j+C_0}] + T_1[a_{1,j+C_1}] + T_2[a_{2,j+C_2}] + T_3[a_{3,j+C_3}], \quad 0 \leq j < \mathrm{N_b}. \tag{4.12}$$

Taking into account that AddRoundKey can be implemented with an additional 32-bit XOR operation per column, we get a look-up table implementation with 4 kB of tables that takes only four table look-ups and four XOR operations per column per round.

Furthermore, the entries $T_0[a], T_1[a], T_2[a]$ and $T_3[a]$ are rotated versions of one another, for all values a. Consequently, at the cost of three additional rotations per round per column, the look-up table implementation can be realized with only one table, i.e. with a total table size of 1 kB. The size of the encryption routine (relevant in applets) can be kept small by including a program to generate the tables instead of the tables themselves.

In the final round, there is no MixColumns operation. This boils down to the fact that $\mathrm{S_{RD}}$ must be used instead of the T-tables. The need for additional tables can be suppressed by extracting the $\mathrm{S_{RD}}$-table from a T-table by masking while executing the final round.

Most operations in the key expansion are 32-bit XOR operations. The additional transformations are the application $\mathrm{S_{RD}}$ and a cyclic shift over 8 bits. This can be implemented very efficiently.

Decryption can be described in terms of the transformations EqRound and EqFinalRound used in the equivalent decryption algorithm. These can be implemented with look-up tables in exactly the same way as the transformations Round and FinalRound. There is no performance degradation compared to encryption. The look-up tables for the decryption are however different. The key expansion to be used in conjunction with the equivalent decryption algorithm is slower, because after the key expansion all but two of the round keys are subject to InvMixColumns (cf. Sect. 3.7).

4.2.2 Bitsliced Software

As discussed in Sect. 10.9.1, the most efficient AES implementations may be vulnerable to timing attacks on processors with cache memory due to table look-ups. This led to AES software that avoids table look-ups altogether.

The technique to avoid table look-ups is called *bitslicing* and was first proposed in [19] for DES. Its basic principle is simple. Each bit of the state and the key is put in a separate CPU word. Taking a bit-level description, e.g. for dedicated hardware, one can now program DES by executing solely bitwise Boolean instructions. This can be made efficient by grouping multiple DES instances in the CPU words. On a CPU supporting 32-bit bitwise Boolean instructions, each CPU word then contains the bits in a specific position of 32 different DES instances. A straightforward implementation of DES in this way would require $64 + 56 = 120$ words. For this bitsliced implementation to be efficient, 32 DES instances must be available that can be computed in parallel. This rules out modes such as CBC or OFB that are strictly serial. Counter mode on the other hand lends itself ideally to this.

In [76] E. Käsper and P. Schwabe applied the bitslice technique to AES. We briefly summarize their approach here and refer the interested reader to the original publication for a more in-depth treatment. To get an efficient bit-level description, they based their approach on the tower field S-box representations of [37]. Additionally, they exploited the large amount of symmetry in AES to reduce the number of parallel instances to eight of the high-end Intel CPUs of the era. As a matter of fact, `SubBytes`, `MixColumns` and `AddRoundKey` treat the 16 bytes of the state in a fully symmetric way, allowing the arrangement of the ith bit of the 16 state bytes in a single CPU word. For `MixColumns` this requires some additional shuffling of bits within a word as a bit depends on bits in all four bytes of its column. `ShiftRows` can be implemented with the same shuffling operations.

4.3 Dedicated Hardware

Rijndael is suited to be implemented in dedicated hardware. Several trade-offs between chip area and speed are possible. Because the implementation in software on general-purpose processors is already very fast, the need for hardware implementations will very probably be limited to two specific cases:

1. Extremely high-speed chip with no area restrictions: the T-tables can be hardwired and the XOR operations can be conducted in parallel.
2. Compact coprocessor on a low-end platform to speed up Rijndael execution: for such platforms typically S_{RD} and the `xtime` (or the complete `MixColumns`) operation can be hardwired.

In dedicated hardware, `xtime` can be implemented with the combination of a hardwired bit transposition and four XOR gates. The `SubBytes` step is the most critical part for a hardware implementation, for two reasons:

1. In order to achieve the highest performance, S_{RD} needs to be instantiated 16 times (disregarding the key schedule). A straightforward implementation with 16 256-byte tables is likely to dominate the chip area requirements or the consumption of logic blocks.
2. Since Rijndael encryption and decryption use different transformations, a circuit that implements Rijndael encryption does not automatically support decryption.

However, when building dedicated hardware for supporting both encryption and decryption, we can limit the required chip area by using parts of the circuit for both transformations. In the following, we explain how S_{RD} and ${S_{RD}}^{-1}$ can be implemented efficiently.

4.3.1 Decomposition of S_{RD}

The Rijndael S-box S_{RD} is constructed from two transformations:

$$S_{RD}[a] = \mathrm{Aff}_8(\mathrm{Inv}_8(a)) \ , \tag{4.13}$$

where $\mathrm{Inv}_8(a)$ is the transformation

$$a \to a^{-1} \text{ in GF}(2^8), \tag{4.14}$$

and $\mathrm{Aff}_8(a)$ is an affine transformation. The transformation $\mathrm{Inv}_8(a)$ is self-inverse and hence

$$S_{RD}{}^{-1}[a] = \mathrm{Inv}_8^{-1}(\mathrm{Aff}_8^{-1}(a)) = \mathrm{Inv}_8(\mathrm{Aff}_8^{-1}(a)) \ . \tag{4.15}$$

Therefore, when we want both S_{RD} and $S_{RD}{}^{-1}$, we need to implement only Inv_8, Aff_8 and Aff_8^{-1}. Since both Aff_8 and Aff_8^{-1} can be implemented with a limited number of XOR gates, the extra hardware can be reduced significantly compared to having to hardwire both S_{RD} and $S_{RD}{}^{-1}$.

The affine transformations Aff_8 and Aff_8^{-1} are defined in Sect. 3.4.1. For ease of reference, we give a tabular description of the functions Aff_8, Aff_8^{-1} and Inv_8 in Appendix A.

4.3.2 Efficient Inversion in GF(2^8)

The problem of designing efficient circuits for inversion in finite fields has been studied extensively before; e.g. by C. Paar and M. Rosner in [124]. We summarize here a possible approach.

Every element of GF(2^8) can be mapped by a linear transformation to an element of $\mathrm{GF}(2^4)^2$, i.e. a polynomial of degree one with coefficients in GF(2^4). In order to define multiplication in $\mathrm{GF}(2^4)^2$, we need a polynomial of degree two that is irreducible over GF(2^4). There exist irreducible polynomials of the form

$$P(x) = x^2 + x + A. \tag{4.16}$$

Here 'A' is a constant element of GF(2^4) that can be chosen to optimize the hardware, as long as $P(x)$ stays irreducible. The inverse of an arbitrary element $(bx + c)$ is then given by the polynomial $(px + q)$ iff

$$\begin{aligned} 1 &= (bx + c) \cdot (px + q) \bmod P(x) && (4.17) \\ &= (cp + bq + bp)x + (cq + bpA). && (4.18) \end{aligned}$$

This gives a set of linear equations in p and q, with the following solution:

$$\begin{cases} p = b(Ab^2 + bc + c^2)^{-1} \\ q = (c + b)(Ab^2 + bc + c^2)^{-1}. \end{cases} \tag{4.19}$$

The problem of generating an inverse in GF(2^8) has been translated into the calculation of an inverse and some operations in GF(2^4). The calculation of an inverse in GF(2^4) can be done with a small table.

The basic approach sketched above has been extended in several ways and optimized [136, 111, 36, 119]. The extensions can be summarized as follows. Firstly, in order to work without any table, one can apply the sketched approach recursively and translate the generation of inverses in GF(2^4) into the calculation of an inverse and some operations in GF(2^2). The calculation of an inverse in GF(2^2), finally, can be done with a few gates. Secondly, there are in total 432 choices of basis in GF(2^8) and GF(2^4) to consider: there are more polynomial bases than described by (4.16), and also normal bases can be considered. A similar optimization effort, with a different outcome, is done for the case of software bitsliced implementations in [37]. Finally, for very compact (but slow) implementations, one can exploit the rotational symmetry of inversion in normal bases [131].

4.3.3 AES-NI

In 2008, Intel and AMD introduced the AES New Instructions (AES-NI) [68]. Later, other processors (SPARC, ARM, IBM, ...) introduced different instructions with similar functionality. The AES-NI instructions combine the steps of the round transformation into a single processor instruction. The set includes instructions to compute one round of AES encryption or decryption, to compute the special last round of AES encryption or decryption, and to perform parts of the AES keyscheduling and `InvMixColumns`.

These instructions can be used to speed up software implementations of AES, but their effect is most prominent when AES is used in a parallel mode of operation. The AES encryption round instruction has a latency of eight cycles. Although a new input can be sent to the instruction in every cycle, the result is available only eight cycles later. (Early versions of the AES-NI instructions had a latency of only six cycles, but could be called only once every two cycles [16, 15]). It follows that the previously popular cipher block chaining (CBC) mode of operation does not enjoy the same speed-up as a parallel mode such as counter mode.

4.4 Multiprocessor Platforms

There is considerable parallelism in the round transformation. All four steps of the round act in a parallel way on bytes, rows or columns of the state. In the look-up table implementation, all table look-ups can in principle be done in parallel. The XOR operations can be done mostly in parallel as well.

The key expansion is clearly of a more sequential nature: the value of $\mathsf{W}[i-1]$ is needed for the computation of $\mathsf{W}[i]$. However, in most applications where speed is critical, the key expansion has to be done only once for a large number of cipher executions. In applications where the cipher key changes often (in extremis, once per application of the block cipher), the key expansion and the cipher rounds can be done in parallel.

A study by C. Clapp [41] indicates that the performance of Rijndael on parallel processors is not constrained by the critical path length. Instead, the limiting factor for Rijndael implementations is the number of memory references that can be done per cycle.

4.5 Conclusions

In this chapter we have shown how Rijndael can be efficiently implemented in dedicated hardware and in software on a wide variety of processors.

5. Design Philosophy

In this chapter we motivate the choices we have made in the process of designing Rijndael and its predecessors. We start with discussing the criteria that are widely considered important for block ciphers, such as security and efficiency. After that, we introduce the criterion of simplicity that plays such an important role in our design approach. We explain what we mean by it and why it is so important. A very effective way to keep things simple is by the introduction of symmetry. After discussing different ways of introducing symmetry, we motivate the choice of operations in Rijndael and its predecessors and our approach to security. This is followed by a discussion of what we think it takes to design a block cipher that satisfies our criteria. We conclude this chapter with a discussion on the generation and usage of round keys.

5.1 Generic Criteria in Cipher Design

In this section we describe a number of design criteria that are adopted by most cryptographers.

5.1.1 Security

The most important criterion for a block cipher is *security*, meaning the absence of cryptanalytic attacks that exploit its internal structure. Among other things, this implies the absence of attacks that have a workload smaller than that of an exhaustive search for the key.

5.1.2 Efficiency

The complementary criterion is that of *efficiency*. Efficiency refers to the amount of resources required to perform an encryption or decryption. In dedicated hardware implementations, energy consumption per encryption or per encrypted bit are relevant, as well as encryption and decryption speed and the required chip area. In software implementations, the encryption/decryption speed and the required amount of working memory and program-storage memory are relevant.

J. Daemen, V. Rijmen, *The Design of Rijndael*, Information Security and Cryptography,
https://doi.org/10.1007/978-3-662-60769-5_5

5.1.3 Key Agility

When quoting the speed of a cipher, one often makes the silent assumption that a large amount of data is encrypted with the same key. In that case the key schedule can be neglected. However, if a cipher key is used to secure messages consisting of a few blocks only, the number of cycles taken by the computation of the key schedule becomes important. The ability to efficiently change keys is called *key agility*.

5.1.4 Versatility

Differences in processor word length and instruction sets may cause the efficiency of a cipher to be very dependent on the processor type. As the AES is implemented on smart cards, smart phones, desktop PCs, workstations, routers, set-top boxes, hardware security modules and probably some other types of devices, we have attempted to design a cipher that is efficient on the widest range of processors possible. Although just a qualifier for efficiency, we call this requirement *versatility*.

5.1.5 Discussion

The criteria of security and efficiency are applied by all cipher designers. There are cases in which efficiency is sacrificed to obtain a higher security margin. The challenge is to come up with a cipher design that offers a reasonable security margin while optimizing efficiency.

The criteria of key agility and versatility are less universal. In some cases these criteria are irrelevant because the cipher is meant for a particular application and will be implemented on a specific platform. For the AES — the successor of the ubiquitous DES — we expected key agility and versatility to be major issues. Still, a large part of the ciphers submitted to the AES focus on efficiency of bulk data encryption on 32-bit processors without much attention to 8-bit processors, multiprocessors or dedicated hardware, or an efficient key schedule.

5.2 Simplicity

A notion that characterizes our design philosophy is *simplicity*. The design process can be broken down into a number of decisions and choices. In each of these decisions and choices, the simplicity criterion has played an important role.

We distinguish *simplicity of specification* and *simplicity of analysis*. A specification is simple if it makes use of a limited number of operations and if the operations by themselves can be easily explained. An obvious advantage of a simple specification is that it facilitates a correct implementation. Another advantage is that a cipher with a simple specification seems a more interesting object to study than a cipher with a complex specification. Moreover, the simplicity of the specification may lead people to believe that it is easier to find a successful attack. In other words, the simplicity of a cipher contributes to the appeal it has for cryptanalysts, and in the absence of successful cryptanalysis, to its cryptographic credibility.

Simplicity of analysis corresponds to the ability to demonstrate and understand in what way the cipher offers protection against known types of cryptanalysis. In this way, resistance against known attacks can be covered in the design phase, thereby providing a certain level of cryptographic credibility from the start. This contributes again to the appeal to cryptanalysts: successful cryptanalysis of a cipher with some credibility gives more prestige than cryptanalysis of an insignificant cipher.

Simplicity of specification does not necessarily imply simplicity of analysis. It is relatively easy to come up with a cipher with a very simple description for which the analysis with respect to known attacks is very hard.

On top of the advantages cited above, we use the criterion of simplicity to obtain a good trade-off between security on the one hand and efficiency and versatility on the other hand. This is explained in Sect. 5.3.

Simplicity can be achieved in a number of ways. In the design of Rijndael and its predecessors, we have mostly realized it through the adoption of symmetry and our choice of operations.

5.3 Symmetry

A very powerful tool for introducing simplicity is symmetry. Symmetry can be applied in several ways. We distinguish symmetry across the rounds, symmetry within the round transformation and symmetry in the steps.

5.3.1 Symmetry Across the Rounds

We design a cipher as the repeated iteration of the same keyed round transformation. This approach has the advantage that in specifications only one round transformation needs to be specified, and in software implementations only one round has to be programmed. Moreover, it allows dedicated hardware implementations that only contain a circuit for the round transformation and the key schedule. In Rijndael, the last round is different from the other ones in order to make the algorithms for decryption and encryption have the same structure (see Chap. 4).

One may wonder whether this symmetry cannot be exploited in cryptanalysis. As a matter of fact, the so-called *slide attacks* as described by A. Biryukov and D. Wagner in [31] exploit this kind of symmetry. However, for slide attacks to work, also the key schedule must exhibit a large degree of symmetry. Hence, protection against known slide attacks can already be achieved with a very simple key schedule, e.g. consisting merely of the XOR of well-chosen constants with the cipher key.

5.3.2 Symmetry Within the Round Transformation

Symmetry within the round transformation implies that it treats all bits of the state in a similar way. In the Feistel round structure, as adopted in the DES (see Chap. 6) this is clearly not the case since the two halves of the state are treated quite differently.

A consequence of our design strategy (see Chap. 9) is that the round transformation consists of a sequence of steps, each with its own particular function. For each of these steps, the symmetry requirement translates easily into some concrete restrictions:

1. **Nonlinear step.** A bricklayer transformation consisting of nonlinear S-boxes operating independently on bytes. The symmetry requirement translates easily into the requirement that the same S-box is used for all byte positions.
2. **Mixing step.** A bricklayer transformation consisting of linear D-boxes operating independently on columns. The symmetry requirement translates into the requirement that the same D-box is used for all column positions. Additionally, alignment between bytes and columns may be imposed: all bits in the same byte are also in the same column.
3. **Transposition step.** The transposition step consist of the mere transposition of bytes. Alignment with the nonlinear step may be imposed: the transposition step is a byte transposition rather than a bit transposition.

These symmetry requirements offer a framework in which only the size of the bytes and the columns, the S-box and the D-box, and the byte transposition need to be specified to fully define the round transformation.

Having a large degree of symmetry in the round transformation may lead to cryptographic weaknesses. An example of such a weakness is the complementation property of the DES [71]. If in Rijndael the key application is not taken into account, there exist a number of byte transpositions π that commute with the round transformation. In other words, we have $\pi \circ \rho = \rho \circ \pi$. If all round keys are 0, this is also valid for the complete cipher. The same property holds if each individual round key is composed of bytes that all have the same value. These symmetry properties can however be eliminated by using even a simple key schedule.

Imposing alignment results in the cipher actually operating on bytes rather than bits. As a matter of fact, this property is exploited in some of the most powerful attacks against reduced-round versions of Rijndael and its relatives to date, the so-called saturation attacks, which are described in Sect. 10.2. The saturation attacks form one of the main motivations behind the number of rounds in Rijndael and its relatives.

Note. Instead of translating symmetry into the requirement for byte alignment, as is done for Rijndael and its relatives, one may choose the opposite: non-alignment. In this case the transposition step moves bits belonging to the same byte to bits in different bytes. This is the approach followed for the bitslice ciphers 3-Way [43], BaseKing [45] and Noekeon [46]. Because of the small size of their S-box, these ciphers are very compact in dedicated hardware. In software they are in general slower than Rijndael and its relatives. Perhaps the best-known bitslice cipher is Serpent, which is the AES candidate submitted by E. Biham et al. [4]. The designers of Serpent have not followed the same simplicity strategy: it has eight different S-boxes giving rise to eight different round transformations, and the mixing step has a substantial amount of asymmetry. These factors make it harder to prove bounds for Serpent than for Rijndael and its relatives and the more symmetric bitslice ciphers mentioned above.

5.3.3 Symmetry in the D-Box

Specifying a D-box with the same size as the one used in the mixing step of Rijndael can in general be done with a binary 32×32 matrix, requiring 128 bytes of storage. By interpreting bytes as elements in a finite field, and restricting ourselves to a matrix multiplication over $\mathrm{GF}(2^8)$, the D-box can be specified with 16 byte values. We have imposed that the matrix is a circulant matrix, imposing on the matrix elements $a_{i,j} = a_{0,j-i \bmod n}$ for all i, j. This reduces the number of bytes required to specify the D-box to 4. Other types of symmetry may be imposed. For example, the mixing step of Anubis [9] makes use of a matrix where the matrix elements are indexed with binary strings i and j instead of integers and the matrix elements satisfy $a_{i,j} = a_{0,i \oplus j}$ for all i, j.

5.3.4 Symmetry and Simplicity in the S-Box

For a discussion on the design underlying the S-box and its predecessors used in Rijndael, we refer to Sect. 3.4.1.

5.3.5 Symmetry Between Encryption and Decryption

In general it is an advantage for a block cipher that encryption and decryption can be performed with the same software program or make use of the same hardware circuit. In Feistel ciphers such as the DES (see Chap. 6) this can easily be achieved by omitting the switching of the two halves in the last round. It suffices to execute the rounds taking the round keys in reverse order.

In ciphers that have a round structure like the one of Rijndael, this is less trivial to achieve. For Rijndael and its predecessors, encryption and decryption are different algorithms. Still, in Sect. 3.7 we derive an equivalent decryption algorithm that has the same structure as the encryption algorithm. By selecting the steps of the round transformation in a careful way, it is possible to design a Rijndael-like block cipher that has encryption and decryption algorithms that are identical with the exception of the key schedule. This is illustrated by the design of Anubis [9].

5.3.6 Additional Benefits of Symmetry

In this section we describe a number of benefits that result from the application of symmetry.

Parallelism. A consequence of the symmetry in the different steps is that they all exhibit a large degree of parallelism. The order in which the S-boxes of the nonlinear step are computed is unimportant, and so they may be all computed in parallel. The same argument is valid for the different D-boxes of the mixing step and for the key application. In dedicated hardware implementations of the round transformation, this gives rise to a critical path consisting only of the S-box, the D-box and the XOR of the key addition. In software implementations, this gives the programmer a lot of flexibility in the order in which the computations are executed. Moreover, it allows the efficient exploitation of parallelism supported by multiprocessors, as C. Clapp demonstrated in [41].

Flexibility in the order of steps. The linearity of three of the four steps and the symmetry of the nonlinear step allow even more freedom in the order in which the steps of the round transformation are executed. The transposition step and the mixing step both commute with the key addition under the condition that the key value is adapted to this changed order. On the other hand, thanks to the fact that the nonlinear step has the same effect on all bytes, it commutes with the transposition step. This gives software implementers even more freedom and in fact allows construction of an equivalent algorithm for decryption that has the same structure as the algorithm for encryption (see Sect. 3.7).

Variable block length. Rijndael shares with the AES candidate RC6 [133] the property that it supports different block lengths. In RC6, the state consists of four 32-bit words and these words appear as arguments in multiplication modulo 2^{32}, XOR and cyclic shift. By adopting another word length, the block length can be varied in steps of 4 bits. For example, adopting a word length of 40 leads to a block length of 160 bits.

The symmetry in the steps of Rijndael similarly facilitates the definition of the round transformation for multiple block lengths. The nonlinear step only requires the block length to be a multiple of the S-box width. The mixing step requires the block length to be a multiple of the column size. The key addition does not impose any condition at all. The only step that must be specified explicitly for each block length supported is the byte transposition.

Changing the block length in RC6 may have a dramatic impact on the efficiency of implementations. For example, implementing 40-bit multiplications and cyclic shifts on a 32-bit processor is not trivial. Changing the block length is easy in the specifications, but costly in implementations. In Rijndael, the basic operations and the components of the state keep their length if the block length changes. This means that the block length of Rijndael can be varied with minimal impact on its computational cost per byte, on any platform.

5.4 Choice of Operations

In the specification of Rijndael and its predecessors, we have limited ourselves to relatively simple operations such as XOR and multiplication with constants in $GF(2^8)$. The S-box makes use of the multiplicative inverse in $GF(2^8)$ and an affine transformation.

With this limitation we have excluded a number of simple and efficient operations that are widely used as components in block ciphers and that appear to have excellent nonlinearity and/or diffusion properties. The first class are arithmetic operations such as addition, subtraction and multiplication, most often performed modulo a number of the form 2^n. The second class are cyclic shifts over an offset that depends on state or key bits. We explain our objections against these operations in the following subsections.

5.4.1 Arithmetic Operations

Addition, subtraction and multiplication seem to be simple operations to describe. Moreover, Boolean transformations based on multiplication seem to perform very well with respect to most common nonlinearity and diffusion criteria. Most processors support arithmetic instructions that execute in as few cycles as a simple bitwise XOR.

Unfortunately, if the word length of the processor does not match the length of the operands, either it becomes hard to take full advantage of the processor power due to carry propagation, or limitations in the processing power become apparent. For example, implementing a 32-bit multiplication modulo 2^{32} on an 8-bit processor smart card requires about 10 multiply instructions and 16 addition instructions.

In dedicated hardware, the number of gates required for addition (or subtraction) is about three times that of a bitwise XOR, and due to the carry propagation the gate delay is much larger and even depends on the word length. Implementing multiplication in dedicated hardware appears to give rise to circuits with a large number of gates and a large gate delay.

Another cost that appears is the protection against power analysis attacks. The carry propagation complicates the implementation of certain protection measures against differential power analysis (DPA) that are trivial for XOR, such as balancing (cf. Sect. 10.9.2).

If arithmetic operations are used that operate on numbers that are represented by more than a single byte, one needs to define in what order these bytes must be interpreted as an integer. In processors there are two architectures: *big endian* and *little endian* [142]. Depending on how the order is defined in the cipher specification, one of the two architectures is typically favored. By not using arithmetic operations, an *endian neutral* cipher can be obtained.

5.4.2 Data-Dependent Shifts

Data-dependent shift operations seem to be simple operations to describe. Moreover, Boolean transformations based on data-dependent shifts seem to perform well with respect to most common nonlinearity and diffusion criteria. Many processors support data-dependent (cyclic) shift instructions that execute in a small fixed number of cycles. Unfortunately, if the word length of the processor does not match the length of the operand that is shifted, it takes several instructions to realize the shift operation.

Protection against implementation attacks (see Sect. 10.9) may be very cumbersome on certain platforms. For example, on a typical 8-bit processor the only shift instructions available are those that shift the content of an 8-bit register over 1 bit. A straightforward implementation of a data-dependent shift would execute in a variable number of cycles, depending on the value of the offset. Providing protection against timing attacks can be achieved by inserting dummy instructions, resulting in a constant number of cycles given by the worst-case offset value. Protecting against DPA on top of that seems a non-trivial exercise and may result in a multiplication of the number of cycles by at least a factor of two.

5.5 Approach to Security

5.5.1 Security Goals

In this section, we present the goals we have set for the security of Rijndael. We introduce two security criteria in order to define the meaning of a successful cryptanalytic attack. Note that we cannot prove that Rijndael satisfies these criteria.

In order to formulate our goals, some security-related concepts need to be defined. A block cipher of block length n_b has 2^{n_b} possible inputs. If the key length is n_k, it defines a set of 2^{n_k} permutations. For a block length of n_b, the number of possible permutations is $2^{n_b}!$. Hence the number of all possible block ciphers of dimensions n_b and n_k is

$$(2^{n_b})!^{2^{(n_k)}} . \tag{5.1}$$

For practical values of the dimensions (e.g. n_b and n_k above 40), the subset of block ciphers with exploitable weaknesses form a negligible minority in this set. We define two security properties *K-secure* and *hermetic* as criteria that are satisfied by the majority of block ciphers for the given dimensions.

Definition 5.5.1. *A block cipher is* K-secure *if all possible attack strategies for it have the same expected work factor and storage requirements as for the majority of possible block ciphers with the same dimensions. This must be the case for all possible modes of access for the adversary (known/chosen/adaptively chosen plaintext/ciphertext, known/chosen/adaptively chosen key relations...) and for any a priori key distribution.*

K-security is a very strong notion of security. If one of the following weaknesses apply to a cipher, it cannot be called K-secure:

1. Existence of key-recovering attacks faster than exhaustive search. These are usually called *shortcut attacks.*
2. Certain symmetry properties in the block cipher (e.g., complementation property).
3. Occurrence of classes of weak keys of non-negligible size (as in IDEA).
4. Related-key attacks.

K-security is essentially a relative measure. It is quite possible to build a K-secure block cipher with a 5-bit block and key length. The lack of security offered by such a primitive is due to its small dimensions, not to the fact that the primitive fails to meet the requirements imposed by these dimensions. Clearly, the longer the key, the higher the targeted security strength.

It is possible to imagine ciphers that have certain weaknesses and still are K-secure. An example of such a weakness would be a block cipher with a block length larger than the key length and a single weak key, for which the permutation is linear. The detection of the usage of the key would take at least a few encryptions, whereas checking whether the key is used would only take a single encryption. If this cipher were to be used for encryption, this single weak key would pose no problem. However, used as a component in a larger scheme, for instance as the compression function of a hash function, this property could introduce a way to efficiently generate collisions. For these reasons we introduce yet another security concept, denoted by the term hermetic.

Definition 5.5.2. *A block cipher is* hermetic *if it does not have weaknesses that are not present for the majority of block ciphers with the same block and key length.*

Informally, a block cipher is hermetic if its internal structure cannot be exploited in any application. For all key and block lengths defined, the security goals are that the Rijndael cipher is K-secure and hermetic. If Rijndael lives up to its goals, the strength against any known or unknown attacks is as good as can be expected from a block cipher with the given dimensions.

5.5.2 Translation of Security Goals into Modern Security Notions

The definitions of K-secure and hermetic are stated in an informal, rather intuitive way. In modern cryptography it is customary to give formal definitions of security of block ciphers. In this section we will discuss the relation between the security definitions we were using at the time when we designed Rijndael and these modern security notions.

We give here an informal description of the modern security notions and refer interested readers to [77] for a more rigorous treatment. Well-established security notions for a block cipher are its *pseudorandom permutation (PRP)* advantage and its *strong pseudorandon permutation (SPRP)* advantage [12]. These are defined by means of a *game* where an algorithm (the *adversary*) must distinguish the block cipher, keyed with a randomly and uniformly selected key not known to the adversary, from a permutation chosen randomly and uniformly from the set of all permutations with the same block size as the block cipher. In the former case we say the adversary is in the *real world* and in the latter case we say it is in the *ideal world*. The adversary is in the real world or in the ideal world, each with probability $1/2$, not knowing which one of the two. In both worlds, the adversary can query the primitive and it is also provided with the specification of the block cipher. In PRP security, the adversary can make only *encryption* queries, while in SPRP security, it can also make *decryption* queries. After analyzing the responses to the queries and performing computations, it must guess whether it is in the ideal or real world. The advantage is basically the adversary's probability of success multiplied by two minus one.

The SPRP advantage of a block cipher with a n_k-bit key that is K-secure is $N2^{-n_k}$ with N the amount of computation where the unit of computation is an execution of the block cipher including the key schedule, and this is irrespective of the number of queries. In other words, the only way to distinguish it from a random permutation is exhaustive key search.

But K-security is actually a much more powerful security notion than can be expressed with an SPRP advantage. Among other things, it also implies resistance against related-key attacks, for a wide range of key relations one can come up with. This is rather a side-effect of the attempt to define a simple to express security goal in the form of K-security than a wish to offer resistance against related-key attacks. Indeed, when dealing with a block cipher with high key agility, there is no need for protocols that allow related-key attacks. The attacks of Biryukov and Khovratovich [72, 149, 82, 24, 29, 28], discussed in Sect. 10.5 have shown that AES-192 and AES-256 do not achieve K-security. In retrospect, if we had been familiar with the concept of SPRP advantage notion at the time, that would have been the appropriate way to express the security goal of Rijndael. Early versions of the concept had already been explored as early as 1988 [97, 11] but only became mainstream after the first edition of this book.

We also presented the security notion of an hermetic cipher to cover unkeyed uses of Rijndael, e.g. as a compression function in hashing. This concept would nowadays be described using the term *ideal cipher*. One could say that a block cipher is hermetic if, when used in an unkeyed mode, it does not allow shortcut attacks. In other words, if the security of the function obtained by applying the unkeyed mode to the block cipher is the one it would have calling an ideal cipher. In the case of collision-resistant hashing this makes sense. One can imagine that it is possible to design a block cipher that, when used as a compression function in a hashing mode, will provide the resistance against collisions that would be obtained with an ideal cipher. However, the notion of hermetic cipher is actually more ambitious: it expresses the property *to behave like an ideal cipher* in all possible circumstances. This is clearly not possible as one can always make up modes or protocols that would be secure in the ideal cipher model but that cannot be secure with any concrete block cipher [35].

5.5.3 Unknown Attacks Versus Known Attacks

'Prediction is very difficult, especially about the future.' (*Niels Bohr*)

Sometimes in cipher design, so-called *resistance against future, as yet unknown, types of cryptanalysis* is used as a rationale to introduce complexity. We prefer to base our ciphers on well-understood components that interact in well-understood ways allowing us to provide bounds that give evidence that the cipher is secure with respect to all known attacks. For ciphers making use of many different operations that interact in hard-to-analyze ways, it is much harder to provide such bounds.

5.5.4 Provable Security Versus Provable Bounds

Often claims are made that a cipher would be *provably secure*. Designing a block cipher that is provably secure in an absolute sense seems for now an unattainable goal. Arguments that have been presented as proofs of security have been shown to be based on (often implicit) assumptions that make these 'proofs of security' irrelevant in the real world. Still, we consider having provable bounds for the workload of known types of cryptanalysis for a block cipher an important feature of the design.

5.6 Approaches to Design

5.6.1 Nonlinearity and Diffusion Criteria

Many papers are devoted to describing nonlinearity and diffusion criteria and counting or characterizing classes of Boolean functions that satisfy them. In most of these papers the Boolean functions are (tacitly) assumed to be (components of) S-boxes located in the F-function of a Feistel structure or in an academic round transformation model such as *so-called* substitution-permutation networks [2, 123]. These networks consist of the alternation of parallel S-boxes and bit permutations, and were proposed in [60, 75]. The S-boxes are considered to be *the* elements in the round transformation that give the cipher its strength against cryptanalysis. Maybe the most important contribution of the wide trail strategy is the demonstration of the importance of the *linear* steps in the round transformation, and quantitative measures for the quality of the linear steps (cf. branch numbers, Sect. 9.3).

Many of the diffusion and nonlinearity criteria described in the cryptology literature are just criteria a block cipher must satisfy in order to be secure. They are necessary conditions for security, but not sufficient. To be of some use in cryptographic design, criteria for the *components* of a cipher are needed rather than criteria for the target *cipher*. Imposing criteria on components in a cipher only makes sense if first a structure of the cipher is defined in which the components have a specific function.

5.6.2 Resistance Against Differential and Linear Cryptanalysis

The discovery of differential and linear cryptanalysis (see Chaps. 6–8 and also Chaps. 12–15) has given rise to a theoretical basis for the design of iterative block ciphers. Nowadays, a new block cipher is only taken seriously if it is accompanied by evidence that it resists differential and linear cryptanalysis. Naturally, differential and linear cryptanalysis are not the only attacks that can be mounted against block ciphers. In Chap. 10 we consider a number of generic types of cryptanalysis and attacks that are specific to the structure of Rijndael and its related ciphers. A block cipher should resist all types of cryptanalysis imaginable. Still, we see that nowadays in most cases resistance against differential and linear cryptanalysis are the criteria that shape a block cipher; the other known attacks are only considered later and resistance against them can be obtained with small modifications in the original design (e.g. the affine transformation in the $\mathrm{S_{RD}}$ to thwart interpolation attacks, cf. Sect. 10.4).

Almost always, an iterative block cipher can be made resistant against differential and linear cryptanalysis by taking enough rounds. Even if a round transformation is used that offers very little nonlinearity or diffusion, repeating it often enough will result in a block cipher that is not breakable by differential or linear cryptanalysis. For an iterated cipher, the workload of an encryption is the workload of the round transformation multiplied by the number of rounds. The engineering challenge is to design a round transformation in such a way that this product is minimized while providing lower bounds for the complexity of differential and linear cryptanalysis that are higher than exhaustive key search.

5.6.3 Local Versus Global Optimization

The engineering challenge can be tackled in different ways. We distinguish two approaches:

1. **local optimization.** The round transformation is designed in such a way that the worst-case behavior of one round is optimized.
2. **global optimization.** The round transformation is designed in such a way that the worst-case behavior of a sequence of rounds is optimized.

In both cases, the worst-case behavior is then used to determine the required number of rounds to offer resistance against differential and linear cryptanalysis. For the actual block cipher, usually some more rounds are taken, to provide a security margin.

In the context of linear cryptanalysis, this worst-case behavior corresponds with the maximum input-output correlation (see Chap. 7) and in the case of differential cryptanalysis it corresponds to the maximum difference propagation probability (see Chap. 8).

In the case of local optimization, the maximum input-output correlation and the maximum difference propagation probability of the round transformation determine the number of rounds required. In Feistel ciphers (see

Chap. 6) it does not make sense to evaluate these criteria over a single round, since part of the state is merely transposed and undergoes no nonlinear operation. Therefore, for Feistel ciphers local optimization is done on a sequence of two rounds.

In Chaps. 7 and 8 we show that to obtain low maximum correlations and difference propagation probabilities, a Boolean transformation must have many input bits. In the local optimization approach the round must thus make use of expensive nonlinear functions such as large S-boxes or modular multiplication. This can be considered to be a *greedy* approach: good nonlinearity is obtained with only few rounds but at a high implementation cost.

The tendency to do local optimization can be found in many ciphers. For example, in [93] X. Lai et al. claim that the maximum difference propagation probability over a single round is an important measure of the resistance that a round transformation offers against differential cryptanalysis. Another example of local optimization is [122] by K. Nyberg and L. Knudsen. All results are obtained in terms of the maximum difference propagation probability of the F-function (see Chap. 6) of a Feistel cipher.

In global optimization, the maximum input-output correlation and difference propagation probability of the round transformation do not play such an important role. Here several approaches are possible. One of the approaches is the *wide trail strategy* that we have adopted for the design of Rijndael and its predecessors. To fully understand the wide trail strategy, we advise reading Chaps. 6–9.

As opposed to local optimization, global optimization allows cheap nonlinear Boolean transformations such as small S-boxes. Global optimization introduces new diffusion criteria. These diffusion criteria no longer specify what the block cipher should satisfy, but give concrete criteria for the design of components of the round transformation. In most cases, the round transformation contains components that realize nonlinearity and components that realize diffusion. The function of the diffusion components is to make sure that the input-output correlation (difference propagation probability) over r rounds is much less than the rth power of the maximum input-output correlation (difference propagation probability) of the round transformation.

For most round transformations, finding the maximum difference propagation probability and the maximum input-output correlation is computationally feasible. Computing the maximum difference propagation probability and the maximum input-output correlation over multiple rounds can, however, become very difficult. In the original differential cryptanalysis and linear cryptanalysis attacks on the DES, finding high difference propagation probabilities and input-output correlations over all but a few rounds of the cipher turned out to be one of the major challenges. In the linear cryptanalysis attack (cf. Chap. 6 and [104, 105]), M. Matsui had to write a sophisticated program that searched for the best linear expression.

In Rijndael and its predecessors, we have made use of symmetry and alignment to easily prove lower bounds for sequences of four rounds (two rounds in SHARK, see Chap. 11). If alignment is not applied, proving bounds becomes more difficult. An illustration of this is the AES finalist Serpent [4], which also applied the principle of global optimization. The Serpent submission contains a table giving the maximum difference propagation probabilities and input-output correlations for 1–7 rounds, clearly illustrating this. Especially the bounds for 5–7 rounds were excellent. Unfortunately, the paper did not give a proof of these bounds nor a description of how they were obtained. Moreover, later the designers of Serpent had to weaken the bounds due to new insights [23]. In our bitslice cipher Noekeon [46], we have provided bounds for four rounds using an exhaustive search program with a relatively simple structure. By exploiting the high level of symmetry in Noekeon, many optimizations were possible in this program, enabling us to *demonstrate* surprisingly good bounds.

5.7 Key-Alternating Cipher Structure

By applying the key with a simple XOR, we simplify the analysis of the cipher and hence make it easier to prove lower bounds in the resistance against particular attacks such as differential and linear cryptanalysis (see Sect. 9.1).

The advantage of the key-alternating structure is that the *quality* of the round transformations in the context of linear or differential cryptanalysis is independent of the round key. By adopting a key-alternating structure, the analysis of linear and differential trails can be conducted without even considering the influence of the key. An example of a radically different approach is the block cipher IDEA [93].

Example 5.7.1. In IDEA the subkeys are applied by means of modular multiplication and addition. The difference propagation probability of the round transformation depends heavily on the value of these subkeys. However, the designers of IDEA have proposed considering alternative notions of *difference* to come to difference propagation probabilities that are independent of the value of the round key. Unfortunately, attacks based on XOR as the difference appear to be more powerful than attacks making use of the alternative notion of difference. Moreover, the existence of weak subkeys and an unfortunate choice in the key schedule give rise to large classes of weak keys for which IDEA exhibits difference propagation probabilities equal to 1. Similar arguments apply for the resistance of IDEA against linear cryptanalysis.

5.8 The Key Schedule

5.8.1 The Function of a Key Schedule

The function of a key schedule is the generation of the round keys from the cipher key. For a key-alternating cipher with a block length of n_b and r rounds, this means $n_b(r+1)$ bits. There is no consensus on the criteria that a key schedule must satisfy. In some design approaches, the key schedule must generate round keys in such a way that they appear to be *mutually independent* and can be considered *random* (see Sects. 7.10.2 and 8.7.2). Moreover, for some ciphers the key schedule is so strong that the knowledge of one round key does not help in finding the cipher key or other round keys. In these ciphers, the key schedule appears to make use of components that can be considered as cryptographic primitives in their own right.

For the key schedule in Rijndael the criteria are less ambitious. Basically, the key schedule is there for three purposes:

1. The first one is the introduction of asymmetry. Asymmetry in the key schedule prevents symmetry in the round transformation and between the rounds, which would lead to weaknesses or allow attacks. Examples of such weaknesses are the complementation property of the DES or weak keys such as in the DES [54]. Examples of attacks that exploit symmetry are slide attacks.
2. The second purpose is resistance against related-key attacks (cf. Sect. 10.5).
3. The third purpose is resistance against attacks in which the cipher key is (partially) known by or can be chosen by the cryptanalyst. This is the case if the cipher is used as the compression function of a hash function [85].

All other attacks are supposed to be prevented by the rounds of the block cipher. The modest criteria we impose can be met by a key schedule that is relatively simple and uses only a small amount of resources. This gives Rijndael its high key agility.

5.8.2 Key Expansion and Key Selection

In Rijndael, the key schedule consists of two parts: the *key expansion*, which maps the n_k-bit cipher key to a so-called *expanded key*, and the *round key selection*, which selects the n_b-bit round keys from the expanded key. This modularity facilitates the definition of a key expansion that is independent of the block length, and a round key selection that is independent of the cipher key length. For Rijndael, round key selection is very simple: the expanded key is seen as the concatenation of n_b-bit round keys starting with the first round key.

5.8.3 The Cost of the Key Expansion

In general, the key schedule is independent of the value of the plaintext or the ciphertext. If a cipher key is used for encrypting (or decrypting) multiple blocks, one may limit the computational cost of the key schedule by performing the key expansion only once and keeping the expanded key in working memory for the complete sequence of encryptions. If cipher keys are used to encrypt large amounts of data, the computational cost of the key schedule can be neglected. Still, in many applications a cipher key is used for the encryption of only a small number of blocks. If the block cipher is used as a one-way function (e.g. in key derivation), or as the compression function in a hash function, each encryption is accompanied by the execution of the key schedule. In these cases, the computational cost of the key schedule is very important. Keeping the expanded key in working memory consumes $n_b(r+1)$ bits. In the case of Rijndael, a block length and key length of 128 bits require 176 bytes of key storage, whereas a block length and key length of 256 bits require 480 bytes of key storage. On some resource-limited platforms such as smart cards there may be not enough working memory available for the storage of the expanded key. To allow efficient implementations on these platforms, the key schedule must allow implementations using a limited amount of working memory in a limited number of processor cycles, in a small program.

5.8.4 A Recursive Key Expansion

We addressed the requirements discussed above by adopting a recursive structure. After the first n_k bits of the expanded key have been initialized with the cipher key, each subsequent bit is computed in terms of bits that have previously been generated. More specifically, if we picture the expanded key as a sequence of 4-byte columns, the column at position i in the expanded key can be computed using the columns at positions from $i - \mathrm{N_k}$ to $i - 1$ only. Let us now consider a *block* consisting of the columns with indices from j to $j + \mathrm{N_k} - 1$. By working out the dependencies, we can show that this block can be computed using columns $j - \mathrm{N_k}$ to $j - 1$. In other words, each $\mathrm{N_k}$-column block is completely determined by the previous $\mathrm{N_k}$-column block. As $\mathrm{N_k}$ columns of the expanded key are sufficient for the computation of all following columns, the key schedule can be implemented using only a working memory that is the size of the cipher key. The round keys are generated on the fly and the key expansion is executed whenever round key bits are required. In the case where the block length is equal to the key length, the blocks described above coincide with round keys, and round key i can be computed from round key $i - 1$ by what can be considered to be one *round* of the key schedule. Additionally, the recursion can be inverted, i.e. column i

can be expressed in terms of columns $i+1$ to $i+N_k$. This implies that the expanded key can be computed backwards, starting from the last N_k columns. This allows on-the-fly round key generation for decryption.

The recursion function must have a low implementation cost while providing sufficient diffusion and asymmetry to thwart the attacks mentioned above. To protect against related-key attacks, nonlinearity can be introduced. More specifically, the nonlinearity should prohibit the full determination of differences in the expanded key from cipher key differences only.

5.9 Conclusions

In this chapter we have tried to make explicit the mindset with which we have designed Rijndael and its predecessors. A large part of this is the formulation of criteria that the result must satisfy. Cipher design is still more engineering than science. In many cases compromises have to be made between conflicting requirements. We are aware that Rijndael is just an attempt to achieve a cipher that satisfies our criteria and that it shows such compromises.

6. The Data Encryption Standard

In this chapter we give a brief description of the block cipher DES [1]. Both differential cryptanalysis and linear cryptanalysis were successfully applied to the DES. Differential cryptanalysis was the first chosen-plaintext attack that was theoretically more efficient than an exhaustive key search for the DES. Linear cryptanalysis was the first attack that achieved the same in a known-plaintext scenario. Resistance against these two attacks is the most important criterion in the design of Rijndael.

We give a summary of the original differential cryptanalysis and linear cryptanalysis attacks using the terminology of their inventors. For a more detailed treatment of the attacks, we refer to the original publications [26, 104]. The only aim of our description is to indicate the aspects of the attacks that determine their expected work factor. For differential cryptanalysis the critical aspect is the maximum *probability* of difference propagations, for linear cryptanalysis it is the maximum *deviation* from 0.5 of the probability that linear expressions hold.

6.1 The DES

The cipher that was the most important object of the attacks to be discussed is the DES [1]. Therefore, we start with a brief description of its structure.

The DES is an iterated block cipher with a block length of 64 bits and a key length of 56 bits. Its main body consists of 16 iterations of a *keyed round function*. The computational graph of the round function is depicted in Fig. 6.1. The state is split into a 32-bit left part L_i and a 32-bit right part R_i. The latter is the argument of the keyed F-function. L_i is modified by combining it with the output of the F-function by means of an XOR operation. Subsequently, the left and the right parts are interchanged. This round function has the so-called Feistel structure: the result of applying a key-dependent function to part of the state is added (using a bitwise XOR operation) to another part of the state, followed by a transposition of parts of the state. A block cipher that has rounds with this structure is called a Feistel cipher.

J. Daemen, V. Rijmen, *The Design of Rijndael*, Information Security and Cryptography,
https://doi.org/10.1007/978-3-662-60769-5_6

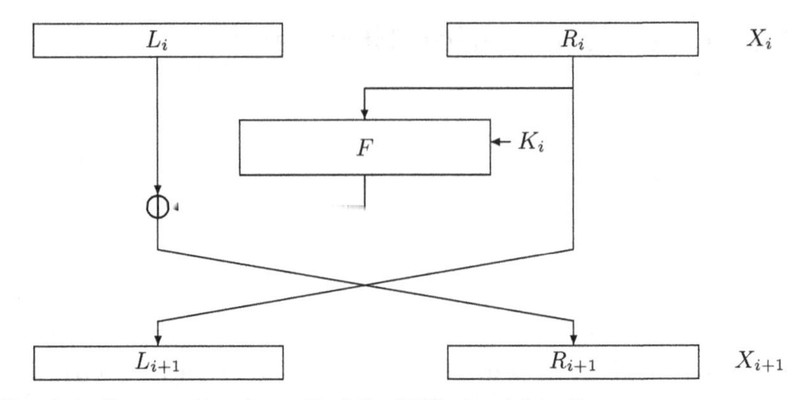

Fig. 6.1. Computational graph of the DES round function

The computational graph of the F-function is depicted in Fig. 6.2. It consists of the succession of four steps:

1. **Expansion** E. The 32 input bits are expanded to a 48-bit vector. In this expansion, the 32-bit vector is split into 4-bit tuples, and the first and last bit of each tuple is duplicated.
2. **Key addition.** The 48-bit vector is modified by combining it with a 48-bit round key using the bitwise XOR operation.
3. **S-boxes.** The resulting 48-bit vector is mapped onto a 32-bit vector by nonlinear S-boxes. The 48-bit vector is split into eight 6-bit tuples, which are converted into eight 4-bit tuples by eight different nonlinear S-boxes that each convert 6 input bits into 4 output bits. As an example, Table 6.1 gives the specification of the second S-box. This table must be read as follows. If the 6-bit input is denoted by $a_1a_2a_3a_4a_5a_6$, the output is given by the entry in row $2a_1 + a_6$ and column $8a_2 + 4a_3 + 2a_4 + a_5$. The 4-bit values are given in hexadecimal notation, e.g. `D` denotes 1101.
4. **Bit permutation** P. The bits of the 32-bit vector are transposed.

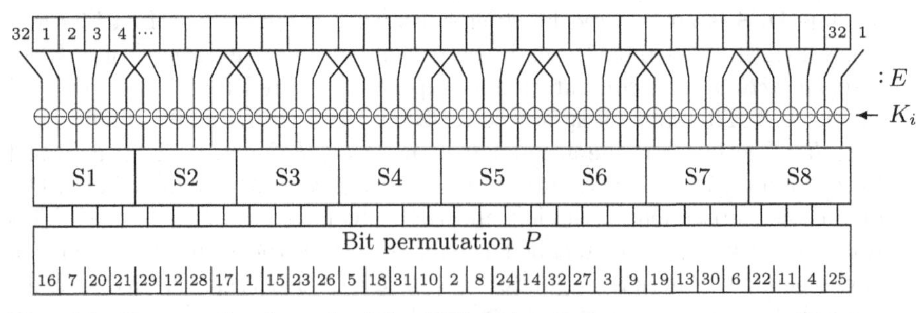

Fig. 6.2. Computational graph of the DES F-function

Observe that the only nonlinear step in the F-function (and also in the round transformation) consists of the S-boxes. The 48-bit round keys are extracted from the 56-bit cipher key by means of a linear key schedule.

	0	1	2	3	4	5	6	7	8	9	A	B	C	D	E	F
0 :	F	1	8	E	6	B	3	4	9	7	2	D	C	0	5	A
1 :	3	D	4	7	F	2	8	E	C	0	1	A	6	9	B	5
2 :	0	E	7	B	A	4	D	1	5	8	C	6	9	3	2	F
3 :	D	8	A	1	3	F	4	2	B	6	7	C	0	5	E	9

Table 6.1. Specification of the DES S-box S_2

6.2 Differential Cryptanalysis

In this section we summarize the most important elements of differential cryptanalysis as E. Biham and A. Shamir described it in [26].

Differential cryptanalysis is a chosen-plaintext (difference) attack in which a large number of plaintext-ciphertext pairs are used to determine the value of key bits. Statistical key information is deduced from ciphertext blocks obtained by encrypting pairs of plaintext blocks with a specific bitwise difference A' under the target key. The work factor of the attack depends critically on the largest probability $\text{Prob}(B'|A')$ with B' being a difference at some fixed intermediate stage of the block cipher, e.g. at the input of the last round.

In a first approximation, the probabilities $\text{Prob}(B'|A')$ for the DES are assumed to be independent of the specific value of the key.

In the basic form of the attack, key information is extracted from the output pairs in the following way. For each pair it is assumed that the intermediate difference is equal to B'. The absolute values of the output pair and the (assumed) intermediate difference B' impose restrictions upon a number ℓ of key bits of the last round key. A pair is said to *suggest* the subkey values that are compatible with these restrictions. While for some pairs many keys are suggested, no keys are found for other pairs, implying that the output values are incompatible with B'. For each suggested subkey value, a corresponding entry in a frequency table is incremented.

The attack is successful if the correct value of the subkey is suggested significantly more often than any other value. Pairs with an intermediate difference not equal to B' are called *wrong pairs*. Subkey values suggested by these pairs are in general wrong. Right pairs, with an intermediate difference equal to B', suggest not only the right subkey value but often also a number of wrong subkey values. For the DES, the wrong suggestions may be considered uniformly distributed among the possible key values if the value $\text{Prob}(B'|A')$ is significantly larger than $\text{Prob}(C'|A')$ for any $C' \neq B'$.

Under these conditions it makes sense to calculate the ratio between the number of times the right value is suggested and the average number of suggestions per entry, the so-called *signal-to-noise* (S/N) *ratio*.

The size of the table of possible values of the ℓ-bit subkey is 2^ℓ. If we denote the average number of suggested subkeys per pair by γ, the S/N ratio is given by

$$\text{S/N} = \text{Prob}(B'|A')2^\ell/\gamma. \tag{6.1}$$

The S/N ratio strongly affects the number of right pairs needed to uniquely identify the correct subkey value. Experimental results [26] showed that for a ratio between 1 and 2 about 20–40 right pairs are enough. For larger ratios only a few right pairs are needed and for ratios that are much smaller than 1 the required number of right pairs makes a practical attack infeasible.

Pairs of differences A' and B' with a large probability $\text{Prob}(B'|A')$ are found by the construction of so-called *characteristics*. An r-round characteristic comprises an $(r+1)$-tuple of difference patterns: $(X'_0, X'_1, \ldots, X'_r)$. The probability of this characteristic is the probability that an initial difference pattern X'_0 propagates to difference patterns $X'_1, X'_2, \ldots, X'_r$ after 1, 2, $\ldots$, r rounds, respectively. Under the so-called *Markov assumption* (cf. also Sect. 8.7.2), i.e. that the propagation probability from X'_{i-1} to X'_i is independent of the propagation probability from X'_0 to X'_{i-1}, this probability is given by

$$\prod_i \text{Prob}(X'_i|X'_{i-1}), \tag{6.2}$$

where $\text{Prob}(X'_i|X'_{i-1})$ is the probability that the difference pattern X'_{i-1} at the input of the round transformation gives rise to X'_i at its output. Hence, the multiple-round characteristic is a sequence of single-round characteristics (X'_{i-1}, X'_i) with probability $\text{Prob}(X'_i|X'_{i-1})$.

In the construction of high-probability characteristics for the DES, advantage is taken of the linearity in the round transformation. Single-round characteristics of the form $(L'_{i-1}\|R'_{i-1}, L'_i\|R'_i)$, where $R'_i = L'_{i-1}$ and $L'_i = R'_{i-1} = 0$ have probability 1 and are called *trivial.* The most probable nontrivial single-round characteristics have an input difference pattern that only affects a small number of the eight S-boxes.

Trivial characteristics have been exploited to construct high-probability *iterative characteristics*. These are characteristics with a periodic sequence of differences. The iterative characteristic with highest probability has a period of two. Of the two involved single-round characteristics, one is trivial. In the other one there is a non-zero difference pattern at the input of three neighboring S-boxes, which propagates to a zero difference pattern at the output of the S-boxes with probability 1/234. Hence, the resulting iterative characteristics have a probability of 1/234 per two rounds.

In the actual differential attacks on the DES, some techniques are used to make the attack more efficient. This involves a special treatment in the first and last rounds. For these techniques we refer to [26].

6.3 Linear Cryptanalysis

In this section we summarize the most important elements of linear cryptanalysis as M. Matsui presented them in [104]. Linear cryptanalysis is a known-plaintext attack in which a large number of plaintext-ciphertext pairs are used to determine the value of key bits.

A condition for applying linear cryptanalysis to a block cipher is to find 'effective' linear expressions. Let $\mathbf{A}[i_1, i_2, \ldots, i_a]$ be the bitwise sum of the bits of $\mathbf{A}$ with indices in a *selection pattern* $\{i_1, i_2, \ldots, i_a\}$; i.e.

$$\mathbf{A}[i_1, i_2, \ldots, i_a] = \mathbf{A}[i_1] \oplus \mathbf{A}[i_2] \oplus \cdots \oplus \mathbf{A}[i_a]. \tag{6.3}$$

Let $\mathbf{P}, \mathbf{C}$ and $\mathbf{K}$ denote the plaintext, the ciphertext and the key, respectively. A linear expression is an expression of the following type:

$$\mathbf{P}[i_1, i_2, \ldots, i_a] \oplus \mathbf{C}[j_1, j_2, \ldots, j_b] = \mathbf{K}[k_1, k_2, \ldots, k_c], \tag{6.4}$$

with $i_1, i_2, \ldots, i_a$, $j_1, j_2, \ldots, j_b$ and $k_1, k_2, \ldots, k_c$ being fixed bit locations. The effectiveness, or deviation, of such a linear expression in linear cryptanalysis is given by $|p - 1/2|$ where p is the probability that the expression holds. By checking the value of the left-hand side of (6.4) for a large number of plaintext-ciphertext pairs, the right-hand side can be guessed by taking the value that occurs most often. In principle, this gives a single bit of information about the key. In [104] it is shown that the probability of making a wrong guess is very small if the number of plaintext-ciphertext pairs is larger than $|p - 1/2|^{-2}$.

In [104] another algorithm is given that determines more than a single bit of key information using a similar linear expression. Instead of (6.4), an expression is used that contains no plaintext or ciphertext bits, but instead contains bits of the intermediate encryption values I_1 and I_{15}, respectively, after exactly one round and after all rounds but one:

$$\mathbf{I}_1[i_1, i_2, \ldots, i_a] \oplus \mathbf{I}_{15}[j_1, j_2, \ldots, j_b] = \mathbf{K}[\ell_1, \ell_2, \ldots, \ell_c]. \tag{6.5}$$

By assuming values for a subset ν_k of the subkey bits of the first and last round, the bits of $\mathbf{I}_1$ and $\mathbf{I}_{15}$ that occur in (6.5) can be calculated. These bits are correct if the values assumed for the key bits with indices in ν_k are correct. Given a large number ℓ of plaintext-ciphertext pairs, the correct values of all bits in ν_k and the value of the right-hand side of (6.5) can be determined in the following way. For all values of the key bits with indices in ν_k, the number of plaintext-ciphertext pairs are counted for which (6.5) holds. For the correct assumption the expected value of this sum is $p\ell$ or $(1-p)\ell$. Thanks to the nonlinear behavior of the round transformation this sum is expected to have significantly less bias for all wrongly assumed subkey values. Given a linear expression (6.5) that holds with probability p, the probability that this algorithm leads to a wrong guess is very small if the number of plaintext-ciphertext pairs is significantly (say more than a factor 8) larger than $|p-1/2|^{-2}$. In an improved version of this attack, this factor 8 is reduced to 1 [105]. Hence, in both variants the value of $|p-1/2|$ is critical for the work factor of the attack.

Effective linear expressions (6.4) and (6.5) are constructed by 'chaining' single-round linear expressions. An $(r-1)$-round linear expression can be turned into an r-round linear expression by appending a single-round linear expression such that all the intermediate bits cancel:

$$\begin{array}{rcl} \mathbf{P}[i_1, i_2, \ldots, i_a] \oplus \mathbf{I}_{r-1}[j_1, j_2, \ldots, j_b] & = & \mathbf{K}[k_1, k_2, \ldots, k_c] \\ & \oplus & \\ \mathbf{I}_{r-1}[j_1, j_2, \ldots, j_b] \oplus \mathbf{I}_r[m_1, m_2, \ldots, m_a] & = & \mathbf{K}[k_2, k_5, \ldots, k_d] \\ & = & \\ \mathbf{P}[i_1, i_2, \ldots, i_a] \oplus \mathbf{I}_r[m_1, m_2, \ldots, m_a] & = & \mathbf{K}[k_1, k_3, \ldots, k_d] \end{array} \quad (6.6)$$

In [104] it is shown that the probability that the resulting linear expression holds can be approximated by $1/2 + 2(p_1 - 1/2)(p_2 - 1/2)$, given that the component linear expressions hold with probabilities p_1 and p_2, respectively.

The DES single-round linear expressions and their probabilities can be studied by observing the dependencies in the computational graph of the round transformation. The selected round output bits completely specify a selection pattern at the output of the S-boxes. If round output bits are selected only from the left half, this involves no S-box output bits at all, resulting in linear expressions that hold with a probability of 1. These are of the following type:

$$\mathbf{I}_{\ell-1}[j_1 + 32, j_2 + 32, \ldots, j_a + 32] = \mathbf{I}_\ell[j_1, j_2, \ldots, j_a]. \quad (6.7)$$

This is called a *trivial* expression. Apparently, the most useful non-trivial single-round linear expressions only select bits coming from a single S-box. For a given S-box, all possible linear expressions and their probabilities can be exhaustively calculated. Together with the key application before the S-boxes, each of these linear expressions can be converted into a single-round linear expression. The most effective multiple-round linear expressions for the DES are constructed by combining single-round trivial expressions with linear expressions involving output bits of only a single S-box. The resulting most effective 14-round linear expression has a probability of $1/2 \pm 1.19 \times 2^{-21}$.

6.4 Conclusions

In this section we have explained the round structure of the DES and have given a summary of the two most important cryptanalytic attacks on the DES using the terminology and formalism of the original publications.

7. Correlation Matrices

In this chapter we consider correlations over Boolean functions and iterated Boolean transformations. Correlations play an important role in cryptanalysis in general and linear cryptanalysis in particular.

We introduce algebraic tools such as *correlation matrices* to adequately describe the properties that make linear cryptanalysis possible. We derive a number of interesting relations and equalities and apply these to iterated Boolean transformations.

7.1 The Walsh-Hadamard Transform

7.1.1 Parities and Masks

A *parity* of a Boolean vector is a binary Boolean function that consists of the XOR of a number of bits. A parity is determined by the positions of the bits of the Boolean vector that are included in the XOR.

The *mask* $\mathbf{w}$ of a parity is a Boolean vector value that has a 1 in the components that are included in the parity and a 0 in all other components. Analogous to the inner product of vectors in linear algebra, we express the parity of vector $\mathbf{a}$ corresponding with mask $\mathbf{w}$ as $\mathbf{w}^{\mathrm{T}}\mathbf{a}$. The concepts of mask and parity are illustrated with an example in Fig. 7.1.

Note that for a vector $\mathbf{a}$ with n bits, there are 2^n different parities. The set of parities of a Boolean vector is in fact the set of all linear binary Boolean functions of that vector.

7.1.2 Correlation

Linear cryptanalysis exploits large correlations over all but a few rounds of a block cipher.

Definition 7.1.1. *The* correlation $\mathrm{C}(f, g)$ *between two binary Boolean functions* $f(\mathbf{a})$ *and* $g(\mathbf{a})$ *is defined as*

$$\mathrm{C}(f, g) = 2 \cdot \mathrm{Prob}(f(\mathbf{a}) = g(\mathbf{a})) - 1. \quad (7.1)$$

J. Daemen, V. Rijmen, *The Design of Rijndael*, Information Security and Cryptography,
https://doi.org/10.1007/978-3-662-60769-5_7

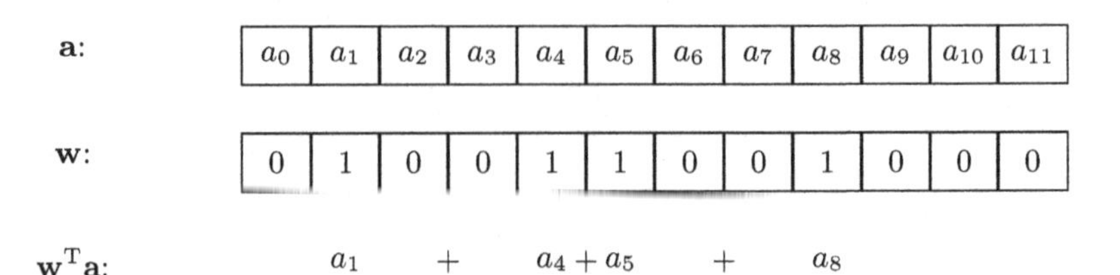

Fig. 7.1. Example of state $\mathbf{a}$, mask $\mathbf{w}$ and parity $\mathbf{w}^\mathrm{T}\mathbf{a}$

From this definition it follows that $\mathrm{C}(f, g) = \mathrm{C}(g, f)$. The correlation between two binary Boolean functions ranges between -1 and 1. If the correlation is different from zero, the binary Boolean functions are said to be *correlated*. If the correlation is 1, the binary Boolean functions are equal; if it is -1, the binary Boolean functions are each other's complement.

7.1.3 Real-Valued Counterpart of a Binary Boolean Function

Let $\hat{f}(\mathbf{a})$ be a real-valued function that is -1 for $f(\mathbf{a}) = 1$ and $+1$ for $f(\mathbf{a}) = 0$. This can be expressed by

$$\hat{f}(\mathbf{a}) = (-1)^{f(\mathbf{a})} \tag{7.2}$$

In this notation the real-valued function corresponding to a parity $\mathbf{w}^\mathrm{T}\mathbf{a}$ becomes $(-1)^{\mathbf{w}^\mathrm{T}\mathbf{a}}$. The real-valued counterpart of the XOR of two binary Boolean functions is the product of their real-valued counterparts, i.e.

$$\widehat{f(\mathbf{a}) + g(\mathbf{a})} = \hat{f}(\mathbf{a})\hat{g}(\mathbf{a}). \tag{7.3}$$

7.1.4 Orthogonality and Correlation

We define the *inner product* of two binary Boolean functions f and g as

$$\langle \hat{f}, \hat{g} \rangle = \sum_{\mathbf{a}} \hat{f}(\mathbf{a})\hat{g}(\mathbf{a}). \tag{7.4}$$

This inner product defines the following *norm*:

$$\|\hat{f}\| = \sqrt{\langle \hat{f}, \hat{f} \rangle}. \tag{7.5}$$

The norm of a binary Boolean function $f(\mathbf{a})$ is equal to the square root of its domain size, i.e. $2^{n/2}$.

From the definition of correlation it follows that

$$C(f,g) = \frac{\langle \hat{f}, \hat{g} \rangle}{\|\hat{f}\| \cdot \|\hat{g}\|}, \tag{7.6}$$

or in words, the correlation between two binary Boolean functions is equal to their inner product divided by their norms. In Fig. 7.2 this is illustrated in a geometrical way.

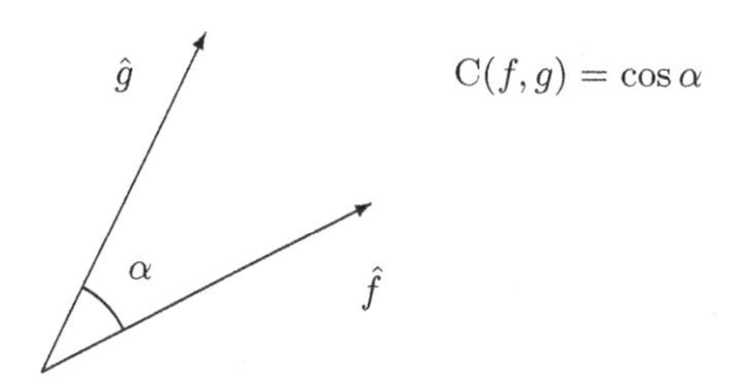

Fig. 7.2. Geometric representation of binary Boolean functions and their correlation

7.1.5 Spectrum of a Binary Boolean Function

The set of binary Boolean functions of an n-bit vector can be seen as elements of a vector space of dimension 2^n. A vector f has 2^n components given by $(-1)^{f(\mathbf{a})}$ for the 2^n values of $\mathbf{a}$. Vector addition corresponds to addition of the components in $\mathbb{R}$, scalar multiplication to multiplication of components with elements of $\mathbb{R}$. This is the ring $(\mathbb{R}^{2^n}, +, \cdot)$.

In $(\mathbb{R}^{2^n}, +, \cdot)$ the parities form an orthogonal basis with respect to the inner product defined by (7.4):

$$\begin{aligned}
\langle (-1)^{\mathbf{u}^T\mathbf{a}}, (-1)^{\mathbf{v}^T\mathbf{a}} \rangle &= \sum_{\mathbf{a}} (-1)^{\mathbf{u}^T\mathbf{a}} (-1)^{\mathbf{v}^T\mathbf{a}} \\
&= \sum_{\mathbf{a}} (-1)^{\mathbf{u}^T\mathbf{a}+\mathbf{v}^T\mathbf{a}} \\
&= \sum_{\mathbf{a}} (-1)^{(\mathbf{u}+\mathbf{v})^T\mathbf{a}} \\
&= 2^n \delta(\mathbf{u}+\mathbf{v}).
\end{aligned}$$

Here $\delta(w)$ is the Kronecker delta function that is equal to 1 if w is the zero vector and 0 otherwise. The representation of a binary Boolean function with respect to the parity basis is called its Walsh-Hadamard spectrum, or simply its *spectrum* [66, 128].

Consider now $\mathrm{C}(f(\mathbf{a}), \mathbf{w}^{\mathrm{T}}\mathbf{a})$, which is the correlation between a binary Boolean function $f(\mathbf{a})$ and the parity $\mathbf{w}^{\mathrm{T}}\mathbf{a}$. If we denote this by $F(\mathbf{w})$, we have

$$\hat{f}(\mathbf{a}) = \sum_{\mathbf{w}} F(\mathbf{w})(-1)^{\mathbf{w}^{\mathrm{T}}\mathbf{a}}, \tag{7.7}$$

where $\mathbf{w}$ ranges over all 2^n possible values. In words, the coordinates of a binary Boolean function in the parity basis are the correlations between the binary Boolean function and the parities. It follows that a Boolean function is completely specified by the set of correlations with all parities.
Dually, we have

$$F(\mathbf{w}) = \mathrm{C}(f(\mathbf{a}), \mathbf{w}^{\mathrm{T}}\mathbf{a}) = 2^{-n} \sum_{\mathbf{a}} \hat{f}(\mathbf{a})(-1)^{\mathbf{w}^{\mathrm{T}}\mathbf{a}}. \tag{7.8}$$

We denote the Walsh-Hadamard transform by the symbol $\mathcal{W}$. We have

$$\mathcal{W} : f(\mathbf{a}) \mapsto F(\mathbf{w}) : \quad F(\mathbf{w}) = \mathcal{W}(f(\mathbf{a})). \tag{7.9}$$

If we take the square of the norm of both sides of (7.7), we obtain:

$$\langle \hat{f}(\mathbf{a}), \hat{f}(\mathbf{a}) \rangle = \langle \sum_{\mathbf{w}} F(\mathbf{w})(-1)^{\mathbf{w}^{\mathrm{T}}\mathbf{a}}, \sum_{\mathbf{v}} F(\mathbf{v})(-1)^{\mathbf{v}^{\mathrm{T}}\mathbf{a}} \rangle. \tag{7.10}$$

Working out both sides gives

$$
\begin{aligned}
2^n &= \sum_{\mathbf{w}} F(\mathbf{w}) \langle (-1)^{\mathbf{w}^{\mathrm{T}}\mathbf{a}}, \sum_{\mathbf{v}} F(\mathbf{v})(-1)^{\mathbf{v}^{\mathrm{T}}\mathbf{a}} \rangle && (7.11)\\
&= \sum_{\mathbf{w}} F(\mathbf{w}) \sum_{\mathbf{v}} F(\mathbf{v}) \langle (-1)^{\mathbf{w}^{\mathrm{T}}\mathbf{a}}, (-1)^{\mathbf{v}^{\mathrm{T}}\mathbf{a}} \rangle && (7.12)\\
&= \sum_{\mathbf{w}} \sum_{\mathbf{v}} F(\mathbf{w}) F(\mathbf{v}) 2^n \delta(\mathbf{w} + \mathbf{v}) && (7.13)\\
&= 2^n \sum_{\mathbf{w}} F^2(\mathbf{w}). && (7.14)
\end{aligned}
$$

Dividing this by 2^n yields the theorem of Parseval [100, p. 416]:

$$\sum_{\mathbf{w}} F^2(\mathbf{w}) = 1. \tag{7.15}$$

This theorem expresses a relation between the number of parities that a given binary Boolean function is correlated with and the amplitude of the correlations. It states that the squares of the correlations corresponding to all input parities sum to 1.

7.2 Composing Binary Boolean Functions

7.2.1 Addition

The spectrum of the sum (XOR) of two binary Boolean functions $f(\mathbf{a})+g(\mathbf{a})$ can be derived using (7.7):

$$\begin{aligned}\hat{f}(\mathbf{a})\hat{g}(\mathbf{a}) &= \sum_{\mathbf{u}} F(\mathbf{u})(-1)^{\mathbf{u}^{\mathrm{T}}\mathbf{a}} \sum_{\mathbf{v}} G(\mathbf{v})(-1)^{\mathbf{v}^{\mathrm{T}}\mathbf{a}} \\ &= \sum_{\mathbf{u}}\sum_{\mathbf{v}} F(\mathbf{u})G(\mathbf{v})(-1)^{(\mathbf{u}+\mathbf{v})^{\mathrm{T}}\mathbf{a}} \\ &= \sum_{\mathbf{w}} \left(\sum_{\mathbf{v}} F(\mathbf{v}+\mathbf{w})G(\mathbf{v}) \right) (-1)^{\mathbf{w}^{\mathrm{T}}\mathbf{a}}. \end{aligned} \tag{7.16}$$

The values of the spectrum $H(\mathbf{w}) = \mathcal{W}(f+g)$ are therefore given by

$$H(\mathbf{w}) = \sum_{\mathbf{v}} F(\mathbf{v}+\mathbf{w})G(\mathbf{v}). \tag{7.17}$$

Hence, the spectrum of the sum of binary Boolean functions is equal to the convolution of the corresponding spectra. We express this as

$$\mathcal{W}(f+g) = \mathcal{W}(f) \otimes \mathcal{W}(g), \tag{7.18}$$

where $\otimes$ denotes the convolution operation. Given this convolution property it is easy to demonstrate some composition properties that are useful in the study of linear cryptanalysis:

1. The spectrum of the complement of a binary Boolean function $g(\mathbf{a}) = f(\mathbf{a}) + 1$ is the negative of the spectrum of $f(\mathbf{a})$: $G(\mathbf{w}) = -F(\mathbf{w})$.
2. The spectrum of the sum of a binary Boolean function and a parity $g(\mathbf{a}) = f(\mathbf{a}) + \mathbf{u}^{\mathrm{T}}\mathbf{a}$ is equal to the spectrum of $f(\mathbf{a})$ transformed by a so-called *dyadic* shift: $G(\mathbf{w}) = F(\mathbf{w}+\mathbf{u})$.

7.2.2 Multiplication

For the multiplication (AND) of two binary Boolean functions we have

$$\widehat{f(\mathbf{a})g(\mathbf{a})} = \frac{1}{2}(1 + \hat{f}(\mathbf{a}) + \hat{g}(\mathbf{a}) - \hat{f}(\mathbf{a})\hat{g}(\mathbf{a})). \tag{7.19}$$

It follows that

$$\mathcal{W}(fg) = \frac{1}{2}(\delta(\mathbf{w}) + \mathcal{W}(f) + \mathcal{W}(g) - \mathcal{W}(f+g)). \tag{7.20}$$

7.2.3 Disjunct Boolean Functions

The subspace of $\mathrm{GF}(2)^n$ generated by the masks $\mathbf{w}$ for which $F(\mathbf{w}) \neq 0$ is called the *support space* of f and is denoted by $\mathcal{V}_f$. The support space of the XOR of two binary Boolean functions is a subspace of the (vector) sum of their corresponding support spaces:

$$\mathcal{V}_{f+g} \subseteq \mathcal{V}_f + \mathcal{V}_g \tag{7.21}$$

This follows directly from the convolution property. Two binary Boolean functions are called *disjunct* if their support spaces are disjunct, i.e. if the intersection of their support spaces only contains the origin. A vector $\mathbf{v} \in \mathcal{V}_{f+g}$ with f and g disjunct has a unique decomposition into a component $\mathbf{u} \in \mathcal{V}_f$ and a component $\mathbf{w} \in \mathcal{V}_g$. In this case the spectrum of $h = f + g$ satisfies

$$H(\mathbf{v}) = F(\mathbf{u})G(\mathbf{w}) \text{ where } \mathbf{v} = \mathbf{u} + \mathbf{w} \text{ and } \mathbf{u} \in \mathcal{V}_f, \mathbf{w} \in \mathcal{V}_g. \tag{7.22}$$

A pair of binary Boolean functions that depend on non-overlapping sets of input bits is a special case of disjunct functions.

7.3 Correlation Matrices

Almost all components in block ciphers are Boolean functions mapping n-bit vectors to m-bit vectors. Examples are S-boxes, round transformations and their steps, and block ciphers themselves. In many cases $m = n$. These functions can be represented by their *correlation matrix*.

A Boolean function $h : \mathrm{GF}(2)^n \rightarrow \mathrm{GF}(2)^m$ can be decomposed into m *component* binary Boolean functions:

$$(h_0, h_1, \ldots, h_{m-1}).$$

Each of these component binary Boolean functions h_i has a spectrum H_i. The vector function H with components H_i can be considered the spectrum of the Boolean function h. As in the case of binary Boolean functions, H completely determines the function h.

The spectrum of any parity of components of h is specified by a simple extension of (7.18):

$$\mathcal{W}(\mathbf{u}^{\mathrm{T}} h) = \bigotimes_{u_i = 1} H_i. \tag{7.23}$$

The correlations between input parities and output parities of a Boolean function h can be arranged in a $2^m \times 2^n$ *correlation matrix* $\mathsf{C}^{(h)}$. The element $\mathsf{C}^{(h)}_{\mathbf{u},\mathbf{w}}$ in row $\mathbf{u}$ and column $\mathbf{w}$ is equal to $\mathsf{C}(\mathbf{u}^{\mathrm{T}} h(\mathbf{a}), \mathbf{w}^{\mathrm{T}} \mathbf{a})$.

Row $\mathbf{u}$ of a correlation matrix can be interpreted as

$$(-1)^{\mathbf{u}^{\mathrm{T}} h(\mathbf{a})} = \sum_{\mathbf{w}} \mathsf{C}^{(h)}_{\mathbf{u},\mathbf{w}} (-1)^{\mathbf{w}^{\mathrm{T}}\mathbf{a}}. \tag{7.24}$$

This expresses an output parity with respect to the basis of input parities.

A binary Boolean function $f(\mathbf{a})$ is a special case of a Boolean function: it has $m = 1$. Its correlation matrix has two rows: row 0 and row 1. Row 1 contains the spectrum of $f(\mathbf{a})$. Row 0 contains the spectrum of the *empty parity*: the binary Boolean function that is equal to 0. This row has a 1 in column 0 and zeroes in all other columns.

7.3.1 Equivalence of a Boolean Function and Its Correlation Matrix

A correlation matrix $cormat^{(h)}$ defines a linear map with domain $\mathbb{R}^{2^n}$ and range $\mathbb{R}^{2^m}$. Let $\mathcal{L}$ be a transformation from the space of binary vectors to the space of real-valued functions that transforms a binary vector of dimension n to a real-valued function of dimension 2^n. $\mathcal{L}$ is defined by

$$\mathcal{L} : \mathrm{GF}(2)^n \to \mathbb{R}^{2^n} : \mathbf{a} \mapsto \mathcal{L}(\mathbf{a}) = \alpha \Leftrightarrow \alpha_u = (-1)^{\mathbf{u}^{\mathrm{T}}\mathbf{a}}. \tag{7.25}$$

Since $\mathcal{L}(\mathbf{a}+\mathbf{b}) = \mathcal{L}(\mathbf{a}) \cdot \mathcal{L}(\mathbf{b})$, $\mathcal{L}$ is a group homomorphism from $(\mathrm{GF}(2)^n, +)$ to $((\mathbb{R}\backslash\{0\})^{2^n}, \cdot)$, where '$\cdot$' denotes the component-wise product. From (7.24) it follows that

$$\mathsf{C}^{(h)} \mathcal{L}(\mathbf{a}) = \mathcal{L}(h(\mathbf{a})). \tag{7.26}$$

In words, applying a Boolean function h to a Boolean vector $\mathbf{a}$ and transforming the corresponding function $(-1)^{\mathbf{u}^{\mathrm{T}}\mathbf{a}}$ with the correlation matrix $\mathsf{C}^{(h)}$ are just different representations of the same operation. This is illustrated in Fig. 7.3.

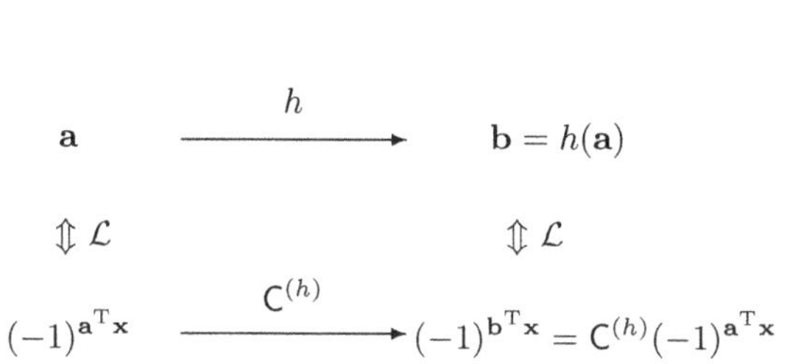

$$\begin{array}{ccc}
\mathbf{a} & \xrightarrow{\quad h \quad} & \mathbf{b} = h(\mathbf{a}) \\
\Updownarrow \mathcal{L} & & \Updownarrow \mathcal{L} \\
(-1)^{\mathbf{a}^{\mathrm{T}}\mathbf{x}} & \xrightarrow{\quad \mathsf{C}^{(h)} \quad} & (-1)^{\mathbf{b}^{\mathrm{T}}\mathbf{x}} = \mathsf{C}^{(h)}(-1)^{\mathbf{a}^{\mathrm{T}}\mathbf{x}}
\end{array}$$

Fig. 7.3. The equivalence of a Boolean function and its correlation matrix

7.3.2 Iterative Boolean Functions

Consider an iterated Boolean function h that is the composition of two Boolean functions $h = h^{(2)} \circ h^{(1)}$ or $h(\mathbf{a}) = h^{(2)}(h^{(1)}(\mathbf{a}))$, where the function $h^{(1)}$ transforms n-bit vectors to p-bit vectors and where the function $h^{(2)}$ transforms p-bit vectors to m-bit vectors. The correlation matrix of h is determined by the correlation matrices of the component functions. We have

$$\begin{aligned}(-1)^{\mathbf{u}^{\mathrm{T}}h(\mathbf{a})} &= \sum_{\mathbf{v}} \mathsf{C}^{(h^{(2)})}_{\mathbf{u},\mathbf{v}}(-1)^{\mathbf{v}^{\mathrm{T}}h^{(1)}(\mathbf{a})} \\ &= \sum_{\mathbf{v}} \mathsf{C}^{(h^{(2)})}_{\mathbf{u},\mathbf{v}} \sum_{\mathbf{w}} \mathsf{C}^{(h^{(1)})}_{\mathbf{v},\mathbf{w}}(-1)^{\mathbf{w}^{\mathrm{T}}\mathbf{a}} \\ &= \sum_{\mathbf{w}} \left(\sum_{\mathbf{v}} \mathsf{C}^{(h^{(2)})}_{\mathbf{u},\mathbf{v}} \mathsf{C}^{(h^{(1)})}_{\mathbf{v},\mathbf{w}} \right) (-1)^{\mathbf{w}^{\mathrm{T}}\mathbf{a}}.\end{aligned}$$

Hence, we have

$$\mathsf{C}^{(h^{(2)} \circ h^{(1)})} = \mathsf{C}^{(h^{(2)})} \times \mathsf{C}^{(h^{(1)})}, \tag{7.27}$$

where $\times$ denotes the matrix product, $\mathsf{C}^{(h^{(1)})}$ is a $2^p \times 2^n$ matrix and $\mathsf{C}^{(h^{(2)})}$ is a $2^m \times 2^p$ matrix. Hence the correlation matrix of the composition of two Boolean functions is the product of their correlation matrices. This is illustrated in Fig. 7.4.

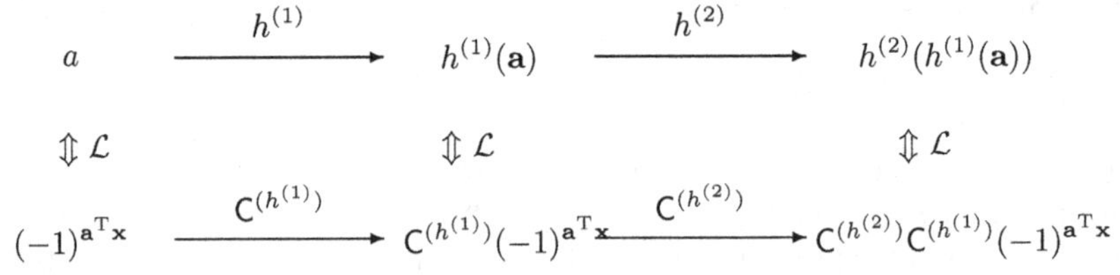

$$\begin{array}{ccccc} a & \xrightarrow{h^{(1)}} & h^{(1)}(\mathbf{a}) & \xrightarrow{h^{(2)}} & h^{(2)}(h^{(1)}(\mathbf{a})) \\ \Updownarrow \mathcal{L} & & \Updownarrow \mathcal{L} & & \Updownarrow \mathcal{L} \\ (-1)^{\mathbf{a}^{\mathrm{T}}\mathbf{x}} & \xrightarrow{\mathsf{C}^{(h^{(1)})}} & \mathsf{C}^{(h^{(1)})}(-1)^{\mathbf{a}^{\mathrm{T}}\mathbf{x}} & \xrightarrow{\mathsf{C}^{(h^{(2)})}} & \mathsf{C}^{(h^{(2)})}\mathsf{C}^{(h^{(1)})}(-1)^{\mathbf{a}^{\mathrm{T}}\mathbf{x}} \end{array}$$

Fig. 7.4. Composition of Boolean functions and multiplication of correlation matrices

The correlations over $h = h^{(2)} \circ h^{(1)}$ are given by

$$\mathsf{C}(\mathbf{u}^{\mathrm{T}}h(\mathbf{a}), \mathbf{w}^{\mathrm{T}}\mathbf{a}) = \sum_{\mathbf{v}} \mathsf{C}(\mathbf{u}^{\mathrm{T}}h^{(2)}(\mathbf{a}), \mathbf{v}^{\mathrm{T}}\mathbf{a})\mathsf{C}(\mathbf{v}^{\mathrm{T}}h^{(1)}(\mathbf{a}), \mathbf{w}^{\mathrm{T}}\mathbf{a}). \tag{7.28}$$

7.3.3 Boolean Permutations

If h is a permutation in $\mathrm{GF}(2)^n$, we have

$$\mathrm{C}(\mathbf{u}^{\mathrm{T}}h^{-1}(\mathbf{a}), \mathbf{w}^{\mathrm{T}}\mathbf{a}) = \mathrm{C}(\mathbf{u}^{\mathrm{T}}\mathbf{b}, \mathbf{w}^{\mathrm{T}}h(\mathbf{b})) = \mathrm{C}(\mathbf{w}^{\mathrm{T}}h(\mathbf{b}), \mathbf{u}^{\mathrm{T}}\mathbf{b}). \tag{7.29}$$

It follows that the correlation matrix of h^{-1} is the transpose of the correlation matrix of h:

$$\mathsf{C}^{(h^{-1})} = \left(\mathsf{C}^{(h)}\right)^{\mathrm{T}}. \tag{7.30}$$

Moreover, from the fact that the composition of a Boolean transformation and its inverse gives the identity function, the product of the corresponding correlation matrices must result in the identity matrix:

$$\mathsf{C}^{(h^{-1})} \times \mathsf{C}^{(h)} = \mathsf{I} = \mathsf{C}^{(h)} \times \mathsf{C}^{(h^{-1})}. \tag{7.31}$$

It follows that

$$\mathsf{C}^{(h^{-1})} = \left(\mathsf{C}^{(h)}\right)^{-1}. \tag{7.32}$$

We can now prove the following theorem.

Theorem 7.3.1. *A Boolean transformation is invertible iff it has an invertible correlation matrix.*

Proof.

$\Rightarrow$ For an invertible Boolean transformation, both (7.30) and (7.32) are valid. Combining these two equations, we have

$$\left(\mathsf{C}^{(h)}\right)^{-1} = \left(\mathsf{C}^{(h)}\right)^{\mathrm{T}}.$$

$\Leftarrow$ Interpreting the rows of the correlation matrix according to (7.24) yields a set of 2^n equations, one for each value of $\mathbf{u}$:

$$(-1)^{\mathbf{u}^{\mathrm{T}}h(\mathbf{a})} = \sum_{\mathbf{w}} \mathsf{C}^{(h)}_{\mathbf{u},\mathbf{w}}(-1)^{\mathbf{w}^{\mathrm{T}}\mathbf{a}}.$$

If we assume that the inverse of $\mathsf{C}^{(h)}$ exists, we can convert this set of 2^n equations, one for each value of $\mathbf{w}$:

$$(-1)^{\mathbf{w}^{\mathrm{T}}\mathbf{a}} = \sum_{\mathbf{u}} \left(\mathsf{C}^{(h)}\right)^{-1}_{\mathbf{w},\mathbf{u}}(-1)^{\mathbf{u}^{\mathrm{T}}h(\mathbf{a})}. \tag{7.33}$$

Assume that we have two Boolean vectors $\mathbf{x}$ and $\mathbf{y}$ for which $h(\mathbf{x}) = h(\mathbf{y})$. By substituting $\mathbf{a}$ in (7.33) by $\mathbf{x}$ and $\mathbf{y}$ respectively, we obtain 2^n equations, one for each value of $\mathbf{w}$:

$$(-1)^{\mathbf{w}^{\mathrm{T}}\mathbf{x}} = (-1)^{\mathbf{w}^{\mathrm{T}}\mathbf{y}}.$$

From this it follows that $\mathbf{x} = \mathbf{y}$ and hence that h is injective. It follows that h is invertible. □

7.4 Special Functions

In the following, the superscript (h) in $\mathsf{C}^{(h)}$ will be omitted.

7.4.1 Addition with a Constant

Consider the function that consists of the addition (XOR) with a constant vector $\mathbf{k}$: $h(\mathbf{a}) = \mathbf{a} + \mathbf{k}$. Since $\mathbf{u}^{\mathrm{T}}h(\mathbf{a}) = \mathbf{u}^{\mathrm{T}}\mathbf{a} + \mathbf{u}^{\mathrm{T}}\mathbf{k}$, the correlation matrix is a diagonal matrix with

$$\mathsf{C}_{\mathbf{u},\mathbf{u}} = (-1)^{\mathbf{u}^{\mathrm{T}}\mathbf{k}}. \tag{7.34}$$

Therefore the effect of the bitwise XOR with a constant vector before (or after) a function h on its correlation matrix is a multiplication of some columns (or rows) by -1.

7.4.2 Linear Functions

Consider a linear function $h(\mathbf{a}) = \mathsf{M}\mathbf{a}$, with M an $m \times n$ binary matrix. We have

$$\mathbf{u}^{\mathrm{T}}h(\mathbf{a}) = \mathbf{u}^{\mathrm{T}}\mathsf{M}\mathbf{a} = (\mathsf{M}^{\mathrm{T}}\mathbf{u})^{\mathrm{T}}\mathbf{a}. \tag{7.35}$$

The elements of the corresponding correlation matrix are given by

$$\mathsf{C}_{\mathbf{u},\mathbf{w}} = \delta(\mathsf{M}^{\mathrm{T}}\mathbf{u} + \mathbf{w}). \tag{7.36}$$

If M is an invertible matrix, the correlation matrix is a permutation matrix. The single non-zero element in row $\mathbf{u}$ is in column $\mathsf{M}^{\mathrm{T}}\mathbf{u}$. The effect of applying an invertible linear function before (or after) a function h on the correlation matrix is only a permutation of its columns (or rows).

7.4.3 Bricklayer Functions

Consider a bricklayer function $\mathbf{b} = h(\mathbf{a})$ that is defined by the following component functions:

$$\mathbf{b}_{(i)} = h_{(i)}\left(\mathbf{a}_{(i)}\right)$$

for $1 \leq i \leq \ell$. For every component function $h_{(i)}$, there is a corresponding correlation matrix denoted by $\mathsf{C}^{(i)}$.

From the fact that the different component functions $h_{(i)}$ operate on non-overlapping sets of input bits and are therefore disjunct, (7.22) can be applied. The elements of the correlation matrix of h are given by

$$\mathsf{C}_{\mathbf{u},\mathbf{w}} = \prod_i \mathsf{C}^{(i)}_{\mathbf{u}_{(i)},\mathbf{w}_{(i)}}, \tag{7.37}$$

where

$$\mathbf{u} = \left(\mathbf{u}_{(1)}, \mathbf{u}_{(2)}, \ldots, \mathbf{u}_{(\ell)}\right)$$

and

$$\mathbf{w} = \left(\mathbf{w}_{(1)}, \mathbf{w}_{(2)}, \ldots, \mathbf{w}_{(\ell)}\right).$$

In words, the correlation between an input parity and an output parity is the product of the correlations of the corresponding input and output parities of the component functions $\mathsf{C}^{(i)}_{\mathbf{u}_{(i)},\mathbf{w}_{(i)}}$.

7.4.4 Keyed Functions

For a keyed function $h[\mathbf{k}]$ the correlations $\mathsf{C}^{(h[\mathbf{k}])}_{\mathbf{u},\mathbf{v}}$ depend on the value of the key $\mathbf{k}$. We define the expected linear potential (ELP) associated with an output mask $\mathbf{u}$ and an input mask $\mathbf{v}$ over a keyed map $h[\mathbf{k}]$ as the average of the square of the correlation $\mathsf{C}^{(h[\mathbf{k}])}_{\mathbf{u},\mathbf{v}}$ over all keys.

Definition 7.4.1 (ELP). *The* ELP *associated with an output mask* $\mathbf{u}$ *and an input mask* $\mathbf{v}$, $\mathrm{ELP}(\mathbf{u},\mathbf{v})$ *over a keyed map* $h[\mathbf{k}]$ *is given by*

$$\mathrm{ELP}(\mathbf{u},\mathbf{v}) = 2^{-|\mathcal{K}|} \sum_{\mathbf{k}\in\mathcal{K}} \left(\mathsf{C}^{(h[\mathbf{k}])}_{\mathbf{u},\mathbf{v}}\right)^2 .$$

7.5 Derived Properties

The concept of the correlation matrix is a valuable tool for demonstrating properties of Boolean functions and their spectra. We will illustrate this with some examples.

Lemma 7.5.1. *The elements of the correlation matrix of a Boolean function satisfy*

$$\mathsf{C}_{(\mathbf{u}+\mathbf{v}),\mathbf{x}} = \sum_{\mathbf{w}} \mathsf{C}_{\mathbf{u},(\mathbf{w}+\mathbf{x})} \mathsf{C}_{\mathbf{v},\mathbf{w}}, \tag{7.38}$$

for all $\mathbf{u},\mathbf{v},\mathbf{x} \in \mathrm{GF}(2)^n$.

Proof. Using the convolution property, we have

$$\mathcal{W}((\mathbf{u}+\mathbf{v})^{\mathrm{T}}h(\mathbf{a})) = \mathcal{W}(\mathbf{u}^{\mathrm{T}}h(\mathbf{a}) + \mathbf{v}^{\mathrm{T}}h(\mathbf{a})) \tag{7.39}$$

$$= \mathcal{W}(\mathbf{u}^{\mathrm{T}}h(\mathbf{a})) \otimes \mathcal{W}(\mathbf{v}^{\mathrm{T}}h(\mathbf{a})). \tag{7.40}$$

Since the components of $\mathcal{W}(\mathbf{u}^{\mathrm{T}}h(\mathbf{a}))$ are given by $\mathsf{C}_{\mathbf{u},\mathbf{w}}$, the projection of (7.40) onto the component with index $\mathbf{x}$ gives rise to (7.38). □

From this lemma we have the following.

Corollary 7.5.1. *The correlation between two output parities defined by* $\mathbf{u}$ *and* $\mathbf{v}$ *is equal to the convolution of columns* $\mathbf{u}$ *and* $\mathbf{v}$ *of the correlation matrix:*

$$\mathsf{C}_{(\mathbf{u}+\mathbf{v}),0} = \sum_{\mathbf{w}} \mathsf{C}_{\mathbf{u},\mathbf{w}} \mathsf{C}_{\mathbf{v},\mathbf{w}}. \tag{7.41}$$

A binary Boolean function is *balanced* if it is 1 (or 0) for exactly half of the elements in the domain. Clearly, being balanced is equivalent to being uncorrelated to the binary Boolean function equal to 0 (or 1). Using the properties of correlation matrices we can now give an elegant proof of the following well-known theorem [2].

Theorem 7.5.1. *A Boolean transformation is invertible iff every output parity is a balanced binary Boolean function of the input bits.*

Proof.

⇒ If h is an invertible transformation, its correlation matrix C is orthogonal. Since $\mathsf{C}_{0,0} = 1$ and all rows and columns have norm 1, it follows that there are no other elements in row 0 or column 0 different from 0. Hence, $\mathsf{C}(\mathbf{u}^{\mathrm{T}}h(\mathbf{a}), 0) = \delta(\mathbf{u})$ or $\mathbf{u}^{\mathrm{T}}h(\mathbf{a})$ is balanced for all $\mathbf{u} \neq 0$.

⇐ The condition that all output parities are balanced binary Boolean functions of input bits corresponds to $\mathsf{C}_{\mathbf{u},0} = 0$ for $\mathbf{u} \neq 0$. If this is the case, we can show that the correlation matrix is orthogonal. The expression $\mathsf{C}^{\mathrm{T}} \times \mathsf{C} = \mathsf{I}$ is equivalent to the following set of conditions:

$$\sum_{\mathbf{w}} \mathsf{C}_{\mathbf{u},\mathbf{w}} \mathsf{C}_{\mathbf{v},\mathbf{w}} = \delta(\mathbf{u}+\mathbf{v}) \text{ for all } \mathbf{u}, \mathbf{v} \in \mathrm{GF}(2)^n. \tag{7.42}$$

Using (7.41), we have

$$\sum_{\mathbf{w}} \mathsf{C}_{\mathbf{u},\mathbf{w}} \mathsf{C}_{\mathbf{v},\mathbf{w}} = \mathsf{C}_{(\mathbf{u}+\mathbf{v}),0}. \tag{7.43}$$

Since $\mathsf{C}_{\mathbf{u},0} = 0$ for all $\mathbf{u} \neq 0$, and $\mathsf{C}_{0,0} = 1$, (7.42) holds for all possible pairs $\mathbf{u}, \mathbf{v}$. It follows that C is an orthogonal matrix, hence h^{-1} exists and is defined by C^{T}. □

Lemma 7.5.2. *The elements of the correlation matrix of a Boolean function with domain* $\mathrm{GF}(2)^n$ *and the spectrum values of a binary Boolean function with domain* $\mathrm{GF}(2)^n$ *are integer multiples of* 2^{1-n}.

Proof. The sum in the right-hand side of (7.8) is always even since its value is of the form $k \cdot (1) + (2^n - k) \cdot (-1) = 2k - 2^n$. It follows that the spectrum values must be integer multiples of 2^{1-n}. □

7.6 Truncating Functions

A function from $\mathrm{GF}(2)^n$ to $\mathrm{GF}(2)^m$ can be converted into a function from $\mathrm{GF}(2)^{n-1}$ to $\mathrm{GF}(2)^m$ by fixing a single bit of the input. More generally, a bit of the input can be set equal to a parity of other input components, possibly complemented. Such a restriction is of the type

$$\mathbf{v}^{\mathrm{T}}\mathbf{a} = \epsilon, \tag{7.44}$$

where $\epsilon \in \mathrm{GF}(2)$. Assume that $v_s \neq 0$.

The restriction can be modeled by a Boolean function $\mathbf{a}' = h^r(\mathbf{a})$ that is applied before h. It maps $\mathrm{GF}(2)^{n-1}$ to $\mathrm{GF}(2)^n$, and is specified by $a'_i = a_i$ for $i \neq s$ and $a'_s = \epsilon + \mathbf{v}^{\mathrm{T}}\mathbf{a} + a_s$. The non-zero elements of the correlation matrix of h^r are

$$\mathsf{C}^{(h^r)}_{\mathbf{w},\mathbf{w}} = 1 \text{ and } \mathsf{C}^{(h^r)}_{(\mathbf{v}+\mathbf{w}),\mathbf{w}} = (-1)^{\epsilon} \text{ for all } \mathbf{w} \text{ where } w_s = 0. \tag{7.45}$$

All columns of this matrix have exactly two non-zero entries with amplitude 1.

The function restricted to the specified subset of inputs is the consecutive application of h^r and the function itself. Hence, its correlation matrix C' is $\mathsf{C} \times \mathsf{C}^{(h^r)}$. The elements of this matrix are

$$\mathsf{C}'_{\mathbf{u},\mathbf{w}} = \mathsf{C}_{\mathbf{u},\mathbf{w}} + (-1)^{\epsilon}\mathsf{C}_{\mathbf{u},(\mathbf{w}+\mathbf{v})} \tag{7.46}$$

if $w_s = 0$, and 0 if $w_s = 1$. The elements in C' are spectrum values of Boolean functions of $(n-1)$-dimensional vectors. Hence, from Lemma 7.5.2 they must be integer multiples of 2^{2-n}.

Applying (7.15) to the rows of the restricted correlation matrices gives additional laws for the spectrum values of Boolean functions. For the single restrictions of the type $\mathbf{v}^{\mathrm{T}}\mathbf{a} = \epsilon$ we have

$$\sum_{\mathbf{w}} (F(\mathbf{w}) + F(\mathbf{w}+\mathbf{v}))^2 = \sum_{\mathbf{w}} (F(\mathbf{w}) - F(\mathbf{w}+\mathbf{v}))^2 = 2. \tag{7.47}$$

Lemma 7.6.1. *The elements of a correlation matrix corresponding to an invertible transformation of n-bit vectors are integer multiples of* 2^{2-n}.

Proof. Let g be the Boolean function from $\mathrm{GF}(2)^{n-1}$ to $\mathrm{GF}(2)^n$ that is obtained by restricting the input of function h. Let the input restriction be specified by the vector $\mathbf{w}$: $\mathbf{w}^T\mathbf{a} = 0$. Then $C^{(g)}_{\mathbf{u},\mathbf{v}} = C^{(h)}_{\mathbf{u},\mathbf{v}} + C^{(h)}_{\mathbf{u},(\mathbf{v}+\mathbf{w})}$ or $C^{(g)}_{\mathbf{u},\mathbf{v}} = 0$. By filling in 0 for $\mathbf{v}$, this yields $C^{(g)}_{\mathbf{u},0} = C^{(h)}_{\mathbf{u},0} + C^{(h)}_{\mathbf{u},\mathbf{w}}$. Now, $C^{(g)}_{\mathbf{u},0}$ must be an integer multiple of 2^{2-n}, and since according to Theorem 7.5.1 $C^{(h)}_{\mathbf{u},0} = 0$, it follows that $C^{(h)}_{\mathbf{u},\mathbf{w}}$ is also an integer multiple of 2^{2-n}. □

7.7 Cross-correlation and Autocorrelation

The cross-correlation function [108, p. 117] of two Boolean functions $f(\mathbf{a})$ and $g(\mathbf{a})$ is denoted by $\hat{c}_{fg}(\mathbf{b})$, and given by

$$\begin{aligned} \hat{c}_{fg}(\mathbf{b}) &= C(f(\mathbf{a}), g(\mathbf{a}+\mathbf{b})) &(7.48)\\ &= 2^{-n}\sum_{\mathbf{a}} \hat{f}(\mathbf{a})\hat{g}(\mathbf{a}+\mathbf{b}) = 2^{-n}\sum_{\mathbf{a}}(-1)^{f(\mathbf{a})+g(\mathbf{a}+\mathbf{b})}. &(7.49)\end{aligned}$$

Now consider FG, the product of the spectra of two binary Boolean functions f and g:

$$\begin{aligned} F(\mathbf{w})G(\mathbf{w}) &= (2^{-n}\sum_{\mathbf{a}} \hat{f}(\mathbf{a})(-1)^{\mathbf{w}^T\mathbf{a}})(2^{-n}\sum_{\mathbf{b}} \hat{g}(\mathbf{b})(-1)^{\mathbf{w}^T\mathbf{b}}) &(7.50)\\ &= 2^{-n}\sum_{\mathbf{a}}(2^{-n}\sum_{\mathbf{b}} \hat{f}(\mathbf{a})\hat{g}(\mathbf{b})(-1)^{\mathbf{w}^T(\mathbf{a}+\mathbf{b})}) &(7.51)\\ &= 2^{-n}\sum_{\mathbf{a}}((2^{-n}\sum_{\mathbf{c}} \hat{f}(\mathbf{a})\hat{g}(\mathbf{a}+\mathbf{c})(-1)^{\mathbf{w}^T\mathbf{c}})) &(7.52)\\ &= \mathcal{W}(2^{-n}\sum_{\mathbf{c}} \hat{f}(\mathbf{a})\hat{g}(\mathbf{a}+\mathbf{c})) &(7.53)\\ &= \mathcal{W}(\hat{c}_{fg}(\mathbf{b})). &(7.54)\end{aligned}$$

Hence the spectrum of the cross-correlation function of two binary Boolean functions equals the product of the spectra of the binary Boolean functions: $\hat{c}_{fg} = \mathcal{W}^{-1}(FG)$.

The cross-correlation function of a binary Boolean function with itself, $\hat{c}_{ff}$, is called the *autocorrelation function* of f and is denoted by $\hat{r}_f$. It follows that the components of the spectrum of the autocorrelation function are the squares of the components of the spectrum of f, i.e.

$$F^2 = \mathcal{W}(\hat{r}_f). \tag{7.55}$$

This equation is generally referred to as the Wiener-Khintchine theorem [128].

7.8 Linear Trails

Let β be an iterative Boolean transformation operating on n-bit vectors:

$$\beta = \rho^{(r)} \circ \rho^{(r-1)} \circ \ldots \circ \rho^{(2)} \circ \rho^{(1)}. \tag{7.56}$$

The correlation matrix of β is the product of the correlation matrices corresponding to the respective Boolean transformations:

$$\mathsf{C}^{(\beta)} = \mathsf{C}^{(\rho^{(r)})} \times \ldots \times \mathsf{C}^{(\rho^{(2)})} \times \mathsf{C}^{(\rho^{(1)})}. \tag{7.57}$$

A *linear trail* U over an iterative Boolean transformation consists of a sequence of $r+1$ masks:

$$\mathsf{U} = \left(\mathbf{u}^{(0)}, \mathbf{u}^{(1)}, \mathbf{u}^{(2)}, \ldots, \mathbf{u}^{(r-1)}, \mathbf{u}^{(r)}\right). \tag{7.58}$$

This linear trail is a sequence of r linear steps $\left(\mathbf{u}^{(i-1)}, \mathbf{u}^{(i)}\right)$ that have a correlation

$$\mathrm{C}\left({\mathbf{u}^{(i)}}^{\mathrm{T}} \rho^{(i)}(\mathbf{a}), {\mathbf{u}^{(i-1)}}^{\mathrm{T}} \mathbf{a}\right).$$

The *correlation contribution* $\mathrm{C_p}$ of a linear trail is the product of the correlation of all its steps:

$$\mathrm{C_p}(\mathsf{U}) = \prod_i \mathsf{C}^{(\rho^{(i)})}_{\mathbf{u}^{(i)}\mathbf{u}^{(i-1)}}. \tag{7.59}$$

As the correlations range from -1 to $+1$, so do correlation contributions. In the following, we will often work with the linear potential (LP) of a trail.

Definition 7.8.1. *The square of the correlation contribution of a linear trail* U *is its* linear potential (LP).

$$\mathrm{LP}(\mathsf{U}) = (\mathrm{C_p}(\mathsf{U}))^2.$$

From the definition of linear trails and (7.57), we can derive the following theorem.

Theorem 7.8.1 (Theorem of Linear Trail Composition). *The correlation between output parity* $\mathbf{u}^{\mathrm{T}}\beta(\mathbf{a})$ *and input parity* $\mathbf{w}^{\mathrm{T}}\mathbf{a}$ *of an iterated Boolean transformation with* r *rounds is the sum of the correlation contributions of all* r*-round linear trails* U *with initial mask* $\mathbf{w}$ *and final mask* $\mathbf{u}$*:*

$$\mathrm{C}(\mathbf{u}^{\mathrm{T}}\beta(\mathbf{a}), \mathbf{w}^{\mathrm{T}}\mathbf{a}) = \sum_{\mathbf{u}^{(0)}=\mathbf{w}, \mathbf{u}^{(r)}=\mathbf{u}} \mathrm{C_p}(\mathsf{U}). \tag{7.60}$$

Both the correlation and the correlation contributions are signed. Some of the correlation contributions will have the same sign as the resulting correlation and contribute positively to its amplitude; the others contribute negatively to its amplitude. We speak of *constructive interference* in the case of two linear trails that have a correlation contribution with the same sign and of *destructive interference* if their correlation contributions have different signs.

7.9 Ciphers

The described formalism and tools can be applied to the calculation of correlations in iterated block ciphers such as the DES and Rijndael.

7.9.1 General Case

In general, an iterative cipher consists of a sequence of keyed rounds, where each round $\rho^{(i)}$ depends on its round key $\mathbf{k}^{(i)}$. In a typical cryptanalytic setting, the round keys are fixed and we can model the cipher as an iterative Boolean transformation.

Linear cryptanalysis requires the knowledge of an output parity and an input parity that have a high correlation over all but a few rounds of the cipher. These correlations are the sum of the correlation contributions of all linear trails that connect the output parity with the input parity.

In general, the correlations over a round depend on the key value, and hence computing the correlation contribution of linear trails requires making assumptions about the round key values. However, in many cases the cipher structure allows the analysis of linear trails without having to make assumptions about the value of the round keys. We will now show that for a key-alternating cipher the amplitude of the correlation contribution is independent of the round key.

7.9.2 Key-Alternating Cipher

We have shown that the Boolean transformation corresponding to a key addition consisting of the XOR with a round key $\mathbf{k}^{(i)}$ has a correlation matrix with only non-zero elements on its diagonal. The element is -1 if $\mathbf{u}^{\mathrm{T}}\mathbf{k}^{(i)} = 1$, and 1 otherwise. If we denote the correlation matrix of ρ by C, the correlation contribution of the linear trail U then becomes

$$\mathsf{C}_{\mathsf{p}}(\mathsf{U}) = \prod_i (-1)^{\mathbf{u}^{(i)\mathrm{T}}\mathbf{k}^{(i)}} \mathsf{C}_{\mathbf{u}^{(i)},\mathbf{u}^{(i-1)}} \quad (7.61)$$

$$= (-1)^{d_{\mathsf{U}}+\sum_i \mathbf{u}^{(i)\mathrm{T}}\mathbf{k}^{(i)}} \mathrm{LP}^{\frac{1}{2}}(\mathsf{U}), \quad (7.62)$$

where $d_{\mathsf{U}} = 1$ if $\prod_i \mathsf{C}_{\mathbf{u}^{(i)}\mathbf{u}^{(i-1)}}$ is negative, and $d_{\mathsf{U}} = 0$ otherwise. $\mathrm{LP}^{\frac{1}{2}}(\mathsf{U})$ is independent of the round keys, and hence only the sign of the correlation contribution is key-dependent. This sign can be expressed as a parity of the expanded key K plus a key-independent constant:

$$s = \mathsf{U}^{\mathrm{T}}\mathsf{K} + d_{\mathsf{U}}, \tag{7.63}$$

where K denotes the expanded key and U denotes the concatenation of the masks $\mathbf{u}^{(i)}$.

The correlation between output parity $\mathbf{u}^{\mathrm{T}}\beta(\mathbf{a})$ and input parity $\mathbf{w}^{\mathrm{T}}\mathbf{a}$ expressed in terms of the correlation contributions of linear trails now becomes

$$\mathrm{C}(\mathbf{v}^{\mathrm{T}}\beta(\mathbf{a}), \mathbf{w}^{\mathrm{T}}\mathbf{a}) = \sum_{\mathbf{u}^{(0)}=\mathbf{w}, \mathbf{u}^{(r)}=\mathbf{v}} (-1)^{d_{\mathsf{U}}+\mathsf{U}^{\mathrm{T}}\mathsf{K}}\mathrm{LP}^{\frac{1}{2}}(\mathsf{U}). \tag{7.64}$$

Even though for a key-alternating cipher the amplitudes of the correlation contribution of the individual linear trails are independent of the round keys, this is not the case for the amplitude of the resulting correlation on the left-hand side of the equation. The terms in the right-hand side of the equation are added or subtracted depending on the value of the round keys. It depends on the value of the round keys whether interference between a pair of linear trails is constructive or destructive.

7.9.3 Averaging over All Round Keys

In Sect. 7.9.2 we have only discussed correlations for cases in which the value of the key is fixed. For key-alternating ciphers we can provide an expression for the expected value of linear potentials, taken over all possible values of the expanded key (i.e. the concatenation of all round keys).

Clearly, the correlation $\mathrm{C}\left(\mathbf{u}^{\mathrm{T}}\beta(\mathbf{a}), \mathbf{w}^{\mathrm{T}}\mathbf{a}\right)$ is determined by the set of linear trails U between $\mathbf{v}$ and $\mathbf{u}$. Due to the fact that the correlation contributions of trails depend on the key, the correlations $\mathrm{C}\left(\mathbf{u}^{\mathrm{T}}\beta(\mathbf{a}), \mathbf{w}^{\mathrm{T}}\mathbf{a}\right)$ also depend on the key.

In the case of a key-alternating cipher, the linear potentials $\mathrm{LP}(\mathsf{U})$ of trails are independent of the key. Let us now express the correlation contribution of trail U as

$$\mathrm{C_p}(\mathsf{U}) = (-1)^{s(\mathsf{U})}\mathrm{LP}^{\frac{1}{2}}(\mathsf{U}),$$

where $s(\mathsf{U})$ is a bit determining the sign. The sign for a trail U is equal to a parity of the expanded key plus a trail-specific bit: $s(\mathsf{U}) = \mathsf{U}^{\mathrm{T}}\mathsf{K} + d_{\mathsf{U}}$. We can now prove the following theorem:

Theorem 7.9.1. *The ELP associated with an input and an output mask is the sum of the linear potentials of all linear trails between the input and output mask:*

$$\mathrm{ELP}(\mathbf{u},\mathbf{v}) = \sum_{\mathsf{U}} \mathrm{LP}(\mathsf{U}). \tag{7.65}$$

Proof. The ELP associated with the input mask $\mathbf{v}$ and output mask $\mathbf{u}$ is given by

$$\begin{aligned}\mathrm{ELP}(\mathbf{u},\mathbf{v}) &= 2^{-n_{\mathsf{K}}} \sum_{\mathsf{K}} \left(\sum_{\mathsf{U}} (-1)^{s(\mathsf{U})} \mathrm{LP}^{\frac{1}{2}}(\mathsf{U}) \right)^2 \\ &= 2^{-n_{\mathsf{K}}} \sum_{\mathsf{K}} \left(\sum_{\mathsf{U}} (-1)^{\mathsf{U}^{\mathrm{T}}\mathsf{K}+d_{\mathsf{U}}} \mathrm{LP}^{\frac{1}{2}}(\mathsf{U}) \right)^2 .\end{aligned} \tag{7.66}$$

Working this out yields

$$\begin{aligned}\mathrm{ELP}(\mathbf{u},\mathbf{v}) &= 2^{-n_{\mathsf{K}}} \sum_{\mathsf{K}} \left(\sum_{\mathsf{U}} (-1)^{\mathsf{U}^{\mathrm{T}}\mathsf{K}+d_{\mathsf{U}}} \mathrm{LP}^{\frac{1}{2}}(\mathsf{U}) \right)^2 \\ &= 2^{-n_{\mathsf{K}}} \sum_{\mathsf{K}} \left(\sum_{\mathsf{U}} (-1)^{\mathsf{U}^{\mathrm{T}}\mathsf{K}+d_{\mathsf{U}}} \mathrm{LP}^{\frac{1}{2}}(\mathsf{U}) \right) \left(\sum_{\mathsf{V}} (-1)^{\mathsf{V}^{\mathrm{T}}\mathsf{K}+d_{\mathsf{V}}} \mathrm{LP}^{\frac{1}{2}}(\mathsf{V}) \right) \\ &= 2^{-n_{\mathsf{K}}} \sum_{K} \sum_{\mathsf{U}} \sum_{\mathsf{V}} \left((-1)^{\mathsf{U}^{\mathrm{T}}\mathsf{K}+d_{\mathsf{U}}} \mathrm{LP}^{\frac{1}{2}}(\mathsf{U}) \right) \left((-1)^{\mathsf{V}^{\mathrm{T}}\mathsf{K}+d_{\mathsf{V}}} \mathrm{LP}^{\frac{1}{2}}(\mathsf{V}) \right) \\ &= 2^{-n_{\mathsf{K}}} \sum_{\mathsf{K}} \sum_{\mathsf{U}} \sum_{\mathsf{V}} (-1)^{(\mathsf{U}+\mathsf{V})^{\mathrm{T}}\mathsf{K}+d_{\mathsf{U}}+d_{\mathsf{V}}} \mathrm{LP}^{\frac{1}{2}}(\mathsf{U}) \mathrm{LP}^{\frac{1}{2}}(\mathsf{V}) \\ &= 2^{-n_{\mathsf{K}}} \sum_{\mathsf{U}} \sum_{\mathsf{V}} (\sum_{\mathsf{K}} (-1)^{(\mathsf{U}+\mathsf{V})^{\mathrm{T}}\mathsf{K}+d_{\mathsf{U}}+d_{\mathsf{V}}}) \mathrm{LP}^{\frac{1}{2}}(\mathsf{U}) \mathrm{LP}^{\frac{1}{2}}(\mathsf{V}).\end{aligned} \tag{7.67}$$

For the factor of $\mathrm{LP}^{\frac{1}{2}}(\mathsf{U})\mathrm{LP}^{\frac{1}{2}}(\mathsf{V})$ in (7.67), we have

$$\sum_{\mathsf{K}} (-1)^{(\mathsf{U}+\mathsf{V})^{\mathrm{T}}\mathsf{K}+d_{\mathsf{U}}+d_{\mathsf{V}}} = 2^{n_{\mathsf{K}}} \delta(\mathsf{U}+\mathsf{V}). \tag{7.68}$$

Clearly, the expression is equal to 0 if $\mathsf{U}+\mathsf{V} \neq 0$: if we sum over all values of K, the exponent of (-1) is 1 for half the terms and 0 for the other half. If $\mathsf{U} = \mathsf{V}$ the exponent of (-1) becomes 0 and the expression is equal to $2^{n_{\mathsf{K}}}$. Substitution in (7.67) yields

$$\begin{aligned}\mathrm{ELP}(\mathbf{u},\mathbf{v})) &= 2^{-n_{\mathsf{K}}} \sum_{\mathsf{U}} \sum_{\mathsf{V}} 2^{n_{\mathsf{K}}} \delta(\mathsf{U}+\mathsf{V}) \mathrm{LP}^{\frac{1}{2}}(\mathsf{U}) \mathrm{LP}^{\frac{1}{2}}(\mathsf{V}) \\ &= \sum_{\mathsf{U}} \mathrm{LP}(\mathsf{U}),\end{aligned}$$

proving our theorem. □

As the ELP associated with a pair of masks $(\mathbf{u},\mathbf{v})$ over a block cipher is determined by the LP values of all linear trails between these masks, we speak of *linear hull* $(\mathbf{u},\mathbf{v})$.

7.9.4 The Effect of the Key Schedule

In the previous section we have taken the average of the linear potentials over all possible values of the expanded key, implying independent round keys. In practice, the values of the round keys are restricted by the key schedule, which computes the round keys from the cipher key. In this section we investigate the effect that the key schedule has on expected values of linear potentials.

First assume that we have a linear or affine key schedule. For the sake of simplicity, we limit ourselves to the linear case, but the conclusions are also valid for the affine case. If the key schedule is linear, the relation between the expanded key K and the cipher key $\mathbf{k}$ can be expressed as multiplication with a binary matrix:

$$\mathsf{K} = \mathsf{M}_{\kappa}\mathbf{k}. \tag{7.69}$$

If we substitute this in (7.66), we obtain

$$\mathrm{ELP}(\mathbf{u},\mathbf{v}) = 2^{-n_{\mathsf{k}}} \sum_{\mathbf{k}} \left(\sum_{\mathsf{U}} (-1)^{\mathsf{U}^{\mathrm{T}}\mathsf{M}_{\kappa}\mathbf{k} + d_{\mathsf{U}}} \mathrm{LP}^{\frac{1}{2}}(\mathsf{U}) \right)^2 . \tag{7.70}$$

Working out the squares in this equation yields

$$\mathrm{ELP}(\mathbf{u},\mathbf{v}) = 2^{-n_{\mathsf{k}}} \sum_{\mathsf{U}} \sum_{\mathsf{V}} \left(\sum_{\mathbf{k}} (-1)^{(\mathsf{U}+\mathsf{V})^{\mathrm{T}}\mathsf{M}_{\kappa}\mathbf{k} + d_{\mathsf{U}} + d_{\mathsf{V}}} \right) \mathrm{LP}^{\frac{1}{2}}(\mathsf{U})\mathrm{LP}^{\frac{1}{2}}(\mathsf{V}). \tag{7.71}$$

For the factor of $\mathrm{LP}^{\frac{1}{2}}(\mathsf{U})\mathrm{LP}^{\frac{1}{2}}(\mathsf{V})$ in (7.67), we have

$$\sum_{\mathbf{k}} (-1)^{(\mathsf{U}+\mathsf{V})^{\mathrm{T}}\mathsf{M}_{\kappa}\mathbf{k} + d_{\mathsf{U}} + d_{\mathsf{V}}} = (-1)^{(d_{\mathsf{U}}+d_{\mathsf{V}})} 2^{n_{\mathsf{k}}} \delta(\mathsf{M}_{\kappa}{}^{\mathrm{T}}(\mathsf{U}+\mathsf{V})). \tag{7.72}$$

As above, the expression is equal to 0 if $(\mathsf{U}+\mathsf{V})^{\mathrm{T}}\mathsf{M}_{\kappa} \neq 0$: if we sum over all values of $\mathbf{k}$, the exponent of (-1) is 1 for half the terms and 0 for the other half. However, if $(\mathsf{U}+\mathsf{V})^{\mathrm{T}}\mathsf{M}_{\kappa} = 0$, all terms have the same sign: $(-1)^{d_{\mathsf{U}}+d_{\mathsf{V}}}$. The condition $\mathsf{M}_{\kappa}{}^{\mathrm{T}}(\mathsf{U}+\mathsf{V}) = 0$ is equivalent to saying that the bitwise difference of the two trails is mapped to 0 by M_{κ}, or equivalently, that the two linear trails depend on the same parities of the cipher key for the sign of their correlation contribution. Let us call such a pair of trails a *key-colliding pair*. The effect of a key-colliding pair on the expression of the expected linear potential in terms of the linear potentials of the trails is the following. Next to the terms $\mathrm{LP}(\mathsf{U})$ and $\mathrm{LP}(\mathsf{V})$ there is twice the term $(-1)^{(d_{\mathsf{U}_i}+d_{\mathsf{U}_j})}\mathrm{LP}^{\frac{1}{2}}(\mathsf{U})\mathrm{LP}^{\frac{1}{2}}(\mathsf{V})$. These four terms can be combined into the expression $(\mathrm{LP}^{\frac{1}{2}}(\mathsf{U})+(-1)^{(d_{\mathsf{U}}+d_{\mathsf{V}})}\mathrm{LP}^{\frac{1}{2}}(\mathsf{V}))^2$. The effect of a trail collision is that their correlation contributions systematically interfere positively if $d_{\mathsf{U}_i} = d_{\mathsf{U}_j}$, and negatively otherwise.

Example 7.9.1. We illustrate the above reasoning on an example in which the key schedule has a dramatic impact. Consider a block cipher $B2[\mathbf{k}^{(1)}, \mathbf{k}^{(2)}](\mathbf{x})$ consisting of two rounds. The first round is encryption with a block cipher B with round key $\mathbf{k}^{(1)}$, and the second round is decryption with that same block cipher B with round key $\mathbf{k}^{(2)}$. Assume we have chosen a state-of-the-art block cipher B. In that case, the expected linear potentials between any pair of input and output masks of $B2$ are the sum of the linear potentials of many linear trails over the composition of B and B^{-1}. The expected value of linear potentials corresponding to any pair of input and output masks is approximately 2^{-n_b}.

Let us now consider a block cipher $C[\mathbf{k}]$ defined by

$$C[\mathbf{k}](\mathbf{x}) = B2[\mathbf{k}, \mathbf{k}](\mathbf{x}).$$

Setting $\mathbf{k}^{(1)} = \mathbf{k}^{(2)} = \mathbf{k}$ can be seen as a very simple key schedule. Clearly, the block cipher $C[\mathbf{k}](\mathbf{x})$ is the identity map, and hence we know that it has linear potentials equal to 1 if the input and output mask are the same, and 0 otherwise. This is the consequence of the fact that in this particular example, the key schedule is such that the round keys can no longer be considered as being independent. We have

$$\mathsf{C}_{\mathbf{u},\mathbf{w}}^{(B2[\mathbf{k}^{(1)},\mathbf{k}^{(2)}])} = \sum_{\mathbf{v}} \mathsf{C}_{\mathbf{u},\mathbf{v}}^{(B[\mathbf{k}^{(1)}])} \mathsf{C}_{\mathbf{v},\mathbf{w}}^{(B^{-1}[\mathbf{k}^{(2)}])} = \sum_{\mathbf{v}} \mathsf{C}_{\mathbf{u},\mathbf{v}}^{(B[\mathbf{k}^{(1)}])} \mathsf{C}_{\mathbf{w},\mathbf{v}}^{(B[\mathbf{k}^{(2)}])}. \quad (7.73)$$

If $\mathbf{k}^{(1)} = \mathbf{k}^{(2)} = \mathbf{k}$, this is reduced to

$$\mathsf{C}_{\mathbf{u},\mathbf{w}}^{(C[\mathbf{k}])} = \sum_{\mathbf{v}} \mathsf{C}_{\mathbf{u},\mathbf{v}}^{(B[\mathbf{k}])} \mathsf{C}_{\mathbf{w},\mathbf{v}}^{(B[\mathbf{k}])} = \delta(\mathbf{u} + \mathbf{w}). \quad (7.74)$$

This follows from the fact that for any given value of $\mathbf{k}$, $\mathsf{C}^{(B[\mathbf{k}])}$ is an orthogonal matrix.

As opposed to this extreme example, key-colliding pairs of linear trails are very rare. The condition is that the two trails depend on the same parity of cipher key bits for their sign. The probability that this is the case for two trails is 2^{-n_k}. If the key schedule is nonlinear, linear trails that always interfere constructively or destructively due to the key schedule cannot ever occur. Instead of $\mathsf{K} = \mathsf{M}_\kappa \mathbf{k}$ we have $\mathsf{K} = f_\kappa(\mathbf{k})$ with f_κ a nonlinear function. The coefficients of the mixed terms are of the form

$$\sum_k (-1)^{(\mathsf{U}+\mathsf{V})^{\mathrm{T}} f_\kappa(\mathbf{k}) + d_{\mathsf{U}} + d_{\mathsf{V}}}. \quad (7.75)$$

It seems hard to come up with a reasonable key schedule for which this expression does not have approximately as many positive as negative terms. If that is the case, the sum of the linear potentials of the linear trails is a very good approximation of the expected linear potentials. Still, taking a nonlinear key schedule to avoid systematic constructive interference seems unnecessary in the light of the rarity of the phenomenon.

7.10 Correlation Matrices and the Linear Cryptanalysis Literature

In this section we make an attempt to position our approach with respect to other formalisms and terminology that are used in some of the cryptographic literature.

7.10.1 Linear Cryptanalysis of the DES

For an overview of the original linear cryptanalysis attack on the DES we refer to Sect. 6.3. The multiple-round linear expressions described in [104] correspond to what we call linear trails. The probability p that such an expression holds corresponds to $\frac{1}{2}(1 + \mathrm{C_p}(\mathsf{U}))$, where $\mathrm{C_p}(\mathsf{U})$ is the correlation contribution of the corresponding linear trail. The usage of probabilities in [104] requires the application of the so-called piling-up lemma in the computation of probabilities of composed transformations. When working with correlations, no such tricks are required: correlations can be simply multiplied.

In [104] the correlation over multiple rounds is approximated by the correlation contribution of a single linear trail. The silent assumption underlying this approximation is that the correlation is dominated by a single linear trail. This seems valid because of the large relative amplitude of the described correlation. There are no linear trails with the same initial and final masks that have a correlation contribution that comes close to the dominant trail.

The amplitude of the correlation of the linear trail is independent of the value of the key, and consists of the product of the correlations of its steps. In general, the elements of the correlation matrix of the DES round function are not independent of the round keys, due to the fact that the inputs of neighboring S-boxes overlap while depending on different key bits. However, in the described linear trails the actual independence is caused by the fact that the steps of the described linear trail only involve bits of a single S-box.

The input-output correlations of the F-function of the DES can be calculated by applying the rules given in Sect. 7.4. The 32-bit mask $\mathbf{b}$ at the output of the bit permutation P is converted into a 32-bit mask $\mathbf{c}$ at the output of the S-boxes by a simple linear function. The 32-bit mask $\mathbf{a}$ at the input of the (linear) expansion E gives rise to a set α of $2^{2\ell}$ 48-bit masks after the expansion, where ℓ is the number of pairwise neighboring S-box pairs that are addressed by $\mathbf{a}$.

On the assumption that the round key is all-zero, the correlation between $\mathbf{c}$ and $\mathbf{a}$ can now be calculated by simply adding the correlations corresponding to $\mathbf{c}$ and all masks in α. Since the S-boxes form a bricklayer function, these correlations can be calculated from the correlation matrices of the individual S-boxes. For $\ell > 0$ the calculations can be greatly simplified by recursively reusing intermediate results in computing these correlations. The total number of calculations can be reduced to less than 16ℓ multiplications and additions of S-box correlations.

The effect of a non-zero round key is the multiplication of some of these correlations by -1. Hence, if $\ell > 0$ the correlation depends on the value of 2ℓ different linear combinations of round key bits. If $\ell = 0$, α only contains a single mask and the correlation is independent of the key.

7.10.2 Linear Hulls

A theorem similar to Theorem 7.9.1 has been proved by K. Nyberg in [121], in her treatment of so-called *linear hulls*. The difference is the following. Theorem 7.9.1 expresses the expected linear potential between an input mask and an output mask, averaged over all values of the expanded keys, as the sum of the linear potentials $\mathsf{LP}(\mathsf{U})$ of the individual trails between these masks. It is valid for key-alternating ciphers. However, the theorem in [121] is proven for DES-like ciphers and does not mention key-alternating ciphers. As the DES is not a key-alternating cipher, the linear potential of a linear trail is in general not independent of the expanded key. In Theorem 7.9.1, the linear potentials $\mathsf{LP}(\mathsf{U})$ of the linear trails must be replaced by the expected linear potentials of the trails $\mathsf{ELP}(\mathsf{U})$, i.e. averaged over all values of the expanded key. In [121] the set of linear trails connecting the same initial mask and final mask is called an *approximate linear hull.*

Unfortunately, the presentation in [121] does not consider the effect of the key schedule and only considers the case of independent round keys. This is often misunderstood by cipher designers as an incentive to design heavy key schedules, in order to make the relations between round keys very complicated, or 'very random'. As we have shown above, linear cryptanalysis does not suggest complicated key schedules as even in the case of a linear key schedule systematic constructive interference of linear trails is very rare.

Extrapolating Theorem 7.9.1 to ciphers that are not key-alternating can be very misleading. First of all, in actual cryptanalysis it is not so much the maximum average linear potential that is relevant but the maximum linear potential corresponding to the given key under attack. We illustrate this with an example.

Example 7.10.1. We consider a cipher B that consists of multiplication with an invertible binary matrix, where the matrix is the key:

$$B[\mathsf{K}](\mathbf{x}) = \mathsf{K}\mathbf{x}.$$

For each given key K, each input parity has a correlation of amplitude 1 with exactly one output parity and no correlation with all other output parities. Averaged over all possible keys K (i.e. invertible matrices), the expected linear potential between any pair of input-output parities as predicted by Theorem 7.9.1 is exactly $(2^{n_b}-1)^{-1}$. Unfortunately, despite this excellent property with respect to average correlation amplitudes, the cipher is linear and trivially breakable.

The following physical metaphor summarizes the problem with the extrapolation. At any given time, on average half of the world's population is asleep. This does not mean that everyone is only half awake all the time.

Even for key-alternating ciphers one must take care in interpreting ELP values. For example, take the case of a high ELP $\mathrm{ELP}(\mathbf{u}, \mathbf{v})$ that is the result of one dominant trail U. We have $\mathrm{ELP}(\mathbf{u}, \mathbf{v}) \approx \mathrm{LP}(\mathsf{U})$ and the required number of plaintext-ciphertext pairs for a given success rate is hence proportional to $\mathrm{LP}^{-1}(\mathsf{U})$. Now consider the case of a high ELP $\mathrm{ELP}(\mathbf{u}, \mathbf{v})$ with two dominant trails U and U' with $\mathrm{LP}^{\frac{1}{2}}(\mathsf{U}') = \mathrm{LP}^{\frac{1}{2}}(\mathsf{U})/2$. Clearly $\mathrm{ELP}(\mathbf{u}, \mathbf{v}) = \mathrm{LP}(\mathsf{U}) + \mathrm{LP}(\mathsf{U}') = 5/4\mathrm{LP}(\mathsf{U})$. One would expect the required number of plaintext-ciphertext pairs in the latter case to be less than in the former case. The magnitude of the correlation of $(\mathbf{u}, \mathbf{v})$ now depends on whether the trails have correlation contributions of equal or opposite signs, something that is determined by a parity of the expanded key. It follows that for half of the keys this correlation has magnitude $3/2\mathrm{LP}^{\frac{1}{2}}(\mathsf{U})$ and for the other half it is $1/2\mathrm{LP}^{\frac{1}{2}}(\mathsf{U})$. The required number of plaintext-ciphertext pairs for a given success rate is hence $4/9\mathrm{LP}^{-1}(\mathsf{U})$ in half of the cases and $2\mathrm{LP}^{-1}(\mathsf{U})$ in the other half. Hence, although the additional trail U' increases the ELP, it also increases the average attack complexity for equal success rates. It follows that ELP values should be interpreted with caution [114]. We think that in design one should focus on worst-case behavior and consider the possibility of constructive interference of (signed) correlation contributions (see Sect. 9.1.1).

7.11 Conclusions

In this chapter we have provided a number of tools for describing and investigating the correlations in Boolean functions, iterated Boolean functions and block ciphers. This includes the concept of the correlation matrix and its properties and the systematic treatment of the silent assumptions made in linear cryptanalysis. We have compared our approach with the formalism usually adopted in the cryptographic literature, and have argued why it is an improvement. An extension of our approach to functions and block ciphers operating on arrays of elements of $\mathrm{GF}(2^n)$ is presented in Chapter 12.

8. Difference Propagation

In this chapter we consider difference propagation in Boolean functions and transformations. Difference propagation plays an important role in cryptanalysis in general and in differential cryptanalysis in particular.

We describe how differences propagate through several types of Boolean functions. We show that the difference propagation probabilities and the correlation potentials of a Boolean function are related by a simple expression. This is followed by a treatment of difference propagation through iterated Boolean transformations and in key-alternating ciphers. Finally we apply our analysis to the differential cryptanalysis of the DES and compare it with the influential concept of Markov ciphers.

8.1 Difference Propagation

Consider a couple of n-bit vectors $\mathbf{a}$ and $\mathbf{a}^*$ with bitwise difference $\mathbf{a}+\mathbf{a}^* = \mathbf{a}'$. Let $\mathbf{b} = h(\mathbf{a}), \mathbf{b}^* = h(\mathbf{a}^*)$ and $\mathbf{b}' = \mathbf{b} + \mathbf{b}^*$. The difference $\mathbf{a}'$ propagates to the difference $\mathbf{b}'$ through h iff

$$h(\mathbf{a}) + h(\mathbf{a} + \mathbf{a}') = \mathbf{b}'. \tag{8.1}$$

In general, $\mathbf{b}'$ is not fully determined by $\mathbf{a}'$ but depends on the value of $\mathbf{a}$ (or equivalently, $\mathbf{a}^*$). The difference propagation probability $\mathrm{DP}^{(h)}(\mathbf{a}', \mathbf{b}')$ is equal to the total number of values $\mathbf{a}$ that satisfy this equation, divided by 2^n.

Definition 8.1.1. *A* difference propagation probability $\mathrm{DP}^{(h)}(\mathbf{a}', \mathbf{b}')$ *is defined as*

$$\mathrm{DP}^{(h)}(\mathbf{a}', \mathbf{b}') = 2^{-n} \sum_{\mathbf{a}} \delta(\mathbf{b}' + h(\mathbf{a} + \mathbf{a}') + h(\mathbf{a})). \tag{8.2}$$

For a pair chosen uniformly from the set of all pairs $(\mathbf{a}, \mathbf{a}^*)$ where $\mathbf{a}+\mathbf{a}^* = \mathbf{a}'$, $\mathrm{DP}^{(h)}(\mathbf{a}', \mathbf{b}')$ is the probability that $h(\mathbf{a}) + h(\mathbf{a}^*) = \mathbf{b}'$.

As explained in Sect. 2.1.9, a function over $\mathrm{GF}(2)^n$ can be mapped to a function over $\mathrm{GF}(2^n)$ and hence described by a polynomial expression (2.36). We can then transform (8.1) into

J. Daemen, V. Rijmen, *The Design of Rijndael*, Information Security and Cryptography,
https://doi.org/10.1007/978-3-662-60769-5_8

$$\sum_i c_i a^i + \sum c_i (a + a')^i + b' = 0. \tag{8.3}$$

This is a polynomial equation in a. For certain special cases, the number of solutions of these polynomials can be analytically determined, providing provable bounds for difference propagation probability. Examples can be found in the paper of K. Nyberg on S-boxes [120] and in Sect. 14.1.

Difference propagation probabilities range between 0 and 1. Since

$$h(\mathbf{a} + \mathbf{a}') + h(\mathbf{a}) = h(\mathbf{a}) + h(\mathbf{a} + \mathbf{a}'), \tag{8.4}$$

their value must be an integer multiple of 2^{1-n}. We have:

$$\sum_{\mathbf{b}'} \mathrm{DP}^{(h)}(\mathbf{a}', \mathbf{b}') = 1. \tag{8.5}$$

The difference propagation from $\mathbf{a}'$ to $\mathbf{b}'$ occurs for a fraction of all possible input values $\mathbf{a}$ (and $\mathbf{a}^*$). This fraction is $\mathrm{DP}^{(h)}(\mathbf{a}', \mathbf{b}')$. If $\mathrm{DP}^{(h)}(\mathbf{a}', \mathbf{b}') = 0$, we say that the input difference $\mathbf{a}'$ and the output difference $\mathbf{b}'$ are *incompatible* through h.

Definition 8.1.2. *The weight of a difference propagation* $(\mathbf{a}', \mathbf{b}')$ *is the negative of the binary logarithm of the* DP, *i.e.*

$$\mathrm{w_r}(\mathbf{a}', \mathbf{b}') = -\log_2 \mathrm{DP}^{(h)}(\mathbf{a}', \mathbf{b}'). \tag{8.6}$$

The weight corresponds to the amount of information (expressed in bits) that the difference propagation gives about $\mathbf{a}$. Equivalently, it is the loss in *entropy* [139] of $\mathbf{a}$ due to the restriction that $\mathbf{a}'$ propagates to $\mathbf{b}'$. The weight ranges between 0 and $n - 1$: in the worst case the difference propagation gives no information on $\mathbf{a}$, and in the best case it fully determines the pair $\{\mathbf{a}, \mathbf{a}^*\}$ without their order, leaving only one bit of uncertainty.

If h is a linear function, the difference pattern at the input completely determines the difference pattern at the output:

$$\mathbf{b}' = \mathbf{b} + \mathbf{b}^* = h(\mathbf{a}) + h(\mathbf{a}^*) = h(\mathbf{a} + \mathbf{a}^*) = h(\mathbf{a}'). \tag{8.7}$$

From $\mathrm{w_r}(\mathbf{a}', \mathbf{b}') = 0$ it follows that this difference propagation does not give any information on $\mathbf{a}$.

8.2 Special Functions

8.2.1 Affine Functions

An *affine* function h from $\mathrm{GF}(2)^n$ to $\mathrm{GF}(2)^m$ is specified by

$$\mathbf{b} = \mathsf{M}\mathbf{a} + \mathbf{k}, \tag{8.8}$$

where M is an $m \times n$ matrix and $\mathbf{k}$ is an m-dimensional vector. The difference propagation for this function is determined by

$$\mathbf{b}' = \mathsf{M}\mathbf{a}'. \tag{8.9}$$

8.2.2 Bricklayer Functions

For a bricklayer function h, the DP is the product of the DP values of the component functions:

$$\mathrm{DP}^{(h)}(\mathbf{a}', \mathbf{b}') = \prod_i \mathrm{DP}^{(h_{(i)})}(\mathbf{a}'_{(i)}, \mathbf{b}'_{(i)}). \tag{8.10}$$

The weight is the sum of the weights of the difference propagation in the component functions:

$$\mathrm{w_r}(\mathbf{a}', \mathbf{b}') = \sum_i \mathrm{w_r}(\mathbf{a}'_{(i)}, \mathbf{b}'_{(i)}), \tag{8.11}$$

where $\mathbf{a}' = (\mathbf{a}'_{(1)}, \mathbf{a}'_{(2)}, \ldots, \mathbf{a}'_{(\ell)})$ and $\mathbf{b}' = (\mathbf{b}'_{(1)}, \mathbf{b}'_{(2)}, \ldots, \mathbf{b}'_{(\ell)})$.

8.2.3 Truncating Functions

A Boolean function h from $\mathrm{GF}(2)^n$ to $\mathrm{GF}(2)^m$ can be converted into a Boolean function h_s from $\mathrm{GF}(2)^n$ to $\mathrm{GF}(2)^{m-1}$ by discarding a single output bit a_s. The difference propagation probabilities of h_s can be expressed in terms of the difference propagation probabilities of h:

$$\mathrm{DP}^{(h_s)}(\mathbf{a}', \mathbf{b}') = \mathrm{DP}^{(h)}(\mathbf{a}', \omega^0) + \mathrm{DP}^{(h)}(\mathbf{a}', \omega^1), \tag{8.12}$$

where $\mathbf{b}'_i = \omega^0_i = \omega^1_i$ for $i \neq s$ and $\omega^1_s = 1$ and $\omega^0_s = 0$. We generalize this to the situation in which only a number of linear combinations of the output are considered. Let λ be a linear function corresponding to an $m \times \ell$ binary matrix M. The difference propagation probabilities of $\theta \circ h$ are given by

$$\mathrm{DP}^{(\lambda \circ h)}(\mathbf{a}', \mathbf{b}') = \sum_{\omega | \mathbf{b}' = \mathsf{M}\omega} \mathrm{DP}^{(h)}(\mathbf{a}', \omega). \tag{8.13}$$

8.2.4 Keyed Functions

For a keyed function $h[\mathbf{k}]$ we define a probability $\mathrm{DP}^{(h[\mathbf{k}])}(\mathbf{a}', \mathbf{b}')$ for each value $\mathbf{k}$ of the key:

$$\mathrm{DP}^{(h[\mathbf{k}])}(\mathbf{a}',\mathbf{b}') = 2^{-n}\sum_{\mathbf{a}}\delta(\mathbf{b}' + h[\mathbf{k}](\mathbf{a}+\mathbf{a}') + h[\mathbf{k}](\mathbf{a})). \tag{8.14}$$

The expected differential probability (EDP) of a difference propagation over a keyed map is the average of the differential propagation $\mathrm{DP}[\mathbf{k}](\mathbf{a},\mathbf{b})$ over all keys.

Definition 8.2.1 (EDP). *The* EDP *of a difference propagation* $(\mathbf{a}',\mathbf{b}')$ *over a keyed map* $h[\mathbf{k}]$ *is given by*

$$\mathrm{EDP}(\mathbf{a}',\mathbf{b}') = 2^{-|\mathcal{K}|}\sum_{\mathbf{k}\in\mathcal{K}}\mathrm{DP}^{(h[\mathbf{k}])}(\mathbf{a}',\mathbf{b}').$$

8.3 Relation Between DP Values and Correlations

The difference propagation probabilities of Boolean functions can be expressed in terms of their spectrum and their correlation matrix elements. The probability of difference propagation $\mathrm{DP}^{(h)}(\mathbf{a}',0)$ is given by

$$\begin{aligned}
\mathrm{DP}^{(h)}(\mathbf{a}',0) &= 2^{-n}\sum_{\mathbf{a}}\delta(h(\mathbf{a}) + h(\mathbf{a}+\mathbf{a}'))\\
&= 2^{-n}\sum_{\mathbf{a}}\frac{1}{2}(1+\hat{h}(\mathbf{a})\hat{h}(\mathbf{a}+\mathbf{a}'))\\
&= 2^{-n}\sum_{\mathbf{a}}\frac{1}{2} + 2^{-n}\sum_{\mathbf{a}}\frac{1}{2}\hat{h}(\mathbf{a})\hat{h}(\mathbf{a}+\mathbf{a}')\\
&= \frac{1}{2}(1+\hat{r}_h(\mathbf{a}'))\\
&= \frac{1}{2}(1+\sum_{\mathbf{w}}(-1)^{\mathbf{w}^{\mathrm{T}}\mathbf{a}'}\hat{H}^2(\mathbf{w})).
\end{aligned} \tag{8.15}$$

The component of the autocorrelation function $\hat{r}_h(\mathbf{a}')$ corresponds to the amount that $\mathrm{DP}^{(h)}(\mathbf{a}',0)$ deviates from 0.5.

For functions from $\mathrm{GF}(2)^n$ to $\mathrm{GF}(2)^m$, we denote the autocorrelation function of $\mathbf{u}^{\mathrm{T}}h(\mathbf{a})$ by $\hat{r}_{\mathbf{u}}(\mathbf{a}')$, i.e.

$$\hat{r}_{\mathbf{u}}(\mathbf{a}') = 2^{-n}\sum_{\mathbf{a}}(-1)^{\mathbf{u}^{\mathrm{T}}h(\mathbf{a})+\mathbf{u}^{\mathrm{T}}h(\mathbf{a}+\mathbf{a}')}. \tag{8.16}$$

Now we can prove the following theorem, which expresses the duality between the difference propagation and the correlation properties of a Boolean function.

Theorem 8.3.1. *The table of DP values and the table of squared correlations of a Boolean function are linked by a (scaled) Walsh-Hadamard transform. We have*

$$\mathrm{DP}(\mathbf{a}', \mathbf{b}') = 2^{-m} \sum_{\mathbf{u},\mathbf{w}} (-1)^{\mathbf{w}^{\mathrm{T}}\mathbf{a}' + \mathbf{u}^{\mathrm{T}}\mathbf{b}'} \mathsf{C}^2_{\mathbf{u},\mathbf{w}}, \tag{8.17}$$

and dually

$$\mathsf{C}^2_{\mathbf{u},\mathbf{w}} = 2^{-n} \sum_{\mathbf{a}',\mathbf{b}'} (-1)^{\mathbf{w}^{\mathrm{T}}\mathbf{a}' + \mathbf{u}^{\mathrm{T}}\mathbf{b}'} \mathrm{DP}(\mathbf{a}', \mathbf{b}'). \tag{8.18}$$

Proof.

$$\begin{aligned}
\mathrm{DP}(\mathbf{a}', \mathbf{b}') &= 2^{-n} \sum_{\mathbf{a}} \delta(h(\mathbf{a}) + h(\mathbf{a} + \mathbf{a}') + \mathbf{b}') \\
&= 2^{-n} \sum_{\mathbf{a}} \prod_{i} \frac{1}{2} \left((-1)^{h_i(\mathbf{a}) + h_i(\mathbf{a}+\mathbf{a}') + \mathbf{b}'_i} + 1 \right) \\
&= 2^{-n} \sum_{\mathbf{a}} 2^{-m} \sum_{\mathbf{u}} \left(\prod_{u_i = 1} (-1)^{h_i(\mathbf{a}) + h_i(\mathbf{a}+\mathbf{a}') + \mathbf{b}'_i} \right) \\
&= 2^{-n} \sum_{\mathbf{a}} 2^{-m} \sum_{\mathbf{u}} (-1)^{\mathbf{u}^{\mathrm{T}}(h(\mathbf{a}) + h(\mathbf{a}+\mathbf{a}') + \mathbf{b}')} \\
&= 2^{-n} \sum_{\mathbf{a}} 2^{-m} \sum_{\mathbf{u}} (-1)^{\mathbf{u}^{\mathrm{T}} h(\mathbf{a}) + \mathbf{u}^{\mathrm{T}} h(\mathbf{a}+\mathbf{a}') + \mathbf{u}^{\mathrm{T}}\mathbf{b}'} \\
&= 2^{-m} \sum_{\mathbf{u}} (-1)^{\mathbf{u}^{\mathrm{T}}\mathbf{b}'} 2^{-n} \sum_{\mathbf{a}} (-1)^{\mathbf{u}^{\mathrm{T}} h(\mathbf{a}) + \mathbf{u}^{\mathrm{T}} h(\mathbf{a}+\mathbf{a}')} \\
&= 2^{-m} \sum_{\mathbf{u}} (-1)^{\mathbf{u}^{\mathrm{T}}\mathbf{b}'} \hat{r}_{\mathbf{u}}(\mathbf{a}') \\
&= 2^{-m} \sum_{\mathbf{u}} (-1)^{\mathbf{u}^{\mathrm{T}}\mathbf{b}'} \sum_{\mathbf{w}} (-1)^{\mathbf{w}^{\mathrm{T}}\mathbf{a}'} \mathsf{C}^2_{\mathbf{u},\mathbf{w}} \\
&= 2^{-m} \sum_{\mathbf{u},\mathbf{w}} (-1)^{\mathbf{w}^{\mathrm{T}}\mathbf{a}' + \mathbf{u}^{\mathrm{T}}\mathbf{b}'} \mathsf{C}^2_{\mathbf{u},\mathbf{w}}.
\end{aligned}$$

□

8.4 Differential Trails

In this section we apply the described formalism and tools to the propagation of differences in iterative Boolean transformations.

8.4.1 General Case

Let β be an iterative Boolean transformation operating on n-bit vectors that is a sequence of r transformations:

$$\beta = \rho^{(r)} \circ \rho^{(r-1)} \circ \ldots \circ \rho^{(2)} \circ \rho^{(1)}. \tag{8.19}$$

A *differential trail* Q over an iterative transformation consists of a sequence of $r+1$ difference patterns:

$$\mathrm{Q} = \left(\mathbf{q}^{(0)}, \mathbf{q}^{(1)}, \mathbf{q}^{(2)}, \ldots, \mathbf{q}^{(r-1)}, \mathbf{q}^{(r)}\right). \tag{8.20}$$

A differential trail has a *probability* that we also denote as its DP. The DP of a differential trail is the number of values $\mathbf{a}^{(0)}$ for which the difference patterns follow the differential trail divided by the number of possible values for $\mathbf{a}^{(0)}$. This differential trail is a sequence of r differential steps $(\mathbf{q}^{(i-1)}, \mathbf{q}^{(i)})$, which have a weight

$$\mathrm{w_r}^{\rho^{(i)}}\left(\mathbf{q}^{(i-1)}, \mathbf{q}^{(i)}\right), \tag{8.21}$$

or $\mathrm{w_r}^{(i)}$ for short.

Definition 8.4.1. *The weight of a differential trail* Q *is the sum of the weights of its differential steps, i.e.*

$$\mathrm{w_r}(\mathrm{Q}) = \sum_i \mathrm{w_r}^{\rho^{(i)}}\left(\mathbf{q}^{(i-1)}, \mathbf{q}^{(i)}\right). \tag{8.22}$$

The significance of the weight of a differential trail is explained in the following section.

8.4.2 Independence of Restrictions

A differential step $(\mathbf{q}^{(i-1)}, \mathbf{q}^{(i)})$ imposes restrictions on the intermediate state $\mathbf{a}^{(i-1)}$ in the following way. The differential step imposes that the value of $\mathbf{a}^{(i-1)}$ is in a set that contains a fraction $2^{-\mathrm{w_r}^{(i)}}$ of all possible values. We denote this set by $\alpha_{i-1}(i)$: the set of possible values of $\mathbf{a}^{(i-1)}$ with the restrictions imposed by the ith step $(\mathbf{q}^{(i-1)}, \mathbf{q}^{(i)})$. As $\mathbf{a}^{(i-1)}$ is completely determined by $\mathbf{a}^{(0)}$, we can consider the set $\alpha_0(i)$ as the set of possible values of $\mathbf{a}^{(0)}$ with the restrictions imposed by the ith step. In the case that β is a permutation, and hence all steps are also permutations, for each element in $\alpha_{i-1}(i)$ there is one element in $\alpha_0(i)$. Both have the same relative size: $2^{-\mathrm{w_r}^{(i)}}$.

Now consider a two-round differential trail. The first step imposes that $\mathbf{a}^{(0)} \in \alpha_0(1)$ and the second step that $\mathbf{a}^{(1)} \in \alpha_1(2)$. We can reduce this second restriction to $\mathbf{a}^{(0)} \in \alpha_0(2)$. The joint restriction imposed by both steps now becomes: $\mathbf{a}^{(0)} \in \alpha_0(1,2)$ where $\alpha_0(1,2) = \alpha_0(1) \cap \alpha_0(2)$. If

$$\Pr\left(\mathbf{x} \in \alpha_0(1) | \mathbf{x} \in \alpha_0(2)\right) = \Pr\left(\mathbf{x} \in \alpha_0(1)\right), \tag{8.23}$$

the restrictions imposed by the first and the second step are independent. In that case, the relative size of $\alpha_0(1,2)$ is equal to $2^{-(\mathrm{w_r}^{(1)}+\mathrm{w_r}^{(2)})}$.

The relative size of the set of values $\mathbf{a}^{(0)}$ that satisfy the restrictions imposed by all the differential steps of a differential trail Q is by definition the DP of Q. While it is easy to compute the weight of a differential trail, computing its DP is in general difficult. If we neglect the correlations between the restrictions of the different steps, the DP of the differential trail is approximated by

$$\mathrm{DP(Q)} \approx 2^{-\mathrm{w_r(Q)}}. \tag{8.24}$$

We refer to Chapter 15 for more details on the accuracy of such an approximation. If $\mathrm{w_r(Q)}$ is of order $n-1$ or larger, (8.24) can no longer be a valid approximation. In this situation, the inevitable (albeit small) correlations between the restrictions come into play. $\mathrm{DP}(Q)$ multiplied by 2^n is the absolute number of inputs $\mathbf{a}^{(0)}$ for which the initial difference pattern propagates along a differential trail Q. For this reason, it must therefore be an (even) integer. Of the differential trails Q with a weight $\mathrm{w_r(Q)}$ above $n-1$, only a fraction $2^{n-1-\mathrm{w_r(Q)}}$ can be expected to actually occur for some $\mathbf{a}^{(0)}$.

Differential cryptanalysis exploits difference propagations $(\mathbf{a}', \mathbf{b}')$ with large probabilities. Since, for a given input value $\mathbf{a}^{(0)}$, exactly one differential trail is followed, the probability of difference propagation $(\mathbf{a}', \mathbf{b}')$ is the sum of the probabilities of all r-round differential trails with initial difference $\mathbf{a}'$ and terminal difference $\mathbf{b}'$. We say that a trail Q is *in* a differential $(\mathbf{a}', \mathbf{b}')$ if $\mathbf{q}^{(0)} = \mathbf{a}'$ and $\mathbf{q}^{(r)} = \mathbf{b}'$. We have

$$\mathrm{DP}(\mathbf{a}', \mathbf{b}') = \sum_{\mathrm{Q} \in (\mathbf{a}', \mathbf{b}')} \mathrm{DP(Q)} \tag{8.25}$$

and

$$\mathrm{EDP}(\mathbf{a}', \mathbf{b}') = \sum_{Q \in (\mathbf{a}', \mathbf{b}')} \mathrm{EDP}(Q)\ . \tag{8.26}$$

8.5 Key-Alternating Cipher

As the round transformation is independent of the key, so is the weight of a differential step over a round. A key addition step has no impact on the difference propagation pattern or the weight. Since the weight of a differential trail is the total of the weight of its differential steps, it is independent of the round keys and hence of the cipher key.

The reduction of the restrictions imposed upon $\mathbf{a}^{(i-1)}$ by $\left(\mathbf{q}^{(i-1)}, \mathbf{q}^{(i)}\right)$, to restrictions on $\mathbf{a}^{(0)}$, involves the round keys. As the signs of the correlations between the different restrictions do depend on the round keys, the probability of a differential trail is in general not independent of the cipher key.

For a differential trail Q with a weight $\mathrm{w_r}(\mathrm{Q})$ above $n-1$, only an expected portion $2^{n-1-\mathrm{w_r}(\mathrm{Q})}$ of the cipher keys will give pairs following it Q.

8.6 The Effect of the Key Schedule

If we use the total weight over all differential steps to predict the difference propagation probability, we make the assumption that the restrictions due to the steps are independent. If we fix the round keys, then we can reduce the restrictions on all differential steps to restrictions on $\mathbf{a}^{(0)}$. It may turn out that the different restrictions are not independent. The reduction of the restrictions from all differential steps to $\mathbf{a}^{(0)}$ involves the round keys, which are in turn derived fromf the cipher key by the key schedule. Hence, the key schedule influences the reduction of restrictions on all differential steps to restrictions on $\mathbf{a}^{(0)}$.

8.7 Differential Trails and the Differential Cryptanalysis Literature

In this section we compare our formalism with the terminology of the original description of differential cryptanalysis and with the concept of Markov ciphers.

8.7.1 Differential Cryptanalysis of the DES Revisited

In this section we match the elements of differential cryptanalysis as described in Sect. 6.2 with those of our framework.

The characteristics with their characteristic probability described in [26] correspond to what we call differential trails and the approximation of their probability based on their weight. In the cryptanalysis of the DES, the difference propagation probability from the initial difference pattern to the final difference pattern is approximated by the probability of the differential trail. This is a valid approximation because of the low relative weight of the differential trail:

1. The odd-round differential steps have a weight equal to 0 and do not impose any restrictions.
2. The even-round differential steps only impose restrictions on few state bits.

3. The state bits after round $i+2$ depend on many state bits after round i. In other words, the correlation between the different restrictions is very weak, if there is any.

For the DES round transformation the distribution of the differential steps and their weight are not independent of the round keys. This dependence was already recognized in [26] where in the analysis the weights of the differential steps are approximated by an average value. The two-round iterative differential trail with approximate probability $1/234$ has in fact a probability that is either $1/146$ or $1/585$, depending on the value of a linear combination of round key bits.

8.7.2 Markov Ciphers

In Sect. 8.4.2 we discussed the determination of the probability of a multiple-round differential trail. This problem has been studied before by X. Lai, J. Massey and S. Murphy in [93]. They were the first to make the distinction between *differentials* and *characteristics*. A differential is a difference propagation from an input difference pattern to an output difference pattern. A characteristic is a differential trail along a number of rounds of a block cipher. In [93] it is shown that the probability of a differential over a sequence of rounds of a block cipher is equal to the sum of the probabilities of all characteristics (differential trails) over those rounds. They also introduce the following concepts:

Markov cipher: A Markov cipher is an iterative cipher whose round transformation satisfies the condition that the differential probability is independent of the choice of one of the component plaintexts under an appropriate definition of difference. All modern ciphers appearing in the mainstream literature are Markov ciphers.

Hypothesis of stochastic equivalence: This hypothesis states that, for virtually all values of the cipher key, the probability of a differential trail can be approximated by the expected value of the probability of the differential trail, averaged over all possible values of the cipher key. We illustrate in Chapter 15 that for many round-reduced variants of modern ciphers, this hypothesis does not hold.

8.8 Conclusions

We have described the propagation of differences in Boolean functions, in iterated Boolean transformations and in block ciphers in general. We have introduced the *differential trail* as the basic building block of difference propagation in block ciphers.

9. The Wide Trail Strategy

In this chapter we explain the strategy that underlies many choices made in the design of Rijndael and its related ciphers.

We start with a detailed explanation of how linear correlations and difference propagation probabilities are built up in key-alternating block ciphers. This is followed by an explanation of the basic principles of the wide trail strategy. Then we introduce an important diffusion measure, the *branch number*, and describe how it is relevant in providing bounds for the DP of differential trails and the LP of linear trails over two rounds. This is followed by a key-alternating cipher structure that combines efficiency with high resistance against linear and differential cryptanalysis. We apply the same principles to the Rijndael cipher structure and prove a theorem that provides a lower bound on the number of active S-boxes in any four-round trail for these ciphers. Finally we provide some concrete constructions for the components used in the described cipher structures, using coding theory and geometrical representations.

9.1 Propagation in Key-Alternating Block Ciphers

In this section we describe the anatomy of correlations and difference propagations in key-alternating block ciphers. This is used to determine the number of rounds required to provide resistance against linear and differential cryptanalysis. In this section we assume that the round transformations do not exhibit correlations with an amplitude of 1 or difference propagations with a DP of 1.

Limiting ourselves to the key-alternating structure allows us to reason more easily about linear and differential trails, since the effect of a key addition on the propagation is quite simple.

9.1.1 Linear Cryptanalysis

For a successful classical linear cryptanalysis attack, the cryptanalyst needs to know a correlation over all but a few rounds of the cipher with an amplitude

J. Daemen, V. Rijmen, *The Design of Rijndael*, Information Security and Cryptography,
https://doi.org/10.1007/978-3-662-60769-5_9

that is significantly larger than $2^{-n_b/2}$. To avoid this, we choose the number of rounds so that there are no such linear trails with a correlation contribution above $n_k^{-1}2^{-n_b/2}$.

This does not guarantee that there are no high correlations over r rounds. In Chap. 7 we have shown that each output parity of a Boolean function is correlated to a number of input parities. Parseval's theorem (7.15) states that the sum of the correlation potentials with all input parities is 1. Under the assumption that the output parity is equally correlated to all 2^{n_b} possible input parities, the correlation to each of these input parities has amplitude $2^{-n_b/2}$. In practice it is very unlikely that such a uniform distribution will be attained, and so correlations will exist that are orders of magnitude higher than $2^{-n_b/2}$. This also applies to the Boolean permutation formed by a cipher for a given value of the cipher key. Hence, the presence of high correlations over (all but a few rounds of) the cipher is a mathematical fact rather than something that may be avoided by design.

However, when we impose an upper bound on the amplitude of the correlation contributions of linear trails, high correlations can only occur as the result of constructive interference of many linear trails that share the same initial and final masks. If this upper bound is $n_k^{-1}2^{-n_b/2}$, any such correlation with an amplitude above $2^{-n_b/2}$ must be the result of at least n_k different linear trails. The condition that a linear trail in this set contributes constructively to the resulting correlation imposes a linear relation on the round key bits. From the point that more than n_k linear trails are combined, it is very unlikely that all such conditions can be satisfied by choosing the appropriate cipher key value.

The strong key-dependence of this interference makes it very unlikely that if a specific output parity has a high correlation with a specific input parity for a given key, this will also be the case for another value of the key. In other words, although it follows from Parseval's theorem that high correlations over the cipher will exist whatever the number of rounds, the strong round key dependence of interference makes locating the input and output masks for which high correlations occur practically infeasible. This is true if the key is known, and even more so if it is unknown.

In the above discussion we have neglected possible linear trail clustering: the fact that sets of linear trails tend to propagate along common intermediate masks. If linear trails tend to cluster, this must be taken into account in the upper bounds for the correlation contributions.

9.1.2 Differential Cryptanalysis

For a successful classical differential cryptanalysis attack, the cryptanalyst needs to know an input difference pattern that propagates to an output difference pattern over all but a few (two or three) rounds of the cipher, with a DP that is significantly larger than 2^{1-n_b}. To avoid this, we choose the number of rounds so that there are no such differential trails with a weight below n_b.

This strategy does not guarantee that there are no such difference propagations with a high DP. For any Boolean function, a difference pattern at the input must propagate to some difference pattern at the output, and the sum of the difference propagation probabilities over all possible output differences is 1. Hence, there must be difference propagations with probabilities equal to or larger than 2^{1-n_b}. This also applies to the Boolean permutation formed by a cipher for a given value of the cipher key. Hence, similar to what we have for correlations, the presence of difference propagations with a high DP over (all but a few rounds of) the cipher is a mathematical fact that cannot be avoided by a careful design.

Let us analyze how, for a given key value, a difference pattern at the input propagates to a difference pattern at the output with some DP y. By definition, there are exactly $y2^{n_b-1}$ pairs with the given input difference pattern and the given output difference pattern. Each of these pairs follows a particular differential trail.

Assuming that the pairs are distributed over the trails according to a Poisson distribution,[1] the expected number of pairs that, for a given key value, follow a differential trail with weight z is 2^{n_b-1-z}. Consider a differential trail with a weight z larger than $n_b - 1$ that is followed by at least one pair. The probability that this trail is followed by more than one pair is approximately 2^{n_b-1-z}. It follows that if there are no differential trails with a weight below $n_b - 1$, the $y2^{n_b-1}$ pairs that have the correct input difference pattern and output difference pattern follow almost $y2^{n_b-1}$ different differential trails.

Hence, if there are no differential trails with a low weight, difference propagations with a large DP are the result of multiple differential trails that happen to be followed by a pair in the given circumstances, i.e. for the given key value. For another key value, each of these individual differential trails may be followed by a pair or may not. This makes predicting the input difference patterns and output difference patterns that have large difference propagation probabilities practically infeasible. This is true if the key is known, and even more so if it is unknown.

In the above discussion we have neglected possible differential trail clustering: the fact that sets of differential trails tend to propagate along common intermediate difference patterns. If differential trails tend to cluster, this must be taken into account in the lower bounds for the weight of the differential trails. Clustering of differential trails in Rijndael and its relatives is treated in Chap. 13.

[1] In Chap. 15 we show that in many ciphers, this assumption does not hold for two-round or four-round trails. However, for trails over more rounds, this still seems a plausible assumption.

9.1.3 Differences Between Linear Trails and Differential Trails

Linear and differential trails propagate in a very similar way. Still, when they are combined to form correlations and difference propagations, respectively, there are a number of very important differences.

The *impact* of a linear trail is its correlation contribution. The correlation contribution can easily be computed and its amplitude is independent of the value of the key. The problem with computing correlations over many rounds is that a correlation may be the result of many linear trails whose interference — constructive or destructive — is strongly key-dependent.

The *impact* of a differential trail is its DP, which is in general infeasible to compute precisely. However, it can be approximated using the *weight* of the differential trail. Unlike its DP, the weight of a differential trail is easy to compute. However, the approximation is only valid for differential trails in which the restrictions imposed by the differential steps are mutually independent and hence that have a weight below $n_b - 1$. If the DP of the individual differential trails were known for a given key, difference propagation probabilities would be easy to compute. For differential trails, destructive interference does not exist.

9.2 The Wide Trail Strategy

The wide trail strategy is an approach used to design the round transformations of key-alternating block ciphers that combine efficiency and resistance against differential and linear cryptanalysis. In this book we describe the strategy for key-alternating block ciphers, but it can also be extended to more general block cipher structures.

We build the round transformations as a sequence of two invertible steps:

1. γ. A local nonlinear transformation. By local, we mean that any output bit depends on only a limited number of input bits and that neighboring output bits depend on neighboring input bits.
2. λ. A linear mixing transformation providing high diffusion. What is meant by high diffusion will be explained in Sect. 9.2.3.

Hence we have a round transformation ρ:

$$\rho = \lambda \circ \gamma. \tag{9.1}$$

We refer to this as a $\gamma\lambda$ round transformation.

9.2.1 The $\gamma\lambda$ Round Structure in Block Ciphers

In block cipher design γ is usually a bricklayer permutation consisting of a number of S-boxes. The state bits of $\mathbf{a}$ are partitioned into n_t $\mathrm{n_s}$-bit *tuples* $\mathbf{a}_i \in Z_2^{\mathrm{n_s}}$ with $i \in \mathcal{I}$ according to the so-called *tuple partition*. $\mathcal{I}$ is called the *index space*. The block size of the cipher is given by $n_b = \mathrm{n_s} n_t$.

Example 9.2.1. Let $\mathcal{X}_1$ be a cipher with a block length of 48 bits. Let the input be divided into six 8-bit tuples:

$$\mathbf{a} = \begin{bmatrix} a_1 \; a_2 \; a_3 \; a_4 \; a_5 \; a_6 \end{bmatrix}.$$

The index space is $\mathcal{I} = \{1, 2, 3, 4, 5, 6\}$.

Figure 9.1 illustrates the different steps of a round and a key addition for a simple example. The block cipher example has a block size of 27 bits. The nonlinear S-boxes operate on $\mathrm{n_s} = 3$ bits at a time. The linear transformation mixes the outputs of the $n_t = 9$ S-boxes. Figure 9.2 gives a more schematic representation, which we will use in the remainder of this chapter.

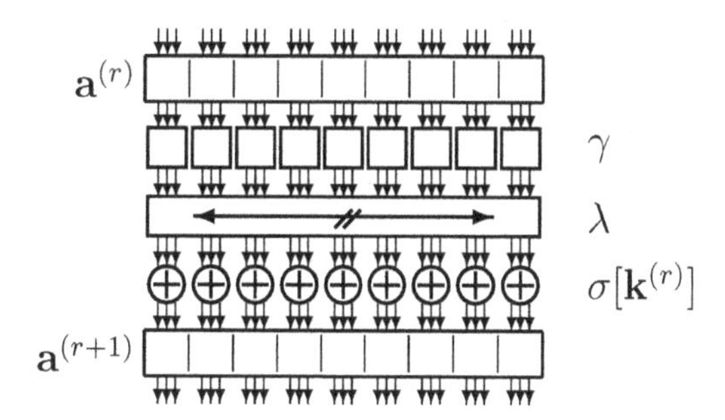

Fig. 9.1. Steps of an example block cipher

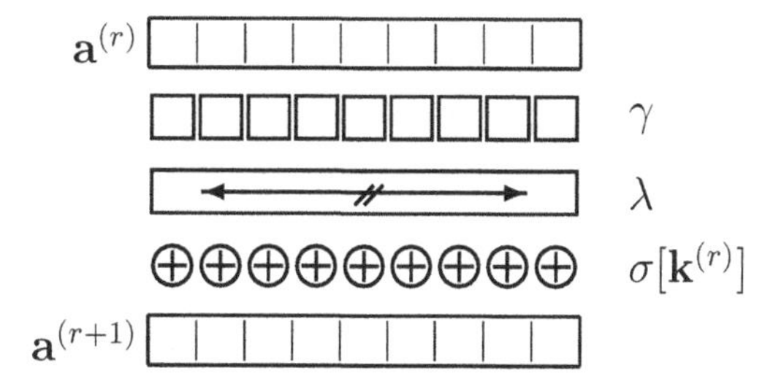

Fig. 9.2. Schematic representation of the different steps of a block cipher

The step γ is a bricklayer permutation composed of S-boxes:

$$\gamma : \mathbf{b} = \gamma(\mathbf{a}) \Leftrightarrow \mathbf{b}_i = S_\gamma(\mathbf{a}_i), \tag{9.2}$$

where S_γ is an invertible nonlinear $\mathrm{n_s}$-bit substitution box. For the purpose of this analysis, S_γ need not be specified. Clearly, the inverse of γ consists of applying the inverse substitution S_γ^{-1} to all bytes. The results of this chapter can easily be generalized to include nonlinear permutations that use different S-boxes for different byte positions. However, this does not result in a plausible improvement of the resistance against known attacks. The use of different S-boxes also increases the program size in software implementations and the required chip area in serial hardware implementations.

The step λ combines the bytes linearly: each byte at its output is a linear function of bytes at its input. λ can be specified at the bit level by a simple $n_\mathsf{b} \times n_\mathsf{b}$ binary matrix M. We have

$$\lambda : \mathbf{b} = \lambda(\mathbf{a}) \Leftrightarrow \mathbf{b} = \mathsf{M}\mathbf{a}. \tag{9.3}$$

λ can also be specified at the byte level. For this purpose the bytes are assumed to code elements in $\mathrm{GF}(2^{\mathrm{n_s}})$ with respect to some basis. In its most general form, we have

$$\lambda : \mathbf{b} = \lambda(\mathbf{a}) \Leftrightarrow b_i = \sum_j \sum_{0 \le \ell < \mathrm{n_s}} M_{i,j,\ell} {a_j}^{2^\ell}. \tag{9.4}$$

In most instances a more simple linear function is chosen that is a special case of (9.4):

$$\lambda : \mathbf{b} = \lambda(\mathbf{a}) \Leftrightarrow b_i = \sum_j M_{i,j} a_j. \tag{9.5}$$

If we consider the state as an array of bytes, this can be expressed as a matrix multiplication:

$$\lambda : \mathbf{b} = \lambda(\mathbf{a}) \Leftrightarrow \mathbf{b} = \mathsf{M} \cdot \mathbf{a} \tag{9.6}$$

where M is an $n_\mathsf{t} \times n_\mathsf{t}$ matrix with elements in $\mathrm{GF}(2^{\mathrm{n_s}})$. The jth column of M is denoted by M_j. The inverse of λ is specified by the matrix M^{-1}.

Example 9.2.2. In $\mathcal{X}_1$, λ could be defined as

$$\lambda\left(\begin{bmatrix} a_1 & a_2 & a_3 & a_4 & a_5 & a_6 \end{bmatrix}\right) = \begin{bmatrix} 2 \cdot a_1 & a_1 + a_2 & a_1 + a_3 + a_4 + a_5 & a_4 + a_5 + a_6 & a_3 + a_5 + a_6 & a_2 + a_3 \end{bmatrix}.$$

The M matrix is then given by

$$\mathsf{M} = \begin{bmatrix} 2 & 0 & 0 & 0 & 0 & 0 \\ 1 & 1 & 0 & 0 & 0 & 0 \\ 1 & 0 & 1 & 1 & 1 & 0 \\ 0 & 0 & 0 & 1 & 1 & 1 \\ 0 & 0 & 1 & 0 & 1 & 1 \\ 0 & 1 & 1 & 0 & 0 & 0 \end{bmatrix}.$$

9.2.2 Weight of a Trail

γ is a bricklayer permutation consisting of S-boxes. Hence, as explained in Sect. 7.4.3, the correlation over γ is the product of the correlations over the different S-box positions for the given input and output masks. We define the *weight* of a correlation as the negative logarithm of its amplitude. The correlation weight for an input mask and output mask is the sum of the correlation weights of the different S-box positions. If the output mask is non-zero for a particular S-box position or byte, we call this S-box or byte *active.*

Similarly, the weight of the difference propagation over γ is the sum of the weights of the difference propagations of the S-box positions for the given input difference pattern and output difference pattern. If the input difference pattern is non-zero for a particular S-box position or byte, we call this S-box or byte *active.*

We take S-boxes that have good nonlinearity properties. For linear cryptanalysis, the relevant property is the maximum amplitude of correlations over the S-box. For differential cryptanalysis, the relevant property is the maximum DP. Once a single S-box has been found with good properties, this can be used for all S-box positions in the nonlinear permutation.

A linear trail is defined by a series of masks. The weight of such a trail is the sum of the weights of the masks of the trail. As the weight of the masks is the sum of the weight of its active S-box positions, the weight of a linear trail is the sum of that of its active S-boxes. An upper bound to the correlation is a lower bound to the weight per S-box. Hence, the weight of a linear trail is equal to or larger than the number of active bytes in all its masks times the minimum (correlation) weight per S-box. We call the number of active bytes in a pattern or a trail its *byte weight.*

A differential trail is defined by a series of difference patterns. The weight of such a trail is the sum of the weights of the difference patterns of the trail. Completely analogous to linear trails, the weight of a differential trail is equal to or larger than the number of active S-boxes times the minimum (differential) weight per S-box.

This suggests two possible mechanisms to eliminate low-weight trails:

1. Choose S-boxes with high minimum differential and correlation weight.
2. Design the round transformation(s) in such a way that there are no relevant trails with a low byte weight.

The maximum correlation amplitude of an $\mathrm{n_s}$-bit invertible S-box is above $2^{-\mathrm{n_s}/2}$, yielding an upper bound for the minimum (correlation) weight of $\mathrm{n_s}/2$. The maximum DP is at least $2^{2-\mathrm{n_s}}$, yielding an upper bound for the minimum (differential) weight of $\mathrm{n_s}-2$. This seems to suggest that one should take large S-boxes.

Instead of spending most of its resources on large S-boxes, the wide trail strategy aims at designing the round transformation(s) such that there are no trails with a low byte weight. In ciphers designed following the wide trail strategy, a relatively large amount of resources is spent in the linear step to provide high multiple-round diffusion.

9.2.3 Diffusion

Diffusion is the term used by C. Shannon to denote the quantitative spreading of information [140]. The exact meaning of the term diffusion depends strongly on the context in which it is used. In this section we will explain what we mean by diffusion in the context of the wide trail strategy.

Inevitably, the nonlinear step γ provides some interaction between the different bits within the bytes that may be referred to as diffusion. However, it does not provide any inter-byte interaction: difference propagation and correlation over γ stay confined within the bytes. In the context of the wide trail strategy, it is not this kind of diffusion we are interested in. We use the term diffusion to indicate properties of a Boolean function that increase the minimum byte weight of linear and differential trails. In this sense, all diffusion is realized by λ; γ does not provide any diffusion at all.

Let us start by considering single-round trails. Obviously, the byte weight of a single-round trail — differential or linear — is equal to the number of active bytes at its input. It follows that the minimum byte weight of a single-round trail is 1, independent of λ.

In two-round trails, the byte weight is the sum of the number of active bytes in the masks or difference patterns in the state at the input of the first and at the second round. The state at the input of the second round is equal to the XOR of the output of the first and a round key. This key addition has no impact on the mask or difference pattern and hence does not impact their byte weight. In this context a relevant diffusion measure of ρ is the minimum number of active bytes at the input and output of ρ. We call this the (byte) *branch number* of ρ. Basically, this branch number provides a lower bound for the minimum byte weight of any two-round trail. The byte branch number ranges between two ('no diffusion at all') and the total number of bytes in the state n_t plus one.

In trails of more than two rounds, the desired diffusion properties of ρ are less trivial. It is clear that any $2n$-round trail is a sequence of n two-round trails and hence that its byte weight is lower bounded by n times the branch number of ρ. One approach would be to design a round transformation with a maximum branch number. However, similar to large S-boxes, transformations that provide high branch numbers have a tendency to have a high implementation cost. More efficient designs can be achieved using a round structure with a limited branch number but with some other particular propagation properties.

For this purpose, λ can be built as a sequence of two steps:

1. θ. A step that provides high local diffusion.
2. π. A step that provides high *dispersion*.

In block cipher design, the mixing step θ is usually a linear bricklayer permutation. Its component permutations operate on a limited number of bytes and have a branch number that is high with respect to their dimensions. The step π takes care of *dispersion*. By dispersion we mean the operation by which bits or bytes that are close to each other in the context of θ are moved to positions that are distant.

Jointly, θ and π have a spectacular effect on patterns with a low Hamming weight: through θ this propagates to a localized pattern with high Hamming weight that is dispersed all over the state by π. There are several approaches to the selection of θ and π. One of these approaches has led to Rijndael and its relatives.

9.3 Branch Numbers and Two-Round Trails

In this section we formally define the *branch number* of a Boolean transformation with respect to a byte partition.

The byte weight of a state is equal to the number of non-zero bytes. This is denoted by $\mathrm{w_t}(\mathbf{a})$. If this is applied to a difference pattern $\mathbf{a}'$, $\mathrm{w_t}(\mathbf{a}')$ is the number of active bytes in $\mathbf{a}'$. Applied to a mask $\mathbf{v}$, $\mathrm{w_t}(\mathbf{v})$ is the number of active bytes in $\mathbf{v}$. We make a distinction between the differential and the linear branch number of a transformation.

Definition 9.3.1. *The* differential branch number *of a transformation ϕ is given by*

$$\mathcal{B}_{\mathrm{d}}(\phi) = \min_{\mathbf{a},\mathbf{b}\neq\mathbf{a}} \{\mathrm{w_t}(\mathbf{a}+\mathbf{b}) + \mathrm{w_t}(\phi(\mathbf{a}) + \phi(\mathbf{b}))\}. \quad (9.7)$$

For a linear transformation $\lambda(\mathbf{a}) + \lambda(\mathbf{b}) = \lambda(\mathbf{a}+\mathbf{b})$, and (9.7) reduces to

$$\mathcal{B}_{\mathrm{d}}(\lambda) = \min_{\mathbf{a}' \neq 0}\{\mathrm{w_t}(\mathbf{a}') + \mathrm{w_t}(\lambda(\mathbf{a}'))\}. \tag{9.8}$$

Analogous to the differential branch number, we can define the linear branch number.

Definition 9.3.2 *The* linear branch number *of a transformation ϕ is given by*

$$\mathcal{B}_{\mathrm{l}}(\phi) = \min_{\alpha,\beta,C(\alpha^{\mathrm{T}}\mathbf{x},\beta^{\mathrm{T}}\phi(\mathbf{x}))\neq 0}\{\mathrm{w_t}(\alpha) + \mathrm{w_t}(\beta)\}. \tag{9.9}$$

If ϕ is a linear transformation denoted by λ, there exists a matrix M such that $\lambda(\mathbf{x}) = \mathsf{M} \cdot \mathbf{x}$. Equation (9.9) can then be simplified to (see Sect. 7.4)

$$\mathcal{B}_{\mathrm{l}}(\lambda) = \min_{\alpha \neq 0}\{\mathrm{w_t}(\alpha) + \mathrm{w_t}(\mathsf{M}^{\mathrm{T}}\alpha)\}. \tag{9.10}$$

It follows that the linear branch number of the linear transformation specified by the matrix M is equal to the differential branch number of the linear transformation specified by the matrix M^{T}. Many of the following discussions are valid both for differential and linear branch numbers, and both $\mathcal{B}_{\mathrm{d}}$ and $\mathcal{B}_{\mathrm{l}}$ are denoted simply by $\mathcal{B}$.

An upper bound for the (differential or linear) branch number of a Boolean transformation ϕ is given by one plus the total number of bytes in the state, denoted by n_α. The output difference pattern or mask corresponding to an input difference pattern or mask with a single non-zero byte can have a maximum weight of n_α. Hence, the branch number of ϕ is upper bounded by

$$\mathcal{B}(\phi) \leq n_\alpha + 1. \tag{9.11}$$

In general, the linear and differential branch numbers of a transformation with respect to a partition are not equal. This is illustrated in Example 9.3.1. However, if the step λ satisfies certain conditions it can be shown that the differential and linear branch numbers are equal. An obvious sufficient condition is the requirement that M be symmetric. Also, if a Boolean transformation has the maximal possible differential or linear branch number, then both branch numbers are equal. This is proven for the case of linear transformations in Sect. 9.6 and for the general case in Chap. 13.

Example 9.3.1. Consider the transformation $\lambda : \mathbf{x} \mapsto \mathsf{A} \cdot \mathbf{x}$ over GF(4), where

$$\mathsf{A} = \begin{bmatrix} 1 & 1 & 1 & 1 \\ 0 & 1 & 0 & 1 \\ 0 & 0 & 1 & 1 \\ 0 & 1 & 1 & 0 \end{bmatrix}. \tag{9.12}$$

A is invertible and hence $\mathcal{B}_{\mathrm{d}}(\theta) \geq 2$ and $\mathcal{B}_{\mathrm{l}}(\theta) \geq 2$. The first column of A has only a single 1 in the first position and hence $\mathsf{A} \cdot (1,0,0,0)^{\mathrm{T}} = (1,0,0,0)^{\mathrm{T}}$;

it follows that $\mathcal{B}_{\mathrm{d}}(\theta) = 2$. A has no rows with a single 1, so A^{T} does not map vectors with weight 1 to weight 1. It does have rows with two non-zero entries; and therefore, $\mathcal{B}_{\mathrm{l}}(\theta) = 3$.

9.3.1 Derived Properties

From the symmetry of Definitions 9.3.1 and 9.3.2 it follows that the branch number of a transformation and that of its inverse are the same. Moreover, we have the following properties:

1. A difference pattern or mask $\mathbf{a}$ is not affected by a key addition and hence its byte weight $\mathrm{w_t}(\mathbf{a})$ is not affected.
2. A bricklayer permutation operating on individual bytes cannot turn an active byte into a non-active byte or vice versa. Hence, it does not affect the byte weight $\mathrm{w_t}$.

Assume that we have a transformation ϕ that is a sequence of a transformation ϕ_1 and a bricklayer transformation ϕ_2 operating on bytes, i.e. $\phi = \phi_2 \circ \phi_1$. As ϕ_2 does not affect the number of active bytes in a propagation pattern, the branch numbers of ϕ and ϕ_1 are the same. More generally, if propagation of patterns is analyzed at the level of bytes, bricklayer transformations operating on individual bytes may be ignored as they leave the difference patterns and masks unchanged.

If we apply this to the byte weight of a $\gamma\lambda$ round transformation ρ, it follows immediately that the (linear or differential) byte branch number of ρ is that of its linear part λ.

9.3.2 A Two-Round Propagation Theorem

The following theorem relates the value of $\mathcal{B}(\lambda)$ to a bound on the number of active bytes in a trail. The proof is valid both for linear and differential trails: in the case of linear trails $\mathcal{B}$ stands for $\mathcal{B}_{\mathrm{l}}$ and in the case of differential trails $\mathcal{B}$ stands for $\mathcal{B}_{\mathrm{d}}$.

Theorem 9.3.1 (Two-Round Propagation Theorem).
For a key-alternating block cipher with a $\gamma\lambda$ round structure, the number of active bytes of any two-round trail is lower bounded by the (byte) branch number of λ.

Proof. Figure 9.3 depicts two rounds. Since the steps γ and $\sigma[\mathbf{k}]$ operate on each byte individually, they do not affect the propagation of patterns. Hence it follows that $\mathrm{w_t}(\mathbf{a}^1)+\mathrm{w_t}(\mathbf{a}^2)$ is only bounded by the properties of the linear step λ of the first round. Definitions 9.3.1 and 9.3.2 imply that the sum of the active bytes before and after λ of the first round is lower bounded by $\mathcal{B}(\lambda)$. □

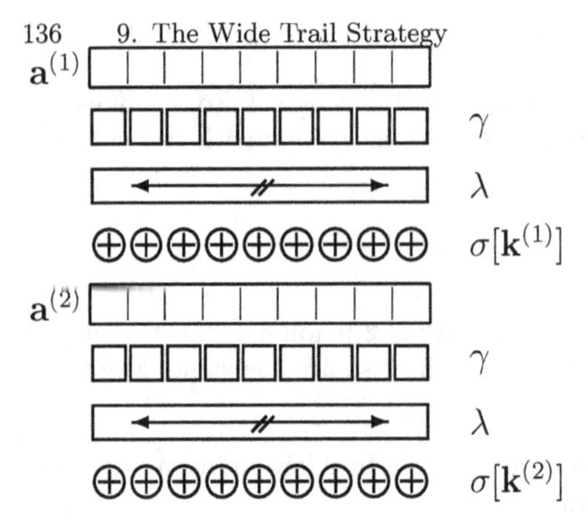

Fig. 9.3. Steps relevant in the proof of Theorem 9.3.1

9.4 An Efficient Key-Alternating Structure

Theorem 9.3.1 seems to suggest that to obtain high lower bounds on the byte weight of multiple-round trails, a transformation λ must be used with a high branch number. However, realizing a high branch number has its computational cost. In this section we elaborate on a cipher structure that is more efficient in providing lower bounds.

We build a key-alternating block cipher that consists of an alternation of two different round transformations defined by

$$\rho^a = \theta \circ \gamma \text{ and} \tag{9.13}$$
$$\rho^b = \Theta \circ \gamma. \tag{9.14}$$

The step γ is defined as before and operates on n_t $\mathrm{n_s}$-bit bytes.

9.4.1 The Diffusion Step θ

With respect to θ, the bytes of the state are grouped into a number of *columns* by a partition Ξ of the index space $\mathcal{I}$. We denote a column by ξ and the number of columns by n_Ξ. The column containing an index i is denoted by $\xi(i)$, and the number of indices in a column ξ by n_ξ. The size of the columns relates to the block length by

$$m \sum_{\xi \in \Xi} n_\xi = m n_t.$$

θ is a bricklayer permutation with component permutations that each operate on a column, as illustrated by Fig. 9.4. Within each column, bytes are linearly combined. We have

$$\theta : \mathbf{b} = \theta(\mathbf{a}) \Leftrightarrow b_i = \sum_{j \in \xi(i)} M_{i,j} a_j. \tag{9.15}$$

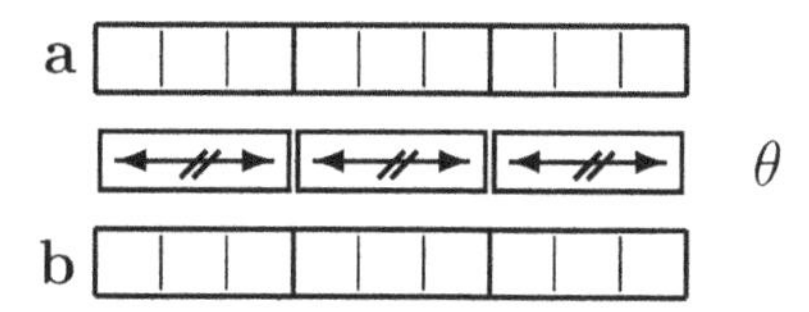

Fig. 9.4. The diffusion step θ

If the array of bytes with indices in ξ is denoted by $\mathbf{a}_\xi$, we have

$$\theta : \mathbf{b} = \theta(\mathbf{a}) \Leftrightarrow \mathbf{b}_\xi = \mathsf{M}_\xi \mathbf{a}_\xi \tag{9.16}$$

where M_ξ is an $n_\xi \times n_\xi$ matrix. The jth column of M_ξ is denoted by $\mathsf{M}_{\xi|j}$. The inverse of θ is specified by the partition Ξ and the matrices M_ξ^{-1}. The bricklayer transformation θ only needs to realize diffusion within the columns and hence has a low implementation cost.

Similar to active bytes, we can speak of active columns. The number of active columns of a propagation pattern $\mathbf{a}$ is denoted by $\mathrm{w_s}(\mathbf{a})$.

The round transformation $\rho^{(a)} = \theta \circ \gamma$ is a bricklayer transformation operating independently on a number of columns. Taking this bricklayer structure into account, we can extend the results of Sect. 9.3 slightly. The branch number of θ is given by the minimum branch number of its component transformations. Applying (9.11) to the component permutations defined by the matrices M_ξ results in the following upper bound:

$$\mathcal{B}(\theta) \leq \min_\xi n_\xi + 1. \tag{9.17}$$

Hence, the smallest column imposes the upper limit for the branch number.

The two-round propagation theorem (Theorem 9.3.1) implies the following lemma.

Lemma 9.4.1. *The byte weight of any two-round trail in which the first round has a $\gamma\theta$ round transformation is lower bounded by $N\mathcal{B}(\theta)$, where N is the number of active columns at the input of the second round.*

Proof. Theorem 9.3.1 can be applied separately to each of the component transformations of the bricklayer transformation $\rho^{(a)}$. For each active column there are at least $\mathcal{B}(\theta)$ active bytes in the two-round trail. If the number of active columns is denoted by N, we obtain the proof. □

Example 9.4.1. In $\mathcal{X}_2$, the partition Ξ has two elements. θ can be defined as

$$\theta\left(\begin{bmatrix} a_1 & a_3 \\ a_2 & a_4 \\ & a_5 \\ & a_6 \end{bmatrix}\right) = \begin{bmatrix} 2a_1 + a_2 & a_3 + a_4 + a_5 \\ a_1 + a_2 & a_4 + a_5 + a_6 \\ & a_3 + a_5 + a_6 \\ & a_3 + a_4 + a_6 \end{bmatrix}.$$

In this case there are two matrices M_ξ:

$$\mathsf{M}_{\xi(0)} = \begin{bmatrix} 2 & 1 \\ 1 & 1 \end{bmatrix}, \qquad \text{and} \qquad \mathsf{M}_{\xi(1)} = \begin{bmatrix} 1 & 1 & 1 & 0 \\ 0 & 1 & 1 & 1 \\ 1 & 0 & 1 & 1 \\ 1 & 1 & 0 & 1 \end{bmatrix}.$$

9.4.2 The Linear Step Θ

Θ mixes bytes across columns:

$$\Theta : \mathbf{b} = \Theta(\mathbf{a}) \Leftrightarrow b_i = \sum_j M_{i,j} a_j. \tag{9.18}$$

The goal of Θ is to provide inter-column diffusion. Its design criterion is to have a high branch number with respect to the column partition. This is denoted by $\mathcal{B}^c(\Theta)$ and called its *column branch number.*

9.4.3 A Lower Bound on the Byte Weight of Four-Round Trails

The combination of the byte branch number of θ and the column branch number of Θ allows us to prove a lower bound on the byte weight of any trail over four rounds starting with $\rho^{(a)}$.

Theorem 9.4.1 (Four-Round Propagation Theorem for $\theta\Theta$ Construction). *For a key-alternating block cipher with round transformations as defined in (9.13) and (9.14), the byte weight of any trail over $\rho^{(b)} \circ \rho^{(a)} \circ \rho^{(b)} \circ \rho^{(a)}$ is lower bounded by $\mathcal{B}(\theta) \times \mathcal{B}^c(\Theta)$.*

Proof. Figure 9.5 depicts four rounds with the key addition steps and the nonlinear steps removed, since these play no role in the trail propagation. It is easy to see that the linear step of the fourth round plays no role. The sum of the number of active columns in $\mathbf{a}^{(2)}$ and $\mathbf{a}^{(3)}$ is lower bounded by $\mathcal{B}^c(\Theta)$. According to Lemma 9.4.1, for each active column in $\mathbf{a}^{(2)}$ there are at least $\mathcal{B}(\theta)$ active bytes in the corresponding columns of $\mathbf{a}^{(1)}$ and $\mathbf{a}^{(2)}$. Similarly, for each active column in $\mathbf{a}^{(3)}$ there are at least $\mathcal{B}(\theta)$ active bytes in the corresponding columns of $\mathbf{a}^{(3)}$ and $\mathbf{a}^{(4)}$. Hence the total number of active bytes is lower bounded by $\mathcal{B}(\theta) \times \mathcal{B}^c(\Theta)$. □

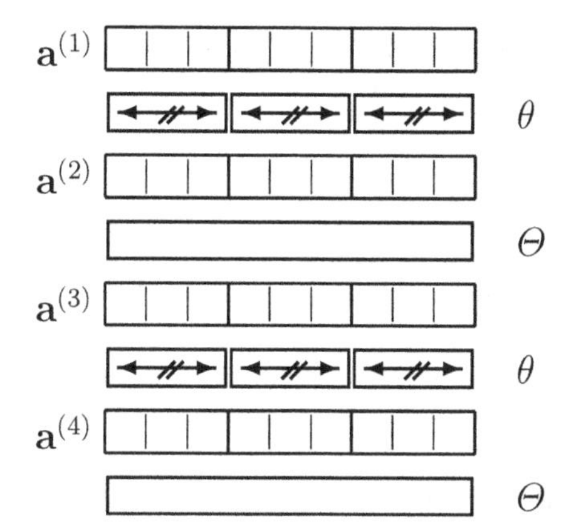

Fig. 9.5. Steps relevant in the proof of Theorem 9.4.1

9.4.4 An Efficient Construction for Θ

As opposed to θ, Θ does not operate on different columns independently and hence may have a much higher implementation cost. In this section we present a construction of Θ in terms of θ and byte transpositions denoted by π. We have

$$\Theta = \pi \circ \theta \circ \pi. \tag{9.19}$$

In the following we will define π, and prove that if π is well chosen the column branch number of Θ can be made equal to the byte branch number of θ.

The byte transposition π. The byte transposition π is defined as

$$\pi : \mathbf{b} = \pi(\mathbf{a}) \Leftrightarrow b_i = a_{p(i)}, \tag{9.20}$$

where $p(i)$ is a permutation of the index space $\mathcal{I}$. The inverse of π is defined by $p^{-1}(i)$.

Example 9.4.2. In the cipher $\mathcal{X}_2$, we define π as the transformation that leaves the first row unchanged and shifts the second row one place to the right:

$$\pi\left(\begin{bmatrix} a_1 & a_3 & a_5 \\ a_2 & a_4 & a_6 \end{bmatrix}\right) = \begin{bmatrix} a_1 & a_3 & a_5 \\ a_6 & a_2 & a_4 \end{bmatrix}.$$

Observe that a byte transposition π does not affect the byte weight of a propagation pattern and hence that the branch number of a transformation is not affected if it is composed with π.

As opposed to θ, π provides inter-column *diffusion*. Intuitively, good diffusion for π would mean that it distributes the different bytes of a column to as many different columns as possible.

We say π is *diffusion optimal* if the different bytes in each column are distributed over all different columns. More formally, we have the following.

Definition 9.4.1. *π is* diffusion optimal *iff*

$$\forall i, j \in \mathcal{I}, i \neq j : (\xi(i) = \xi(j)) \Rightarrow (\xi(p(i)) \neq \xi(p(j))). \tag{9.21}$$

It is easy to see that this implies the same condition for π^{-1}. A diffusion optimal byte transposition π implies $\mathrm{w_s}(\pi(\mathbf{a})) \geq \max_\xi(\mathrm{w_t}(a_\xi))$. Therefore a diffusion optimal transformation can only exist if $n_\Xi \geq \max_i(n_{\xi_i})$. In words, π can only be diffusion optimal if there are at least as many columns as there are bytes in the largest column.

If π is diffusion optimal, we can prove that the column branch number of the transformation Θ is equal to the branch number of θ.

Lemma 9.4.2. *If π is a diffusion optimal transposition of bytes, the column branch number of $\pi \circ \phi \circ \pi$ is equal to the byte branch number of ϕ.*

Proof. We refer to Fig. 9.6 for the notation used in this proof. Firstly, we demonstrate that

$$\mathrm{w_s}(\mathbf{a}) + \mathrm{w_s}(\mathbf{d}) \geq \mathcal{B}(\phi). \tag{9.22}$$

For any active column in $\mathbf{b}$, the number of active bytes in that column and the corresponding column of $\mathbf{c}$ is at least $\mathcal{B}(\phi)$. π moves all active bytes in an active column of $\mathbf{c}$ to different columns in $\mathbf{d}$, and π^{-1} moves all active bytes in an active column of $\mathbf{b}$ to different columns in $\mathbf{a}$. It follows that the sum of the number of active columns in $\mathbf{a}$ and in $\mathbf{d}$ is lower bounded by the byte branch number of ϕ.
Now we only have to prove that the sum of the number of active columns in $\mathbf{a}$ and in $\mathbf{d}$ is upper bounded by the byte branch number of ϕ. Assume that $\mathbf{b}$, and equivalently $\mathbf{c}$, only have one active column and that ϕ restricted to this column has branch number $\mathcal{B}(\phi)$. In that case, there exists a configuration in which the sum of the number of active bytes in $\mathbf{b}$ and $\mathbf{c}$ is equal to $\mathcal{B}(\phi)$. π moves the active bytes in the active column of $\mathbf{c}$ to different columns in $\mathbf{d}$, and π^{-1} moves the active bytes in the active column of $\mathbf{b}$ to different columns in $\mathbf{a}$, and hence the total number of active columns in $\mathbf{a}$ and $\mathbf{d}$ is equal to $\mathcal{B}(\phi)$. □

9.5 The Round Structure of Rijndael

9.5.1 A Key-Iterated Structure

The efficient structure described in Sect. 9.4 uses two different round transformations. It is possible to define a block cipher structure with only one round

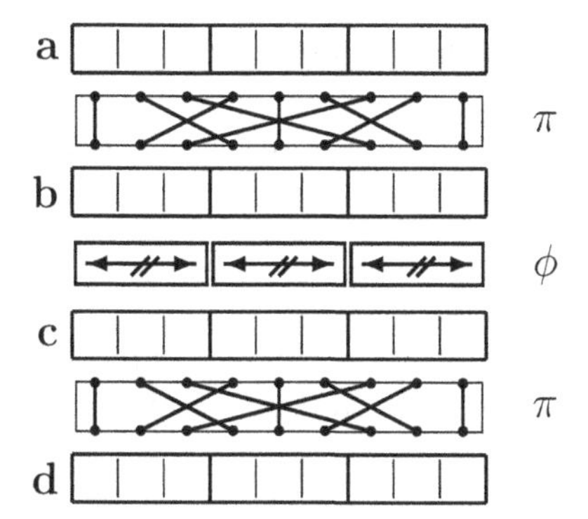

Fig. 9.6. Steps relevant in the proof of Lemma 9.4.2

transformation that achieves the same bound. This is the round structure used in Rijndael and most of the related ciphers. The advantage of having a single round transformation is a reduction in program size in software implementations and chip area in dedicated hardware implementations. For this purpose, λ can be built as the sequence of two steps:

1. θ. The linear bricklayer transformation that provides high local diffusion, as defined in Sect. 9.4.1.
2. π. The byte transposition that provides high *dispersion*, as defined in Sect. 9.4.4.

Hence we have the following for the round transformation:

$$\rho^c = \theta \circ \pi \circ \gamma. \tag{9.23}$$

Fig. 9.7 gives a schematic representation of the different steps of a round. The steps of the round transformation are defined in such a way that they

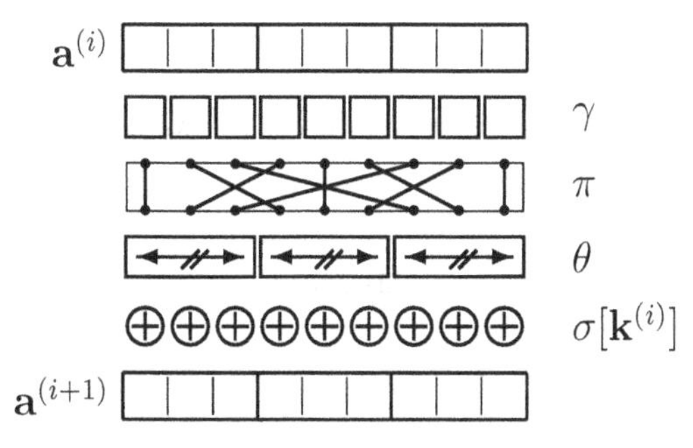

Fig. 9.7. Sequence of steps of a $\gamma\pi\theta$ round transformation, followed by a key addition

impose strict lower bounds on the number of active S-boxes in four-round trails.

For two-round trails the number of active bytes is lower bounded by $\mathcal{B}(\rho) = \mathcal{B}(\lambda) = \mathcal{B}(\theta)$. For four rounds, we can prove the following important theorem.

Theorem 9.5.1 (Four-Round Propagation Theorem).
For a key-iterated block cipher with a $\gamma\pi\theta$ round transformation and diffusion optimal π, the number of active S-boxes in a four-round trail is lower bounded by $(\mathcal{B}(\theta))^2$.

Proof. Firstly, we show that the transformation consisting of four rounds ρ^c as defined in (9.23) is equivalent to four rounds of the construction with ρ^a and ρ^b as defined in (9.13) and (9.14). For simplicity, we leave out the key addition steps, but the proof works in the same way if the key addition steps are present. Let $\mathcal{A}$ be defined as

$$\begin{aligned}\mathcal{A} &= \rho^c \circ \rho^c \circ \rho^c \circ \rho^c \\ &= (\theta \circ \pi \circ \gamma) \circ (\theta \circ \pi \circ \gamma) \circ (\theta \circ \pi \circ \gamma) \circ (\theta \circ \pi \circ \gamma).\end{aligned}$$

γ is a bricklayer permutation, operating on every byte separately and operating independently of the byte's position. Therefore γ commutes with π, which only moves the bytes to different positions. We get

$$\begin{aligned}\mathcal{A} &= (\theta \circ \gamma) \circ (\pi \circ \theta \circ \pi \circ \gamma) \circ (\theta \circ \gamma) \circ (\pi \circ \theta \circ \pi \circ \gamma) \\ &= \rho^a \circ \rho^b \circ \rho^a \circ \rho^b,\end{aligned}$$

where Θ of ρ^b is defined exactly as in (9.19). Now we can apply Lemma 9.4.2 and Theorem 9.4.1 to finish the proof. □

In a four-round trail there can be only $4n_t$ active bytes. One may wonder how the lower bound of Theorem 9.5.1 relates to this upper bound. From (9.11) we have that $\mathcal{B}^2 \leq \min(n_\xi + 1)^2 = \min n_\xi^2 + 2 \min n_\xi + 1$. Diffusion-optimality implies that $\min(n_\xi + 1)^2 \leq \min n_\xi n_\Xi + 2 \min n_\xi + 1 \leq n_t + 2n_t + n_t = 4n_t$. Hence, the lower bound of Theorem 9.5.1 is always below the upper bound of $4n_t$.

9.5.2 Applying the Wide Trail Strategy to Rijndael

To provide resistance against differential and linear cryptanalysis, Rijndael has been designed according to the wide trail strategy: the four-round propagation theorem is applicable to Rijndael. It exhibits the key-iterated round structure described above:

1. `SubBytes`: the nonlinear step γ, operating on the state bytes in parallel.

2. `ShiftRows`: the transposition step π.
3. `MixColumns`: the mixing step θ, operating on columns of four bytes each.

The coefficients of `MixColumns` have been selected in such a way that both the differential branch number and the linear branch number (see Definitions 9.3.1 and 9.3.2) of `MixColumns` are equal to 5. Since `ShiftRows` moves the bytes of each column to four different columns, it is diffusion optimal (see Definition 9.4.1). Hence, the four-round propagation theorem (Theorem 9.5.1) proves that the number of active S-boxes in a four-round differential trail or linear trail is lower bounded by 25.

S_{RD} has been selected in such a way that the maximum correlation over it is at most 2^{-3}, and the DP is at most 2^{-6}, in other words, that the weight of any difference propagation is at least 6.

This gives a minimum weight of 150 for any four-round differential or linear trail. These results hold for all block lengths of Rijndael and are independent of the value of the round keys. Hence there are no eight-round trails with a weight below 300 or a correlation contribution above 2^{-150}.

9.6 Constructions for θ

For memory-efficient implementations, all columns preferably have the same size. The fact that the branch number is upper bounded by the smallest column (see Eq. (9.11)) points in the same direction. Hence we will consider in the following only the case where all columns have the same size.

Additionally we can reduce program and chip size by imposing the requirement that θ acts in the same way on each column. In this case the same matrix M_ξ is used for all columns.

Imposing additional symmetry conditions on the matrix M_ξ can lead to even more compact implementations. For instance, M_ξ can be defined as a circulant matrix.

Definition 9.6.1. *An $n \times n$ matrix A is* circulant *if there exist n constants $a_1, \ldots, a_n$ and a 'step' $c \neq 0$ such that for all i, j $(0 \leq i, j < n)$*

$$a_{i,j} = a_{i+cj \bmod n} \,. \tag{9.24}$$

If $\gcd(c, n) = 1$ it can be proven that $\mathcal{B}_\mathrm{l}(\lambda) = \mathcal{B}_\mathrm{d}(\lambda)$.

The branch numbers of linear functions can be studied using the framework of linear codes over $\mathrm{GF}(2^p)$. Codes can be associated with Boolean transformations in the following way.

Definition 9.6.2. *Let θ be a transformation from $GF(2^{\mathrm{n_s}})^n$ to $GF(2^{\mathrm{n_s}})^n$. The* associated code *of θ, $\mathcal{C}_\theta$, is the code that has codewords given by the vectors $(\mathbf{x}, \theta(\mathbf{x}))^\mathrm{T}$. The code $\mathcal{C}_\theta$ has $2^{\mathrm{n_s}n}$ codewords and has length $2n$.*

If θ is defined as $\theta(\mathbf{x}) = \mathsf{A} \cdot \mathbf{x}$, then $\mathcal{C}_\theta$ is a linear $[2n, n, d]$ code. Code $\mathcal{C}_\theta$ consists of the vectors $(\mathbf{x}, \mathsf{A} \cdot \mathbf{x})^{\mathrm{T}}$, where $\mathbf{x}$ takes all possible input values. Equivalently, the generator matrix G_θ of $\mathcal{C}_\theta$ is given by

$$\mathsf{G}_\theta = \left[\mathsf{I}\ \mathsf{A}^{\mathrm{T}}\right] , \tag{9.25}$$

and the parity-check matrix H_θ is given by

$$\mathsf{H}_\theta = [-\mathsf{A}\ \mathsf{I}] = [\mathsf{A}\ \mathsf{I}] . \tag{9.26}$$

We can construct matrices M_ξ giving rise to a maximum branch number from an MDS code. Because of this connection, M_ξ matrices are often called *MDS matrices.*

It follows from Definition 9.3.1 that the differential branch number of a transformation θ equals the minimal distance between two different codewords of its associated code $\mathcal{C}_\theta$. The theory of linear codes addresses the problems of determining the distance of a linear code and the construction of linear codes with a given distance. The relations between linear transformations and linear codes allow us to construct efficiently transformations with high branch numbers. As a first example, the upper bound on the differential branch number given in (9.11) corresponds to the Singleton bound for codes (Theorem 2.2.1). Theorem 2.2.2 states that a linear code has distance d if and only if every $d - 1$ columns of the parity-check matrix H are linearly independent and there exists some set of d columns that are linearly dependent. Reconsidering the matrix A of Example 9.3.1, all columns in $\mathsf{H} = [-\mathsf{A}\ \mathsf{I}]$ are non-zero, hence every set of one column is linearly independent. Since two columns are equal, there is a set of two columns that is not linearly independent. Therefore the differential branch number equals two. Theorem 2.2.3 states that a linear code with maximal distance requires that every square submatrix of A is nonsingular. An immediate consequence is that transformations can have maximal branch numbers only if they are invertible. Furthermore, a transformation with maximal linear branch number has also maximal differential branch number, and vice versa. Indeed, if all submatrices of A are nonsingular, then this holds also for A^{T}.

The following theorem relates the linear branch number of a linear transformation to the dual of the associated code.

Theorem 9.6.1. *If $\mathcal{C}_\theta$ is the associated code of the linear transformation θ, then the linear branch number of θ is equal to the distance of the dual code of $\mathcal{C}_\theta$.*

Proof. We give the proof for binary codes only. If θ is specified by the matrix A, then $\left[\mathsf{I}\ \mathsf{A}^{\mathrm{T}}\right]$ is a generator matrix for $\mathcal{C}_\theta$, and $[\mathsf{A}\ \mathsf{I}]$ is a generator matrix for the dual of $\mathcal{C}_\theta$. It follows from (9.10) that the minimal distance of the code generated by $[\mathsf{A}\ \mathsf{I}]$ equals the linear branch number of θ. □

It follows that transformations that have an associated code that is MDS have equal differential and linear branch numbers.

9.7 Choices for the Structure of $\mathcal{I}$ and π

In this section we present several concrete constructions for π and the implications with respect to trails.

We present two general structures for $\mathcal{I}$ and π. In the first structure the different bytes of a state are arranged in a multidimensional regular array or hypercube of dimension d and side n_ξ. Ciphers constructed in this way have a block size of $\mathrm{n_s} n_\xi^d$. In the second structure the bytes of a state are arranged in a rectangle with one side equal to n_ξ. This gives more freedom for the choice of the block size of the cipher.

9.7.1 The Hypercube Structure

In this construction the columns ξ are arranged in a hypercube. The step π corresponds to a rotation of the hypercube around a diagonal axis (called the p-axis).

The indices $i \in \mathcal{I}$ are represented by a vector of length d and elements i_j between 0 and $n_\xi - 1$. We have

$$\mathbf{i} = (i_1, i_2, \ldots, i_d). \tag{9.27}$$

The columns ξ are given by

$$\mathbf{j} \in \xi(\mathbf{i}) \text{ if } j_1 = i_1, j_2 = i_2, \ldots \text{ and } j_{d-1} = i_{d-1}. \tag{9.28}$$

$p(i)$, defining π, is given by

$$p : \mathbf{j} = p(\mathbf{i}) \Leftrightarrow (j_1, j_2, \ldots, j_{d-1}, j_d) = (i_2, i_3, \ldots, i_d, i_1). \tag{9.29}$$

Clearly, π is diffusion optimal (if $d > 1$). We will briefly illustrate this for d equal to 1, 2 and 3.

Dimension 1. Dimension 1 is a degenerate case because the partition counts only one column, and π cannot be diffusion optimal. SHARK [132] is an example where $n_t = n_\xi = 8$ and $\mathrm{n_s} = 8$, resulting in a block size of 64 bits.

Dimension 2. Figure 9.8 shows the two-dimensional array, the transposition π and the partition Ξ.

The two-dimensional structure is adopted in Square [44], with $\mathrm{n_s} = 8$ and $n_\xi = 4$, resulting in a block cipher with a block size of 128 bits in which every four-round trail has at least $\mathcal{B}^2 = 25$ active S-boxes.

Crypton [96] has the same structure and transposition π as SQUARE, but it uses a different step θ. Since for Crypton $\mathcal{B}(\theta) = 4$, there are at least 16 active S-boxes in every four-round trail.

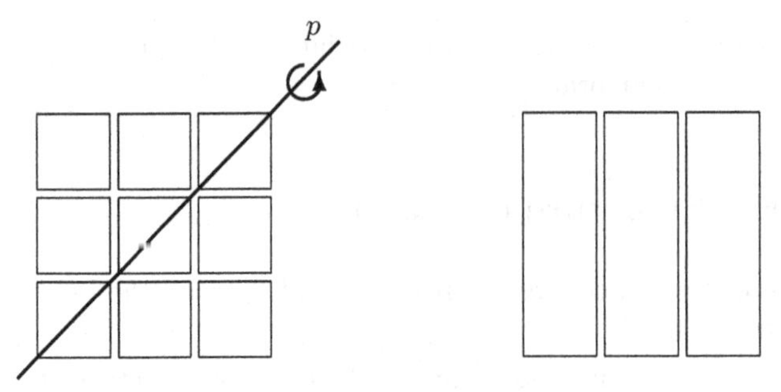

Fig. 9.8. Example of the hypercube structure with $d = 2$ and $n_\xi = 3$. The p-axis is indicated on the left

Dimension 3. For dimension three, with $n_\xi = 2$ and $\mathrm{n_s} = 8$, we get a 64-bit cipher that has some similarity to the block cipher SAFER designed by J. Massey [103], however the round transformation of SAFER actually looks more like a triple application of $\theta \circ \pi$ for every application of γ. Therefore SAFER also can (almost) be seen as an example of a cipher with a diffusion layer of dimension 1.

Theorem 9.5.1 guarantees for our constructions a lower bound on the number of active S-boxes per four rounds of 9. For trails of more than four rounds, the minimum number of active S-boxes per round rises significantly: after six rounds for instance there are already a minimum of 18 active S-boxes. Figure 9.9 shows an example for the arrangement of the bytes and the columns.

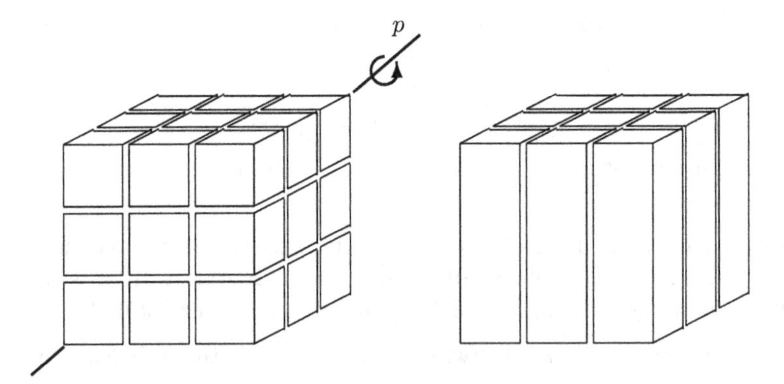

Fig. 9.9. Example for the hypercube structure with $d = 3$ and $n_\xi = 3$. The bytes are shown on the left and the colunms are shown on the right

9.7.2 The Rectangular Structure

In this construction the columns ξ are arranged in a rectangle. The other dimensions of the array are determined from the required block size of the cipher. Figure 9.10 shows the arrangement of the bytes and the columns for an example where $n_\xi = 3$ and $n_\Xi = 5$. The step π leaves the first row invariant, shifts the second row by one position, and the third row by two positions.

Generally, if the step π shifts every row by a different number of bytes, the diffusion of π is optimal. (Note that this is only possible if $n_\Xi \geq n_\xi$, i.e. if the number of rows is at most the number of columns.)

If a byte has $\mathrm{n_s} = 8$ bits and every column contains $n_\xi = 4$ bytes, then setting the number of columns n_Ξ to 4, 5, 6, 7 or 8 gives a block size of 128, 160, 192, 224 or 256 bits, respectively. This is exactly the structure adopted in Rijndael [47]. BKSQ [48] is a cipher tailored for smart cards. Therefore its dimensions are kept small: $\mathrm{n_s} = 8, n_\xi = 3$ and $n_\Xi = 4$ to give a block length of 96 bits.

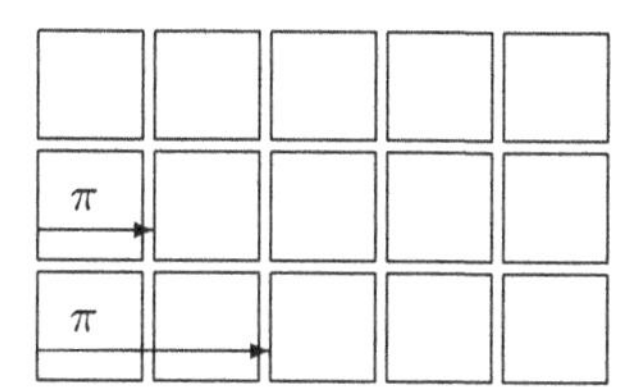

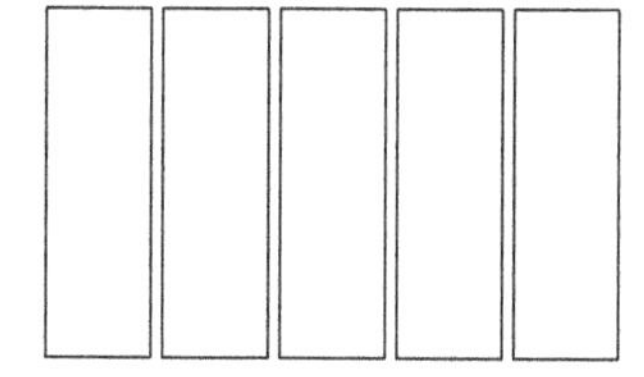

Fig. 9.10. Example of the rectangular structure. The bytes are shown on the left and the columns are on the right

9.8 Conclusions

In this chapter we have given the design strategy that is the foundation of the Rijndael structure. The proposed cipher structure allows us to give provable bounds on the correlation of linear trails and the weight of differential trails, while at the same time allowing efficient implementations.

Finally, we show that Rijndael and its related ciphers are instances of a cipher family that allows a large flexibility in block length without losing the properties of efficiency and high resistance against cryptanalysis.

10. Cryptanalysis

The resistance of Rijndael against linear and differential cryptanalysis has been treated extensively in Chaps. 7 to 9. In this chapter we discuss the resistance of Rijndael against various other cryptanalytic attacks. None of these attacks poses a threat to Rijndael in a practical sense. We also touch briefly on the topic of implementation attacks.

10.1 Truncated Differentials

The concept of truncated differentials was described by L. Knudsen in [84]. The corresponding class of attacks exploit the fact that in some ciphers, differential trails (see Chap. 8) tend to cluster. We refer to Chap. 13 for a treatment in depth. In short, clustering takes place if for certain sets of input difference patterns and output difference patterns, the number of differential trails is exceedingly large. The expected probability that a differential trail stays within the boundaries of the cluster can be computed independently of the probabilities of the individual differential trails. Ciphers in which all steps operate on the state in bytes are prone to be susceptible to this type of attack. Since this is the case for Rijndael, with all steps operating on bytes rather than individual bits, we investigated its resistance against truncated differentials.

10.2 Saturation Attacks

In the paper presenting the block cipher Square [44], a dedicated attack by L. Knudsen on reduced versions of Square is described. The attack is often referred to as the 'Square' attack. The attack exploits the byte-oriented structure of Square, and is also applicable to reduced versions of Rijndael. N. Ferguson et al. [61] proposed some optimizations that reduce the work factor of the attack. In [99], S. Lucks proposes the name *'saturation attack'* for this type of attack. Later, these attacks have been called *Structural attacks* [30] and *Integral attacks* [88].

J. Daemen, V. Rijmen, *The Design of Rijndael*, Information Security and Cryptography,
https://doi.org/10.1007/978-3-662-60769-5_10

The saturation attack is a chosen-plaintext attack on ciphers with the Rijndael round structure. It can be mounted independently of the choice of the S-box in the nonlinear step and the key schedule. The version we describe here is for the case in which the columns of mixing step `MixColumns` have a maximum branch number and the byte transposition `ShiftRows` is diffusion optimal. If one of these two conditions is not fulfilled, the attack is slightly different but has comparable complexity. In this section we describe the attack on a cipher in which the tuples are 8-bit values. Generalizing the attack to other tuple sizes is trivial.

Applied to Rijndael, the saturation attack is faster than an exhaustive key search for reduced-round versions of up to six rounds. After describing the basic attack on four rounds, we will show how it can be extended to five and six rounds.

10.2.1 Preliminaries

Let a Λ-set be a set of 256 states with some relation to one another. We define four relations:

Constant: A Λ-set is *constant* in position i, j if

$$\forall \mathbf{x}, \mathbf{y} \in \Lambda : x_{i,j} = y_{i,j}.$$

Active: A Λ-set is *active* in position i, j if

$$\forall \mathbf{x}, \mathbf{y} \in \Lambda : x_{i,j} \neq y_{i,j}.$$

Balanced: A Λ-set is *balanced* in position i, j if

$$\sum_{l \in \Lambda} x^l_{i,j} = 0.$$

Unknown: A Λ-set is *unknown* in position i, j if we cannot prove that it has one of the three previous properties.

Observe that if a Λ-set is active or constant in a position i, j, then it is also balanced in that position. Observe further that if a Λ-set contains only active or constant positions, then the application of the steps `SubBytes` or `AddRoundKey` on the states of a Λ-set results in a different Λ-set with the positions of the active bytes unchanged. Application of the step `ShiftRows` results in a Λ-set in which the active bytes are transposed by `ShiftRows`. Application of the step `MixColumns` does not conserve active positions. However, since every output byte of `MixColumns` is a linear combination with invertible coefficients of the four input bytes in the same column, an input column with a single active byte gives rise to an output column with all four bytes active.

10.2.2 The Basic Attack

Consider a Λ-set in which only one position is active and all other positions are constant. We will now trace the evolution of the properties through three rounds. In the first round, `MixColumns` converts the active position to a complete column of active positions. In the second round, the four active positions of this column are spread over four distinct columns by `ShiftRows`. Subsequently, `MixColumns` of the second round converts this to four columns of only active positions. The `SubBytes` and `ShiftRows` of the third round do not change the properties. Let the inputs of `MixColumns` in the third round be denoted by $\mathbf{a}^l$, and the outputs by $\mathbf{b}^l$. Then we have for all positions i, j that

$$\begin{aligned}\sum_l b^l_{i,j} &= \sum_l \texttt{MixColumns}(a^l_{i,j}) \\ &= \sum_l \left(\texttt{02} \cdot a^l_{i,j} + \texttt{03} \cdot a^l_{i+1,j} + a^l_{i+2,j} + a^l_{i+3,j}\right) \\ &= \texttt{02} \cdot \sum_l a^l_{i,j} + \texttt{03} \cdot \sum_l a^l_{i+1,j} + \sum_l a^l_{i+2,j} + \sum_l a^l_{i+3,j} \\ &= 0 + 0 + 0 + 0 = 0.\end{aligned}$$

Hence, all positions at the input of the fourth round are balanced. This property is in general destroyed by the subsequent application of `SubBytes`.

We assume that the fourth round is a `FinalRound`, i.e. it does not include a `MixColumns` operation. Every output byte of the fourth round depends on only one input byte of the fourth round. Let the input of the fourth round be denoted by $\mathbf{c}$, the output by $\mathbf{d}$ and the round key of the fourth round by $\mathbf{k}$. We have

$$\mathbf{d} = \texttt{AddRoundKey}\left(\texttt{ShiftRows}\left(\texttt{SubBytes}(\mathbf{c})\right), \mathbf{k}\right) \tag{10.1}$$

$$d_{i,j} = \mathrm{S_{RD}}[c_{i,j+C_i}] + k_{i,j}, \ \forall i, j \tag{10.2}$$

$$c_{i,j} = \mathrm{S_{RD}}^{-1}[d_{i,j-C_i} + k_{i,j-C_i}], \ \forall i, j, \tag{10.3}$$

where the operations on the column index are, as always, performed in modulo $\mathrm{N_b}$. Using (10.3), the value of $c_{i,j}$ can be calculated from the ciphertexts for all elements of the Λ-set by assuming a value for $k_{i,j-C_i}$. If the assumed value for $k_{i,j-C_i}$ is equal to the correct round key byte, the following equations must hold:

$$\sum_l c^l_{i,j} = 0, \ \forall i, j. \tag{10.4}$$

If (10.4) does not hold, the assumed value for the key byte must be wrong. This is expected to eliminate all but approximately one key value. This can be repeated for the other positions of $\mathbf{k}$. Since checking (10.4) for a single

Λ-set leaves only 1/256 of the wrong key assumptions as possible candidates, the cipher key can be found with overwhelming probability with only two Λ-sets. The work factor of the attack is determined by the processing of the first set of 2^8 plaintexts. For all possible values of one round key byte, (10.4) has to be evaluated. This means 2^{16} XOR operations and S-box look-ups. This corresponds to roughly 2^{10} executions of the four-round cipher. A negligible number of possible values for the round key byte have to be checked against the second set of plaintexts. In order to recover a full round key, the attack needs to be repeated 16 times. This results in a total complexity of 2^{14} cipher executions.

10.2.3 Influence of the Final Round

At first sight, it seems that the removal of the operation `MixColumns` in the final round of Rijndael makes the cipher weaker against the saturation attack. We will now show that adding a `MixColumns` operation in the last round would not increase the resistance against this attack. Let the input of the fourth round still be denoted by $\mathbf{c}$, and the output of a 'full' fourth round (including `MixColumns`) by $\mathbf{e}$. We have

$$\mathbf{e} = \texttt{AddRoundKey}\left(\texttt{MixColumns}\left(\texttt{ShiftRows}\left(\texttt{SubBytes}(\mathbf{c})\right)\right), \mathbf{k}\right) \quad (10.5)$$

$$\begin{aligned} e_{i,j} &= \texttt{02} \cdot \mathrm{S_{RD}}[c_{i,j+C_i}] + \texttt{03} \cdot \mathrm{S_{RD}}[c_{i+1,j+C_{i+1}}] \\ &\quad + \mathrm{S_{RD}}[c_{i+2,j+C_{i+2}}] + \mathrm{S_{RD}}[c_{i+3,j+C_{i+3}}] + k_{i,j}, \ \forall i,j. \end{aligned} \quad (10.6)$$

There are $4\mathrm{N_b}$ equations (10.6): one for each value of i,j. The equations can be solved for the bytes of $\mathbf{c}$, e.g. for $c_{0,0}$:

$$\begin{aligned} c_{0,0} &= \mathrm{S_{RD}}^{-1}[\texttt{0E} \cdot (e_{0,0} + k_{0,0}) + \texttt{0B} \cdot (e_{1,-C_1} + k_{1,-C_1}) \\ &\quad + \texttt{0D} \cdot (e_{2,-C_2} + k_{2,-C_2}) + \texttt{09} \cdot (e_{3,-C_3} + k_{3,-C_3})] \quad (10.7) \\ &= \mathrm{S_{RD}}^{-1}[\texttt{0E} \cdot e_{0,0} + \texttt{0B} \cdot e_{1,-C_1} + \texttt{0D} \cdot e_{2,-C_2} + \texttt{09} \cdot e_{3,-C_3} \\ &\quad + k'_{0,0}], \quad (10.8) \end{aligned}$$

where the equivalent key $\mathbf{k}'$ is defined as

$$\mathbf{k}' = \texttt{InvMixColumns}(\texttt{InvShiftRows}(\mathbf{k})). \quad (10.9)$$

Similar equations hold for the other bytes of $\mathbf{c}$. The value of $c_{0,0}$ in all elements of the Λ-set can be calculated from the value of the ciphertext by assuming a value for one byte of the equivalent key $\mathbf{k}'$, and the same attack as before can be applied in order to recover the bytes of the equivalent key $\mathbf{k}'$. When all bytes of $\mathbf{k}'$ have been determined, (10.9) can be used to determine $\mathbf{k}$.

We conclude that the removal of the `MixColumns` step in the final round does *not* weaken Rijndael with respect to the four-round saturation attack. This conclusion agrees with the results of Sect. 3.7.2. Since the order of the steps `MixColumns` and `AddRoundKey` in the final round can be inverted, `MixColumns` can be moved after the last key addition and thus a cryptanalyst can easily factor it out, even if he does not know the round key.

10.2.4 Extension at the End

If a round is added, we have to calculate the value of $c_{i,j+C_i}$ from (10.3) using the output of the fifth round instead of the fourth round. This can be done by additionally assuming a value for a set of 4 bytes of the fifth round key. As in the case of the four-round attack, wrong key assumptions are eliminated by verifying (10.4).

In this five-round attack, 2^{40} key values must be checked, and this must be repeated four times. Since checking (10.4) for a single Λ-set leaves only 1/256 of the wrong key assumptions as possible candidates, the cipher key can be found with overwhelming probability with only five Λ-sets. The complexity of the attack can be estimated at four runs $\times$ 2^{40} possible values for five key bytes $\times$ 2^8 ciphertexts in the set $\times$ five S-box look-ups per test, or 2^{46} five-round cipher executions.

10.2.5 Extension at the Beginning

The basic idea of this extension is to work with sets of plaintexts that result in a Λ-set with a single active position and 15 constant positions at the output of the first round.

We consider a set of 2^{32} plaintexts, such that one column at the input of `MixColumns` in the first round ranges over all possible values and all other positions are constant. Since `MixColumns` and `AddRoundKey` are invertible and work independently on the four columns, this property will be conserved: at the output of the first round the states will have constant values for three columns, and the value of the fourth column will range over all 2^{32} possibilities. This set of 2^{32} plaintexts can be considered as a union of 2^{24} Λ-sets, where each Λ-set has one active position at the output of the first round, and all other positions are constant. It is not possible to separate the plaintexts of the different Λ-sets, but evidently, since (10.4) must hold for every individual Λ-set, it must also hold when the sum goes over all 2^{32} values. Therefore, the round key of the final round can be recovered byte by byte, in the same way as for the four-round attack. This five-round attack requires two structures of 2^{32} chosen plaintexts. The work factor of this attack can be estimated at 16 runs $\times$ 2^{32} plaintexts in the set $\times 2^8$ possible values for the key byte $\times$ one S-box look-up per test, or 2^{38} five-round cipher executions.

10.2.6 Attacks on Six Rounds

Combining both extensions results in a six-round attack. The work factor can be estimated at four runs $\times$ 2^{32} plaintexts in the set $\times$ 2^{40} possible values for 5 key bytes $\times$ 5 S-box look-ups per test, or 2^{70} six-round cipher executions. N. Ferguson et al. explain in [61] a way to do the required calculations more efficiently. In this way, the work factor of the six-round attack can be further reduced to 2^{46} six-round cipher executions. The work factor and memory requirements are summarized in Table 10.1. S. Lucks observed that for Rijndael with key lengths of 192 or 256, the six-round attack can be extended by one more round by guessing an additional round key [98]. The work factor of the attack increases accordingly.

Table 10.1. Complexity of saturation attacks applied to Rijndael

Attack	No. of plaintexts	No. of cipher executions	Memory
Basic (four rounds)	2^9	2^{14}	small
Extension at end	2^{11}	2^{46}	small
Extension at beginning	2^{33}	2^{38}	2^{32}
Both extensions	2^{35}	2^{46}	2^{32}

10.2.7 The Herds Attack and Other Extensions

N. Ferguson et al. describe in [61] a further extension of the saturation attack, known as the herds attack. The authors describe a seven-round attack that requires $2^{128} - 2^{119}$ chosen plaintexts and 2^{64} bits of memory. The attack has a workload comparable to 2^{120} encryptions.

The herds attack can be extended into an eight-round attack with the same plaintext requirements and using 2^{104} bits of memory. The workload is too large to be applicable to the case of 128-bit keys. For 192-bit keys, the workload is comparable to 2^{188} encryptions. For 256-bit keys, this becomes 2^{204} encryptions.

J. Nakahara et al. apply the saturation attack to versions of Rijndael with larger block sizes, i.e. versions that are not included in AES [116]. They describe attacks on up to seven rounds of Rijndael versions with block lengths of 160, 192, 224 and 256 bits. All these attacks have a data complexity, memory complexity or time complexity above 2^{128}. Minier et al. provide an overview of saturation attacks and their underlying principles for AES and all versions of Rijndael [113].

Sun et al. present a saturation property like the one discussed in Sect. 10.2.2, extending over five rounds instead of four [141].

10.2.8 Division Cryptanalysis

Division cryptanalysis was introduced in [143]. For any $\mathbf{x}, \mathbf{u} \in \mathrm{GF}(2)^n$ we define $\mathbf{x}^{\mathbf{u}}$ as the product of the coordinates x_i where the corresponding coordinate $u_i = 1$. A set $X \subseteq \mathrm{GF}(2)^n$ has *division property* $\mathcal{D}_k^n$ if

$$\sum_{\mathbf{x} \in X} \mathbf{x}^{\mathbf{u}} = 0, \forall \mathbf{u} \text{ with } \mathrm{w_h}(\mathbf{u}) < k \ . \tag{10.10}$$

Usually it is assumed that k is the largest integer with this property. Every set with an even number of elements trivially satisfies at least $\mathcal{D}_1^n$.

It can easily be verified that a Λ-set is balanced iff it has division property $\mathcal{D}_2^n$; a Λ-set is active iff it has division property $\mathcal{D}_n^n$. Note that division properties can easily be defined for multi-sets, and also the concepts of active, balanced and constant sets can be extended to multi-sets. For these extensions, $\mathcal{D}_2^n$ is still equivalent to balanced multi-sets; active sets still have division property $\mathcal{D}_n^n$, but the reverse may no longer be true, depending on the way an active multi-set is defined.

The intuition behind division cryptanalysis is that a saturation attack may possibly be improved by considering the properties $\mathcal{D}_k^n$ for $2 < k < n$. The propagation of $\mathcal{D}_k^n$ properties through an S-box depends to a first approximation on the algebraic degree of the S-box. A more detailed analysis is provided in [34]. The propagation of $\mathcal{D}_k^n$ properties through a linear diffusion layer is currently determined on an ad hoc basis.

Division cryptanalysis has been used to construct new attacks on (reduced variants of) several Rijndael-like ciphers, but not on Rijndael itself.

10.3 Gilbert-Minier and Demirci-Selçuk Attack

The saturation attack on Rijndael reduced to six rounds is based on the fact that three rounds of Rijndael can be distinguished from a random permutation. H. Gilbert and M. Minier developed a four-round distinguisher that allows an attack on Rijndael that is reduced to seven rounds [64, 112]. Due to the increased work factor of the attack, it is more efficient than exhaustive key search for only some of the key lengths. H. Demirci and A.A. Selçuk extended this work and developed an attack on Rijndael reduced to eight rounds [55].

10.3.1 The Four-Round Distinguisher

Let $R4_{\mathbf{k}}$, $R5_{\mathbf{k}}$ denote the action of Rijndael, reduced to four, respectively five rounds, under the unknown key $\mathbf{k}$. Let $\mathbf{a}$ denote the input, $\mathbf{e}$ the output after four rounds and $\mathbf{f}$ the output after five rounds of Rijndael.

Gilbert and Minier show in [64] that the structure in the transformation `Round` can be used to derive relations between $\mathbf{a}$ and $\mathbf{e} = R4_{\mathbf{k}}(\mathbf{a})$. More particularly, one can define a family of functions derived from $R4_{\mathbf{k}}$:

$$f_{\mathbf{c}} : \mathrm{GF}(2^8) \rightarrow \mathrm{GF}(2^8) : x \mapsto f_{\mathbf{c}}(x) = y, \tag{10.11}$$

where $x = a_{0,0}$, $y = \texttt{0E}e_{0,0} + \texttt{0B}e_{1,0} + \texttt{0D}e_{2,0} + \texttt{09}e_{3,0}$ and $\mathbf{c}$ is a vector containing only 10 coefficients from $\mathrm{GF}(2^8)$, which depend on the key $\mathbf{k}$ and the 15 remaining elements of $\mathbf{a}$. Gilbert and Minier further show that four of the 10 coefficients depend only on $\mathbf{k}$ and on $a_{1,0}, a_{2,0}, a_{3,0}$. It follows that if for a fixed key, 2^{16} different values for $a_{1,0}, a_{2,0}, a_{3,0}$ are selected, with large probability at least two choices will result in the same function f. This property holds for all values of the key $\mathbf{k}$ and can be used to distinguish $R4_{\mathbf{k}}$ from a random permutation. Note that the distinguisher does not work with probability 1. More information on the construction of this four-round distinguisher can be found in [64]. In [112], Minier presents a stronger distinguisher, which however doesn't lead to a stronger or more efficient attack.

10.3.2 The Attack on Seven Rounds

In the same way as the six-round saturation attack, the seven-round attack is mounted by adding one round before the distinguisher and two rounds after it.

By assuming a value for 4 key bytes of the first round key, it is possible to determine a set of plaintexts such that the inputs of the second round are constant in three columns. This set is divided into subsets with constant values for the 'parameters' u, v and w at the input of the second round. There should be 2^{16} subsets, with 16 plaintexts in each subset. 16 values for x suffices to determine whether two sets of parameters result in identical functions, with negligible false-alarm probability. It can be shown that the required plaintexts for all 2^{32} values of the 4 bytes of the first round key can be drawn from a set of 2^{32} plaintexts.

Each of the bytes $e_{i,j}$ can be expressed as a function of 4 ciphertext bytes and 5 key bytes. Hence the y values depend on 20 key bytes, which have to be guessed in order to perform the attack. The work complexity of the attack can be estimated at about 2^{192} executions of the round transformation, which is below the complexity of an exhaustive search for a 256-bit key, and approximately equal to the complexity of an exhaustive search for a 192-bit key.

A variant of this attack works only for the 128-bit key schedule, and is claimed to be marginally faster than an exhaustive search for a 128-bit key.

10.3.3 The Demirci-Selçuk Attack

Demirci and Selçuk extend in [55] the Gilbert-Minier attack and derive relations between $\mathbf{a}$ and $\mathbf{f} = R5_{\mathbf{k}}(\mathbf{a})$. They obtain the following family of functions:

$$g_{\mathbf{d}} : \mathrm{GF}(2^8) \rightarrow \mathrm{GF}(2^8) : x \mapsto g_{\mathbf{d}}(x) = y, \tag{10.12}$$

where $x = a_{0,0}$, $y = \mathtt{0E}f_{0,0} + \mathtt{0B}f_{1,0} + \mathtt{0D}f_{2,0} + \mathtt{09}f_{3,0}$ and $\mathbf{d}$ is a vector containing 26 coefficients from $\mathrm{GF}(2^8)$, which depend on the key $\mathbf{k}$ and the 15 remaining elements of $\mathbf{a}$.

26 key-dependent coefficients cannot be determined more efficiently than searching for a 128-bit key, but even the naive method is faster than searching for a 256-bit key. Demirci and Selçuk employ a time-memory trade-off to reduce the complexity of the coefficient-determination step below the complexity of the search for a 192-bit key. Subsequently, they use the relations to mount meet-in-the-middle attacks on Rijndael reduced to seven and eight rounds. For further details of these attacks, we refer to [55].

10.4 Interpolation Attacks

In [73] T. Jakobsen and L. Knudsen introduced a new algebraic attack on block ciphers. The attack is feasible if the components in the cipher have a compact algebraic expression and can be combined to give expressions with manageable complexity. The basis of the attack is that if the constructed polynomials have a small degree, only a few cipher input/output pairs are necessary to solve for the (key-dependent) coefficients of the polynomials.

$\mathrm{S_{RD}}$ takes bytes as input and produces bytes as output. Like any other transformation with this input size and output size, it can be expressed as a polynomial over $\mathrm{GF}(2^8)$. The polynomial expression of $\mathrm{S_{RD}}$ can, for example, be found by means of the Lagrange interpolation technique. The polynomial expression for $\mathrm{S_{RD}}$ is given by

$$\begin{aligned}\mathrm{S_{RD}}[x] = \mathtt{63} + \mathtt{8F}x^{127} + \mathtt{B5}x^{191} + \mathtt{01}x^{223} + \mathtt{F4}x^{239} \\ + \mathtt{25}x^{247} + \mathtt{F9}x^{251} + \mathtt{09}x^{253} + \mathtt{05}x^{254}.\end{aligned} \tag{10.13}$$

This complicated expression of $\mathrm{S_{RD}}$ in $\mathrm{GF}(2^8)$, in combination with the effect of the mixing and transposition steps, prohibits interpolation attacks on more than a few rounds of Rijndael.

The techniques in [73] can be extended to use rational expressions or in fact any other type of expression. We found no simple rational expression for $\mathrm{S_{RD}}$ but it seems impossible to prove that no usable expression can be found. A second possible extension of this attack is the use of approximate expressions, as proposed by T. Jakobsen in [74]. It remains an open problem whether any useful expression can be derived in this way.

N. Ferguson et al. describe how one can derive an algebraic expression for ten-round Rijndael [62]. The expression would count 2^{50} terms. Although this is certainly an interesting result, the authors are not aware of ways to use this expression in an actual attack. Another interesting and as yet unanswered question is how this compares with other block ciphers. Cid et al. present an overview of algebraic aspects of the AES in [40]. In [33] Bouillaguet et al. use algebraic techniques to construct equations describing reduced-round versions of Rijndael and to solve them for the key.

10.5 Related-Key Attacks

This section is based on material that appeared earlier in [53]. In [18], E. Biham introduced a related-key attack. Later it was demonstrated by J. Kelsey et al. that several ciphers have related-key weaknesses [81]. In a related-key attack, the attacker submits queries containing an input x or an output y for the block cipher. Additionally, the adversary specifies for each query a function G, which needs to be applied to the secret key k. The *Key Access Scheme (KAS)* of a related-key attack defines the relations between the keys under which the attacker can query the block cipher. A KAS consists of a set of functions Γ and a set of domains α. The attacker can query the block cipher under all keys l for which there is a function $G_i \in \Gamma$ and a constant a_j in a domain $A_j \in \alpha$ such that $l = G_i(k, a_j)$.

A KAS may lead to the observation of properties in block ciphers that at first sight appear to be weaknesses, but on closer inspection turn out not to be so. For example, consider the KAS containing the two following functions: $G_1(k, a) = k \oplus a$ and with some abuse of notation $G_2(k, a) = k + a \bmod 2^n$, with k and a interpreted as integers and a an n_{k}-bit string/integer. For any block cipher B, if the least significant bit of k equals 0, then for all text inputs x

$$B[G_1(k, 1)](x) = B[G_2(k, 1)](x). \tag{10.14}$$

On the other hand, if the least significant bit of k equals 1, then for most block ciphers (10.14) will hold for a very low fraction of the inputs x. By repeating the queries for $a_j = 2, 4, 8, \ldots$ the attacker recovers k. However, this 'attack' does not allow us to distinguish any block cipher from the ideal cipher, because the property holds also for the latter.

A more subtle characteristic of related-key security follows from time/data trade-off considerations. If an attacker can obtain for an arbitrary input x the values $B[k_1](x), B[k_2](x), \ldots, B[k_n](x)$, then an exhaustive search for the key can be accelerated with a factor d. By setting $k_j = G(k, a_j)$ we can convert this result to the related-key scenario [13]. Hence all KAS result in a seeming erosion of security proportional to the number of functions and constants.

10.5.1 The Key Schedule of Rijndael-256

We first introduce a new notation for the key schedule of Rijndael with 256-bit keys. This key schedule produces 15 round keys, which we will denote here by $\mathsf{K}[0],\ldots,\mathsf{K}[14]$. The first two round keys are simple copies of the two halves of the key: $[\mathsf{K}[0]\ \mathsf{K}[1]] = \mathsf{K}$. The 13 following keys are derived iteratively by the repeated application of a transformation that we denote by φ. The transformation φ is defined as follows:

$$\varphi\left(\mathsf{K}[2t],\mathsf{K}[2t+1]\right) = \left(\mathsf{K}[2t+2],\mathsf{K}[2t+3]\right)$$
$$\Leftrightarrow \begin{cases} \mathsf{K}[2t+2]_{i,0} = \mathrm{S_{RD}}\left[\mathsf{K}[2t+1]_{j+1,3}\right] + \mathsf{K}[2t]_{j,0} \\ \mathsf{K}[2t+2]_{i,1} = \mathsf{K}[2t+2]_{i,0} + \mathsf{K}[2t]_{j,1} \\ \mathsf{K}[2t+2]_{i,2} = \mathsf{K}[2t+2]_{i,1} + \mathsf{K}[2t]_{j,2} \\ \mathsf{K}[2t+2]_{i,3} = \mathsf{K}[2t+2]_{i,2} + \mathsf{K}[2t]_{j,3} \\ \mathsf{K}[2t+3]_{i,0} = \mathrm{S_{RD}}\left[\mathsf{K}[2t+2]_{i,3}\right] + \mathsf{K}[2t+1]j,0 \\ \mathsf{K}[2t+3]_{i,1} = \mathsf{K}[2t+3]_{i,0} + \mathsf{K}[2t+1]_{j,1} \\ \mathsf{K}[2t+3]_{i,2} = \mathsf{K}[2t+3]_{i,1} + \mathsf{K}[2t+1]_{j,2} \\ \mathsf{K}[2t+3]_{i,3} = \mathsf{K}[2t+3]_{i,2} + \mathsf{K}[2t+1]_{j,3} \end{cases}, \quad \text{for } j = 0,1,2,3.$$

We denote by φ^t the transformation constructed by iterating t times φ. Finally, we denote by $\varphi^{0.5}$ the transformation that computes only one new round key:

$$\varphi\left(\mathsf{K}[2t],\mathsf{K}[2t+1]\right) = \left(\mathsf{K}[2t+2],\mathsf{K}[2t+3]\right)$$
$$\Rightarrow \varphi^{0.5}\left(\mathsf{K}[2t],\mathsf{K}[2t+1]\right) = \left(\mathsf{K}[2t+1],\mathsf{K}[2t+2]\right).$$

10.5.2 The Biryukov-Khovratovich Attack

Related-key differential attacks on Rijndael or reduced variants of Rijndael have been studied in several papers [72, 149, 82, 24, 29, 28]. We briefly present here the attack by Biryukov and Khovratovich on Rijndael-256 [28]. Their attack requires $2^{99.5}$ chosen plaintexts/ciphertexts. It has a computational complexity of $2^{99.5}$ encryptions and requires a memory of 2^{77} states.

The attack is a *Boomerang attack*, which uses quartets instead of pairs [145]. An important feature of the Biryukov-Khovratovich attack is that they allow the attacker to query the block cipher under related keys, where the relation is expressed in terms of the *round keys*. Hence, this attack could also be called a *related round key attack*. Denote the four related keys by K_A, K_B, K_C, K_D, and the round keys generated from them by $\mathsf{K}_A[t]$, $\mathsf{K}_B[t]$, $\mathsf{K}_C[t]$, $\mathsf{K}_D[t]$ with $0 \le t \le 14$. The attack imposes eight relations between round keys, described by four state differences, denoted by M_1, M_2, M_3, M_4:

$$\mathsf{M}_1 = \mathsf{K}_A[2] + \mathsf{K}_B[2] \qquad = \mathsf{K}_C[2] + \mathsf{K}_D[2] = \begin{bmatrix} \texttt{00 00 00 00} \\ \texttt{00 01 00 01} \\ \texttt{00 00 00 00} \\ \texttt{00 00 00 00} \end{bmatrix} \tag{10.15}$$

$$\mathsf{M}_2 = \mathsf{K}_A[3] + \mathsf{K}_B[3] \qquad = \mathsf{K}_C[3] + \mathsf{K}_D[3] = \begin{bmatrix} \texttt{3E 00 3E 00} \\ \texttt{21 00 21 00} \\ \texttt{1F 00 1F 00} \\ \texttt{1F 00 1F 00} \end{bmatrix} \tag{10.16}$$

$$\mathsf{M}_3 = \mathsf{K}_A[7] + \mathsf{K}_C[7] \qquad = \mathsf{K}_B[7] + \mathsf{K}_D[7] = \begin{bmatrix} \texttt{3E 3E 3E 3E} \\ \texttt{21 21 21 21} \\ \texttt{1F 1F 1F 1F} \\ \texttt{1F 1F 1F 1F} \end{bmatrix} \tag{10.17}$$

$$\mathsf{M}_4 = \mathsf{K}_A[8] + \mathsf{K}_C[8] \qquad = \mathsf{K}_B[8] + \mathsf{K}_D[8] = \begin{bmatrix} \texttt{01 00 01 00} \\ \texttt{00 00 00 00} \\ \texttt{00 00 00 00} \\ \texttt{00 00 00 00} \end{bmatrix} \tag{10.18}$$

These round key relations can also be described as nonlinear relations between the keys:

$$\begin{aligned}
\mathsf{K}_B &= \varphi^{-1}\left(\varphi\left(\mathsf{K}_A\right) + \left[\mathsf{M}_1\ \mathsf{M}_2\right]\right)\\
\mathsf{K}_D &= \varphi^{-1}\left(\varphi\left(\mathsf{K}_C\right) + \left[\mathsf{M}_1\ \mathsf{M}_2\right]\right)\\
\mathsf{K}_C &= \varphi^{-3.5}\left(\varphi^{3.5}\left(\mathsf{K}_A\right) + \left[\mathsf{M}_3\ \mathsf{M}_4\right]\right)\\
\mathsf{K}_D &= \varphi^{-3.5}\left(\varphi^{3.5}\left(\mathsf{K}_B\right) + \left[\mathsf{M}_3\ \mathsf{M}_4\right]\right).
\end{aligned}$$

A related-key differential attack on Rijndael performs better than a standard differential attack because the differences introduced in the keys will cause differences in the round keys, which in turn will partially cancel out the differences in the state during the encryption (with some probability). A related round key attack performs better than a related-key attack because the attacker can directly choose the differences in round keys, thereby making a part of the attack deterministic instead of probabilistic.

10.5.3 The KAS of the Biryukov-Khovratovich Attack

Biryukov and Khovratovich did not define the Key Access Scheme of their attack. In order to assess the impact of their attack, we try to define one here. As a first attempt to define a KAS, we propose the following function:

$$G(\mathsf{K}, \mathsf{A}_1, \mathsf{A}_2) = \varphi^{-1}\left(\varphi^{-2.5}\left(\varphi^{3.5}(\mathsf{K}) + \mathsf{A}_2\right) + \mathsf{A}_1\right), \tag{10.19}$$

where A_1, A_2 can be any couple of 256-bit constants. Indeed, in order to construct a quartet of keys as needed for the Biryukov-Khovratovich attack, we can set $\mathsf{K}_A = \mathsf{K}$ and compute the three remaining keys using (10.19) as follows:

$$\begin{aligned}
\mathsf{K}_B &= G(\mathsf{K}_A, \left[\mathsf{M}_1\ \mathsf{M}_2\right], \mathsf{0}) \\
\mathsf{K}_C &= G(\mathsf{K}_A, \mathsf{0}, \left[\mathsf{M}_3\ \mathsf{M}_4\right]) \\
\mathsf{K}_D &= G(\mathsf{K}_A, \left[\mathsf{M}_1\ \mathsf{M}_2\right], \left[\mathsf{M}_3\ \mathsf{M}_4\right]).
\end{aligned}$$

However, we show now that we can use (10.19) to recover the key of any block cipher. Let a, b denote two values in GF(2^8) with as condition that the differential (a, b) has DP 2^{-7} over the $\mathrm{S_{RD}}$. We define the following four states:

$$\mathsf{D}_1 = \begin{bmatrix} 00 & 00 & 00 & 00 \\ 00 & 00 & 00 & 00 \\ 00 & 00 & 00 & 00 \\ 00 & 00 & 00 & 00 \end{bmatrix}, \qquad \mathsf{D}_2 = \begin{bmatrix} 00 & 00 & 00 & 00 \\ 00 & a & a & 00 \\ 00 & 00 & 00 & 00 \\ 00 & 00 & 00 & 00 \end{bmatrix},$$

$$\mathsf{D}_3 = \begin{bmatrix} b & b & b & b \\ 00 & 00 & 00 & 00 \\ 00 & 00 & 00 & 00 \\ 00 & 00 & 00 & 00 \end{bmatrix}, \qquad \mathsf{D}_4 = \begin{bmatrix} 00 & 00 & 00 & 00 \\ 00 & a & a & a \\ 00 & 00 & 00 & 00 \\ 00 & 00 & 00 & 00 \end{bmatrix}.$$

With these definitions, the differential $\left(\left[\mathsf{D}_1\ \mathsf{D}_2\right], \left[\mathsf{D}_3\ \mathsf{D}_4\right]\right)$ has DP 2^{-7} over $\varphi^{2.5}$, since only one S-box gets a non-zero input difference (equal to a). It follows that for all keys K the relation

$$G(\mathsf{K}, \mathsf{X}, \mathsf{Y}) = G\left(\mathsf{K}, \mathsf{X} + \left[\mathsf{D}_1\ \mathsf{D}_2\right], \mathsf{Y} + \left[\mathsf{D}_3\ \mathsf{D}_4\right]\right)$$

holds for a fraction of 2^{-7} of the states X, Y. It follows that for each key, it is easy to find solutions X, Y, i.e. to create collisions for the Key Access Scheme. A collision in the Key Access Scheme is trivially detectable by the attacker and can be used to recover one byte of the round key (cf. the example with XOR and modular addition). Note that the number of queries needed to find with high probability a collision can be reduced to 25 by working with several choices of (a, b) simultaneously. The attack can be repeated 19 times with slightly different choices for D_1, D_2, D_3, D_4, allowing the attacker to recover 160 bits of key material using only 500 related-key queries.

Although the attack uses $\mathrm{S_{RD}}$ and the Rijndael key schedule, it works for any block cipher: φ and the S-box are used only to describe the KAS; they do not have to be used in the block cipher that is being attacked. It follows that this first attempt to define a Key Access Scheme is not sound.

In order to repair the Key Access Scheme, we can modify it as follows. We forbid the use of values A_1, A_2 that result in a differential over $\varphi^{2.5}$ with non-zero DP, while at the same time not excluding the values that are needed for the Biryukov-Khovratovich attack. Although this restriction repairs the KAS, it has to be designed specially for this attack and could be called contrived.

10.6 Biclique Attacks

Biclique attacks can be described as extensions of meet-in-the-middle attacks. A *biclique* of dimension d for a key-dependent function $f[k]$ is a structure consisting of 2^d ciphertexts c_i, 2^d internal states s_j and 2^{2d} keys $k_{i,j}$ such that

$$\forall i, j \in \{0, 1, \ldots, 2^n - 1\} : \; c_i = f[k_{i,j}](s_j).$$

In order to perform a biclique attack on a cipher $B[k] = f[k] \circ g[k]$, an attacker first builds a biclique structure for $f[k]$. Subsequently, the attacker obtains the 2^d plaintexts p_i corresponding to the ciphertexts c_i. Finally, the attacker checks whether $\exists i, j \in \{0, 1, \ldots, 2^n - 1\} : \; g[k_{i,j}](p_i) = s_j$. Each $k_{i,j}$ satisfying this relation is a valid candidate for the key.

An efficient way to construct bicliques is by using differential trails satisfying some extra conditions [32]. Bogdanov et al. claim that they can perform a key recovery on Rijndael at 1/5 to 1/3 of the complexity of an exhaustive search for the key, depending on the length of the key.

10.7 Rebound Attacks

The rebound attack was proposed in [109] for the cryptanalysis of Rijndael-based hash functions. It can be described as an improved version of a differential attack, applicable to hash functions only. From a perspective of differential cryptanalysis, finding a collision for a hash function corresponds to finding a pair that follows a trail through that hash function with output difference zero. It follows that differential cryptanalysis of hash functions is intuitively very similar to differential cryptanalysis of block ciphers. However, there are also important differences between these two cases, which can be observed also in the rebound attack.

In the case of block ciphers, an adversary that wants to find a pair following a trail can usually do little better than simply trying out pairs. The effort is proportional to the inverse of the DP of the trail. Since hash functions do not have a secret key, an adversary can do better than that. In principle, an adversary could simply write out the equations that determine whether a pair follows a trail and solve them. In practice, these equations are highly nonlinear and difficult to solve. However, it is often possible to determine some of the message bits, thereby increasing the probability that a random guess for the remaining part of the solution will be correct. Typically, the equations arising from the first steps of the hash function are easier to solve, because they do not yet depend on all message words. These techniques are known in the literature under the name message modification techniques.

The rebound attack consists of two phases, called the inbound and outbound phases. According to these phases, the compression function, internal block cipher or permutation of a hash function is split into three sub-parts. Let W be the permutation, then we get $W = W_{\text{forward}} \circ W_{\text{inbound}} \circ W_{\text{backward}}$. The part of the inbound phase is placed in the middle of the permutation and the two parts of the outbound phase (forward and backward) are placed next to the inbound part. Two high-probability (truncated) differential trails are constructed through the two parts of the outbound phase. Subsequently, these trails are connected by a third trail, which runs through the inbound phase. Similar to the message modification technique of Wang and Yu [147], the freedom in the choice of inputs or (internal) state variables is used to efficiently fulfill most conditions of the differential trail through the inbound phase, hence its probability is of less importance in the attack complexity.

10.8 Impossible-Differential Attacks

10.8.1 Principle of the Attack

As explained in Sect. 6.2, in a classical differential attack (partial) information about the key is derived from the ciphertexts and the difference $\mathbf{b}'$ of some intermediate values. Since the difference $\mathbf{b}'$ is known only with probability $\mathrm{DP}^{(h)}(\mathbf{a}', \mathbf{b}')$, where $\mathbf{a}'$ is the plaintext difference, the information about the key is probabilistic. Since the work factor of the classical attack depends on the largest probability $\mathrm{DP}^{(h)}(\mathbf{a}', \mathbf{b}')$ over all choices of $\mathbf{a}'$ and $\mathbf{b}'$, a designer might try to ensure that $\mathrm{DP}^{(h)}(\mathbf{a}', \mathbf{b}')$ is small for all $\mathbf{a}'$, $\mathbf{b}'$. In fact, this is one of the objectives of the Wide Trail Strategy; see Sect. 9.2.

Impossible-differential attacks [22, 83], on the contrary, exploit difference propagations $(\mathbf{a}', \mathbf{b}')$ with $\mathrm{DP}^{(h)}(\mathbf{a}', \mathbf{b}') = 0$. Note that the difference propagation probability equals the sum of the DP of all trails Q with $\mathbf{q}^{(0)} = \mathbf{a}', \mathbf{q}^{(r)} = \mathbf{b}'$, see (8.25), so in order to have $\mathrm{DP}^{(h)}(\mathbf{a}', \mathbf{b}') = 0$ it is required that $\mathrm{DP}(\mathrm{Q}) = 0$ for all Q. If a difference propagation with $\mathrm{DP} = 0$ is known, then it can be exploited as follows.

Obtain the ciphertexts $\mathbf{c}, \mathbf{c}^*$ corresponding to a pair of plaintexts with difference $\mathbf{a}'$. Let $g[\mathbf{k}]$ denote the operation that partially decrypts the ciphertexts to the intermediate states $\mathbf{b}, \mathbf{b}^*$. In other words, the block cipher $B[\mathbf{k}] = g^{-1}[\mathbf{k}] \circ h[\mathbf{k}]$. By definition, any key $\mathbf{k}$ that satisfies

$$g[\mathbf{k}](\mathbf{c}) + g[\mathbf{k}](\mathbf{c}^*) = \mathbf{b}' \tag{10.20}$$

must be different from the key used to compute $\mathbf{c}, \mathbf{c}^*$. By using multiple pairs of plaintexts, the probability for a wrong key to survive this test can be made very small. This attack can be more efficient than exhaustive search for the key only if the function $h[\mathbf{k}]$ can be computed without full knowledge of the key.

The impossible-differential attack can be generalized to difference propagations with $\mathrm{DP}^{(h)}(\mathbf{a}', \mathbf{b}') \approx 0$, provided that the probability that the correct key satisfies (10.20) is significantly smaller than the probability that an incorrect key satisfies (10.20) [87]. One can again define an S/N ratio, similarly to the approach in Sect. 6.2.

10.8.2 Application to Rijndael

From the analysis in Sect. 10.2 it follows easily that there are many difference propagations over three rounds of Rijndael with $\mathrm{DP}^{(h)}(\mathbf{a}', \mathbf{b}') = 0$. For example, if $\mathrm{w_t}(\mathbf{a}') = \mathrm{w_t}(\mathbf{b}') = 1$, then we always have $\mathrm{DP}^{(h)}(\mathbf{a}', \mathbf{b}') = 0$. An impossible differential over four rounds of Rijndael was first presented in [25]. An impossible-differential attack for seven rounds of Rijndael with block length 128 bits is described in [101]. Impossible-differential attacks for up to 10 rounds of Rijndael with larger block lengths are described in [146].

10.9 Implementation Attacks

Implementation attacks are based not only on mathematical properties of the cipher, but also on physical characteristics of the implementation. Typical examples are timing attacks [90], introduced by P. Kocher, and power analysis [91], introduced by P. Kocher et al. In timing attacks, key information is derived from the total execution time of the encryption algorithm. In power analysis attacks, key information is derived from the power consumption of the device executing the encryption algorithm. Power analysis attacks can be generalized to other measurable quantities such as electro-magnetic emanation or heat dissipation from the device.

10.9.1 Timing Attacks

A timing attack can be mounted if the execution time of the encryption algorithm depends on the value of the key. Let us illustrate this by an example. Assume that we have a cipher implementation in which an instruction is executed on the condition that a certain key-dependent intermediate result $\mathbf{b}$ takes a specific value. If no special precautions are taken, the total execution time of the cipher will vary depending on whether or not the conditional instruction is executed. Hence, it is possible to deduce the value of $\mathbf{b}$ by carefully measuring the execution time. It suffices to compare the encryption time for different values of $\mathbf{b}$, while taking care that all other parameters influencing the encryption time are kept constant or averaged out.

An implementation can be protected against timing attacks by ensuring that the encryption time is independent of the value of the key. For conditional instructions, this can be done by inserting dummy instructions in the shortest path until all paths take the same time. However, this solution might leave the cipher unprotected against power analysis attacks.

In Rijndael on low-end CPUs, a possible weakness with respect to timing attacks is the implementation of the finite-field multiplications in `MixColumns`, namely the subroutine `xtime`. The weakness in `xtime` can easily be eliminated by defining a 256-byte table and using a look-up table to implement `xtime` (see Sect. 4.1.1). This reduces a Rijndael program to a fixed sequence of table look-ups and XORs.

However, it soon turned out that these table look-ups form a potential weakness with respect to timing attacks. Namely, in 2002, D. Page suggested that cache mechanisms in modern processors could be used to mount side channel attacks [125]. In 2005, G. Bertoni et al. published the first cache miss attacks against Rijndael implementations with T-tables [17]. This was followed by a long series of papers improving and refining the attacks and applying them to other ciphers. Cache miss attacks exploit the fact that the latency of a table look-up depends on whether the addressed table entry is in cache memory or not, and that this entry, or rather its offset, depends on the key. T-table Rijndael implementations on high-end CPUs lend themselves well to cache miss attacks due to the relatively large T-tables. To thwart these attacks, Rijndael implementations were developed that have no table look-ups; see Sect. 4.2.2. Moreover they led Intel, and later also other CPU manufacturers, to hardwire dedicated Rijndael instructions in their processors; see Sect. 4.3.3.

10.9.2 Power Analysis

Simple power analysis (SPA) is an attack where the attacker obtains measurements of the power consumption of the device during the execution of one (or a few) encryptions. Typically, this type of attack is applicable to devices that depend on external power supplies, e.g. smart cards. If the power consumption pattern of the hardware depends on the *instruction* being executed, the attacker can deduce the sequence of instructions. If the sequence or the type of instructions depends on the value of the key, then the power consumption pattern leaks information about the key. Rijndael can easily be implemented with a fixed sequence of instructions, which prevents this type of attack.

In most processors, the power consumption pattern of an instruction depends on the value of the *operands*. For example, setting a bit in a register might consume more power than clearing it. Usually, the variation in the power consumption due to the difference in operand value is so small that it is buried in noise and is not revealed in power consumption measurements. However, by combining measurements of many encryptions, the attacker can average out the noise and obtain information about the value of the operand. This class of attack is called *differential power analysis* (DPA). Protecting implementations against these sophisticated attacks is much harder than for timing attacks and SPA, especially if the signal-to-noise ratio is high. Proposed countermeasures can be divided into three classes:

Balancing: It is possible to reduce the vulnerability of each individual instruction against power analysis by a redesign of the hardware to minimize or eliminate completely the dependency of the power consumption on the value of the operands. This redesign can also be simulated by changing the software in such a way that all data words contain at all times the complement of each of the data bits as well as the data bits themselves. In this way, the correlation between power consumption and input values can be diminished. It seems unlikely that the dependency can be eliminated completely since there will always be small physical variations in the devices.

Masking: In this approach, instructions on a sensitive variable $\mathbf{x}$ are replaced by instructions on operands $\mathbf{x}^{(1)}, \mathbf{x}^{(2)}, \ldots, \mathbf{x}^{(s)}$ with $\mathbf{x} = \mathbf{x}^{(1)} + \mathbf{x}^{(2)} + \cdots + \mathbf{x}^{(s)}$. The approach can be described as the application of Multi-Party Computation techniques at micro-level.

Leakage-resilient cryptography: In this approach the mode of operation of the block cipher and the key management are redesigned such that each secret key is used only a small number of times. The goal of this approach is to ensure that before an attacker has collected enough measurements of the power consumption of the device to be able to recover the key, the key is already put out of use. One has to ensure that measurements obtained for one key cannot be used to accelerate an attack on a subsequent key.

10.10 Conclusions

Resistance against linear and differential attacks was a design criterion of Rijndael. From the number of publications alone, we can conclude that during the AES selection process, Rijndael attracted a significant amount of attention from the cryptographic community. Square, the direct predecessor of Rijndael, has also been scrutinized vigorously for weaknesses. The complexity of the published attacks on reduced versions of Rijndael indicate that with the current state-of-the-art cryptographic techniques, no practical attacks can be mounted on a full version of Rijndael.

In order to resist implementation attacks, care has to be taken when implementing the algorithm. Because of its simplicity, Rijndael has a number of advantages when it comes to protecting its implementation against this kind of attack.

The selection of cryptanalytic results that we reviewed in this chapter is by no means complete. The decisions on what material we discussed here at length, what material we only cited and what material we left out are based on our personal tastes and bound by the incompleteness of our knowledge. Whether a certain paper was included in this chapter or not should not be seen as a judgment on its quality. We mention here a few attack models that were introduced after the AES process.

Known-key and chosen-key attacks look at the security of the block cipher when some of the bits of the key are known to, or can be chosen by the cryptanalyst. This type of security can be relevant for example when the block cipher is used as a building block for a cryptographic hash function. Chosen-key attacks on Rijndael are described in [63].

Subspace attacks can be described as an extension of differential cryptanalysis, where the cryptanalyst tracks the propagation not of pairs of texts, but of a larger set of texts that form a subspace. Its application to Rijndael is described in [67]. Another extension of differential cryptanalysis is the Yo-yo attack, invented specially to cryptanalyze Rijndael and similar ciphers [135].

11. The Road to Rijndael

We did not design Rijndael from scratch. In fact, prior to its design, we had already published three block ciphers that are similar to Rijndael. Each of these ciphers inherits properties from its predecessor and enriches them with new ideas. Hence, Rijndael can be seen as a step in an evolution process. In this chapter, we discuss the similarities and differences between Rijndael and its predecessors.

11.1 Overview

The design of Rijndael is only one step in a long process of our research on the design of secure and efficient block ciphers using the wide trail design strategy. In this section, we briefly present the different ciphers that we designed along the way. We also discuss common elements of the round transformation structure, and the differences in the first or the last round.

11.1.1 Evolution

SHARK. The first cipher in the series was SHARK [132]. In this cipher, we first used MDS codes to build a mixing step. The mixing step of SHARK has the one-dimensional structure described in Sect. 9.7.1. The round transformation of SHARK is modular and in principle easily extendible to any block length that is a multiple of 8. However, for a block length of $8n$ bits, an efficient implementation of the round transformation uses tables that require $n^2 \times 256$ bytes of memory. For block lengths of 128 bits, this becomes inefficient on most common processors.

Square. The cipher Square was published in [44]. It has a block length of 128 bits, yet requires only sixteen 8-bit to 32-bit table look-ups per round, whereas an extension of SHARK to this block length would require sixteen 8-bit to 128-bit table look-ups per round. The increased efficiency is achieved by using a two-dimensional structure, as discussed in Sect. 9.7.1, and the introduction of a transposition step. The round transformation of Square

J. Daemen, V. Rijmen, *The Design of Rijndael*, Information Security and Cryptography,
https://doi.org/10.1007/978-3-662-60769-5_11

uses tables that require $n \times 256$ bytes in total, for a block length of $8n$. Note that n has to be a square number. For Square, n has been fixed to 16.

Another improvement in Square concerns the implementations on processors with limited RAM. These processors typically have no space for the extended tables. By restricting the coefficients in the mixing step to small values, the performance on these limited processors becomes acceptable for practical applications.

A fourth improvement in Square is the introduction of an efficient and elegant key schedule.

BKSQ. The cipher BKSQ was published in [48]. In this cipher, the round transformation structure of Square is further generalized. The state is no longer 'square', but can become 'rectangular'. This allows ciphers with block lengths of $8n_1n_2$ bits to be defined.

A second modification with respect to Square is the introduction of nonlinearity in the key schedule.

11.1.2 The Round Transformation

SHARK, Square, BKSQ and Rijndael are key-iterated block ciphers: they consist of the alternation of a key-independent round transformation ρ with a key addition, here denoted by $\sigma[\mathbf{k}]$. The round transformation is the sequence of a nonlinear bricklayer permutation, here denoted by γ, and a linear step, here denoted by λ. The three operations $\sigma[\mathbf{k}]$, γ and λ can be ordered in six different ways in the round transformation. However, we will show that with respect to security, all the orderings are equivalent.

Equivalence of orderings. Firstly, we recall from Sect. 3.7.2 that

$$\sigma[\lambda(\mathbf{k})] \circ \lambda \equiv \lambda \circ \sigma[\mathbf{k}]. \tag{11.1}$$

Both orderings can be chosen in the definition of the cipher's round transformation, without making a difference in the security analysis or performance of the cipher.

Secondly, consider the following key-dependent round transformations that are rotated versions of one another:

$$\rho_1 = \sigma[\mathbf{k}] \circ \lambda \circ \gamma \tag{11.2}$$
$$\rho_2 = \lambda \circ \gamma \circ \sigma[\mathbf{k}]. \tag{11.3}$$

A cipher defined as the iteration of R ρ_1 round transformations can also be described as an iteration of ρ_2 round transformations, with a special definition for the first round and the last round.

We conclude that the same ordering of operations in the cipher can follow from different definitions of the round transformation. In fact, from the previous arguments it follows that all six orderings of the operations in the round transformation result in equivalent ciphers, except for the definition of the key schedule and the definition of the first and the last round.

Boundary effects. The first and/or the last round of the ciphers can differ from the other rounds in several ways. Firstly, operations performed before the first key application or after the last key application can usually be factored out by the cryptanalyst and hence do not contribute to the security of the cipher. The only exceptions to this rule are the modes of operation where only a part of the state may be output. Therefore, if in the definition of the cipher, the round transformation does not start (end) with a round key application, an extra round key application has to be added to the beginning (end) of the cipher.

Secondly, because of (11.1) it is usually possible to leave out one application of λ in the first or the last round, since it does not improve the security of the cipher. Removing one application of λ usually helps to give the inverse of the cipher the same structure as the cipher.

11.2 SHARK

Both the block length and the key length of SHARK can easily be varied. In [132] it is proposed to use a block length of 8 bytes, or 64 bits. Let the number of bytes in the input be denoted by n. For a block length of 64 bits, $n = 8$.

The structure. The round transformation of SHARK has the simple $\gamma\lambda$ structure, as defined in Sect. 9.2.1. The elements of a state $\mathbf{a}$ are denoted by a_i, $0 \leq i < n$. The cipher consists of eight rounds.

The linear transformation. The mixing step of SHARK is derived from a linear code over GF(2^8) with length $2n$, dimension n and minimal distance $n+1$. This construction corresponds to the one-dimensional structure discussed in Sect. 9.7.1. The transformation is denoted by λ. For $n = 8$, we have

$$\lambda(\mathbf{a}) = \begin{bmatrix} a_0 \; a_1 \; a_2 \; a_3 \; a_4 \; a_5 \; a_6 \; a_7 \end{bmatrix} \times \begin{bmatrix} \texttt{CE 95 57 82 8A 19 B0 01} \\ \texttt{E7 FE 05 D2 52 C1 88 F1} \\ \texttt{B9 DA 4D D1 9E 17 83 86} \\ \texttt{D0 9D 26 2C 5D 9F 6D 75} \\ \texttt{52 A9 07 6C B9 8F 70 17} \\ \texttt{87 28 3A 5A F4 33 0B 6C} \\ \texttt{74 51 15 CF 09 A4 62 09} \\ \texttt{0B 31 7F 86 BE 05 83 34} \end{bmatrix} . \quad (11.4)$$

The branch number of λ is 9 $(= n + 1)$.

As explained in Chap. 4, λ can be implemented efficiently by extending the tables that specify the substitution boxes. In SHARK, there are n tables, requiring $n \times 256$ bytes of memory each. When $n = 8$, this gives a total of 16 kB.

The nonlinear transformation. The nonlinear transformation is a bricklayer permutation of S-boxes operating on bytes, denoted by γ. The same S-box is used for all byte positions. We have

$$\gamma : \mathbf{b} = \gamma(\mathbf{a}) \Leftrightarrow b_i = \mathrm{S}_\gamma(a_i), \tag{11.5}$$

where S_γ is an invertible 8-bit substitution table or S-box.

As in Rijndael, the S-box of SHARK is based on the function $F(x) = x^{-1}$ over GF(2^8), as proposed by K. Nyberg in [120]. An affine transformation is added in order to make the description of the S-boxes less simple. This transformation is not equivalent to the transformation that is applied in the S-boxes of Rijndael.

The round key application. In [132], two alternative ways to introduce the round key in the round transformation are proposed. The first is a key addition in the form of a bitwise XOR of the state with a round key; the second version uses a key-controlled affine transform.

XOR. In the first alternative, the 64 state bits are modified by means of an XOR with a 64-bit round key. This operation is denoted $\sigma_+[\mathbf{k}^{(r)}]$. The resulting cipher is a key-iterated cipher with all its advantages; see Chap. 9. A limitation of the simple scheme is that the entropy of the round key is 'only' 64 bits.

Affine transformation. Let $\kappa^{(t)}$ be a key-dependent invertible 8×8 matrix over GF(2^8). The second alternative for the key application is then denoted by $\sigma_{\mathrm{AT}}[\kappa^{(t)}, \mathbf{k}^{(t)}]$ and defined as

$$\sigma_{\mathrm{AT}}[\kappa^{(t)}, \mathbf{k}^{(t)}] : \mathbf{b} = \sigma_{\mathrm{AT}}[\kappa^{(t)}, \mathbf{k}^{(t)}](\mathbf{a}) \Leftrightarrow \mathbf{b} = \kappa^{(t)} \times \mathbf{a} + \mathbf{k}^{(t)}. \tag{11.6}$$

The resulting operation on the state is linear. Since the operation has to be invertible, it must be ensured that all $\kappa^{(t)}$ are invertible matrices. Each round now introduces more key material, increasing the number of round key bits introduced in the key application to 9×64 bits. The computational overhead of this operation is very high. We can restrict the $\kappa^{(t)}$ to a certain subspace, for instance by letting the $\kappa^{(t)}$ be diagonal matrices. The number of round key bits introduced in the key application then becomes close to 2×64 bits.

The cipher. The round transformation, denoted by ρ, consists of a sequence of two steps:

$$\rho = \lambda \circ \gamma. \tag{11.7}$$

SHARK is defined with seven rounds, followed by a final round where the mixing step is absent. The applications of the round transformation are interleaved with nine round key applications.

The key schedule. The key schedule expands the key $\mathbf{K}$ to the round keys $\mathbf{K}^{(t)}$. The key schedule of SHARK operates in the following way. The cipher key is concatenated with itself until it has a length of 9×64 bits, or 9×128 bits for the extended version. This string is encrypted with SHARK in CFB mode, using a fixed key. The first 448 bits of the output form the round keys $\mathbf{k}^{(t)}$. For the extended version, the next 448 bits are used to form the diagonal elements of the matrices $\kappa^{(t)}$. If one of these elements is zero, then it is discarded and all the following values are shifted down one place. An extra encryption of the all-zero string is added at the end to provide the extra diagonal elements. The fixed key used during the key schedule is formed in the following way. The matrices $\kappa^{(t)}$ are equal to the identity matrix. The vectors $\mathbf{k}^{(t)}$ are taken from an expanded substitution table, which is used in the combined implementation of the nonlinear step and the mixing step.

While this mechanism for round key generation in principle makes it possible to use a key of $64 \times 9(\times 2)$ bits, it is suggested that the key length should not exceed 128 bits.

11.3 Square

Square can be considered as an extension of the simple SHARK variant where the mixing step is changed, a byte transposition step has been introduced, and an efficient and elegant key schedule has been introduced. Square has a block length of 128 bits and a key length of 128 bits.

The structure. The round transformation of Square is almost identical to the round transformation of Rijndael when the block length equals 128 bits. The round transformation consists of a sequence of three distinct steps that operate on the *state*: a 4×4 array of bytes. The element of a state $\mathbf{a}$ in row i and column j is specified as $a_{i,j}$. Both indices start from 0. The steps are illustrated in Fig. 11.1.

The mixing step. The mixing step θ is similar to `MixColumns` in Rijndael, except that it operates on the rows of the state instead of the columns. We have

$$\theta(a) = \begin{bmatrix} a_{0,0} & a_{0,1} & a_{0,2} & a_{0,3} \\ a_{1,0} & a_{1,1} & a_{1,2} & a_{1,3} \\ a_{2,0} & a_{2,1} & a_{2,2} & a_{2,3} \\ a_{3,0} & a_{3,1} & a_{3,2} & a_{3,3} \end{bmatrix} \times \begin{bmatrix} \mathtt{2} & \mathtt{1} & \mathtt{1} & \mathtt{3} \\ \mathtt{3} & \mathtt{2} & \mathtt{1} & \mathtt{1} \\ \mathtt{1} & \mathtt{3} & \mathtt{2} & \mathtt{1} \\ \mathtt{1} & \mathtt{1} & \mathtt{3} & \mathtt{2} \end{bmatrix}, \tag{11.8}$$

where the multiplication is in $\mathrm{GF}(2^8)$. The coefficients have been chosen to maximize the branch number of θ, and to facilitate the implementation on 8-bit processors.

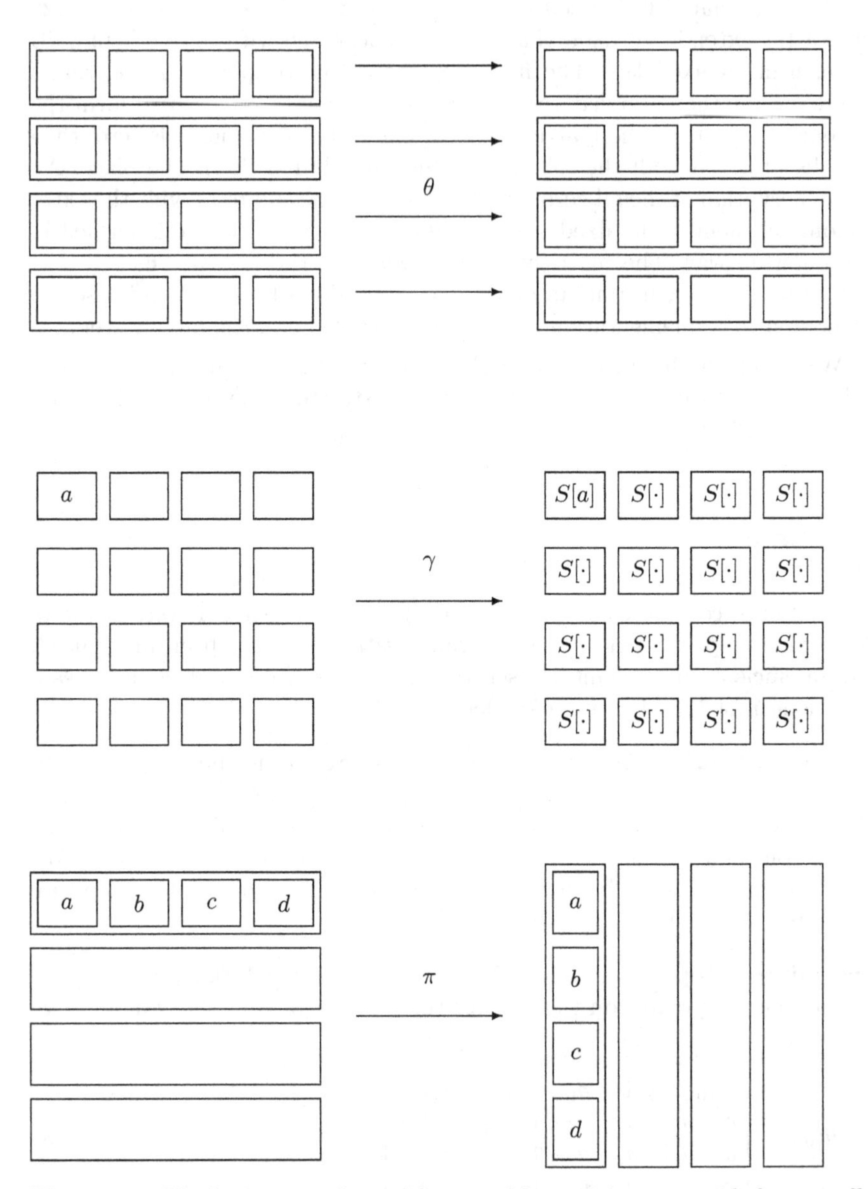

Fig. 11.1. The basic operations of Square. θ is a mixing step with four parallel linear transformations. γ consists of 16 separate substitutions. π is a transposition

The byte transposition. The byte transposition π interchanges rows and columns of a state. If the state is considered as a matrix, it corresponds to the matrix transposition operation. We have

$$\pi : \mathbf{b} = \pi(\mathbf{a}) \Leftrightarrow b_{i,j} = a_{j,i}. \tag{11.9}$$

π is an involution, hence $\pi^{-1} = \pi$.

The nonlinear step. The nonlinear step γ is a bricklayer permutation operating on bytes. We have

$$\gamma : \mathbf{b} = \gamma(\mathbf{a}) \Leftrightarrow b_{i,j} = \mathrm{S}_\gamma(a_{i,j}), \tag{11.10}$$

where S_γ is an invertible 8-bit substitution table or S-box. The S-box of Square is exactly the same as the S-box of SHARK.

The key addition. The key addition with round key $\mathbf{k}^{(t)}$ is denoted by $\sigma[\mathbf{k}^{(t)}]$. It is identical to the key addition in Rijndael, and the simple key application of SHARK.

The cipher. The round transformation ρ is a sequence of three steps:

$$\rho = \pi \circ \gamma \circ \theta. \tag{11.11}$$

Square is defined as eight rounds interleaved with nine key addition steps. These transformations are preceded by an initial application of θ^{-1}. Note that the θ^{-1} before $\sigma[\mathbf{k}^{(0)}]$ can be incorporated in the first round. The initial θ^{-1} can be discarded by omitting θ in the first round and applying $\theta(\mathbf{k}^{(0)})$ instead of $\mathbf{k}^{(0)}$. The same simplification can be applied to the algorithm for decryption.

The key schedule. The key schedule is linear. It is defined in terms of the rows of the key. We can define a left byte-rotation operation $\mathsf{rotl}(a_i)$ on a row as

$$\mathsf{rotl}[a_{i,0}a_{i,1}a_{i,2}a_{i,3}] = [a_{i,1}a_{i,2}a_{i,3}a_{i,0}] \tag{11.12}$$

and a right byte-rotation $\mathsf{rotr}(a_i)$ as its inverse.

The round keys $\mathbf{k}^{(t)}$ are derived from the cipher key $\mathbf{K}$ in the following way. $\mathbf{k}^{(0)}$ equals the cipher key $\mathbf{K}$. The other round keys are derived iteratively by means of an invertible affine transformation, called 'the round key evolution' and denoted by ψ:

$$\psi : \mathbf{k}^{(t)} = \psi(\mathbf{k}^{(t-1)}). \tag{11.13}$$

The round key evolution ψ is defined by

$$\begin{aligned}
\mathbf{k}_0^{(t+1)} &= \mathbf{k}_0^{(t)} + \mathsf{rotl}(\mathbf{k}_3^{(t)}) + C^{(t)} \\
\mathbf{k}_1^{(t+1)} &= \mathbf{k}_1^{(t)} + \mathbf{k}_0^{(t+1)} \\
\mathbf{k}_2^{(t+1)} &= \mathbf{k}_2^{(t)} + \mathbf{k}_1^{(t+1)} \\
\mathbf{k}_3^{(t+1)} &= \mathbf{k}_3^{(t)} + \mathbf{k}_2^{(t+1)} .
\end{aligned} \tag{11.14}$$

The round constants $C^{(t)}$ are also defined iteratively. We have $C^{(0)} = 1$ and $C^{(t)} = 2 \cdot C^{(t-1)}$.

11.4 BKSQ

BKSQ is an iterated block cipher with a block length of 96 bits and a key length of 96, 144 or 192 bits. Its intended use case was that of a (second) pre-image resistant one-way function in a lightweight hash signature scheme offering about 96 bits of security. Most available block ciphers have block lengths of 64 or 128 bits. A block length of 64 bits would only offer 64 bits of security. A block length of 128 bits would be overkill. BKSQ is tailored towards these applications. Still, it can also be used for efficient MACing and encryption on a smart card.

The structure. Let the input of the cipher be denoted by a string of 12 bytes: $p_0 p_1 \ldots p_{11}$. These bytes can be rearranged into a 3×4 array, or state $\mathbf{a}$:

$$\mathbf{a} = \begin{bmatrix} a_{0,0} & a_{0,1} & a_{0,2} & a_{0,3} \\ a_{1,0} & a_{1,1} & a_{1,2} & a_{1,3} \\ a_{2,0} & a_{2,1} & a_{2,2} & a_{2,3} \end{bmatrix} = \begin{bmatrix} p_0 & p_3 & p_6 & p_9 \\ p_1 & p_4 & p_7 & p_{10} \\ p_2 & p_5 & p_8 & p_{11} \end{bmatrix} . \tag{11.15}$$

The basic building blocks of the cipher operate on this array. Figure 11.2 gives a graphical illustration of the building blocks.

The linear transformations. BKSQ uses two linear transformations. The first transformation is similar to `MixColumns` in Rijndael, except that it operates on columns of length 3 instead of length 4. This transformation is denoted by θ. We have

$$\theta(\mathbf{a}) = \begin{bmatrix} 3 & 2 & 2 \\ 2 & 3 & 2 \\ 2 & 2 & 3 \end{bmatrix} \times \begin{bmatrix} a_{0,0} & a_{0,1} & a_{0,2} & a_{0,3} \\ a_{1,0} & a_{1,1} & a_{1,2} & a_{1,3} \\ a_{2,0} & a_{2,1} & a_{2,2} & a_{2,3} \end{bmatrix} . \tag{11.16}$$

This choice for the coefficients makes it possible to implement θ very efficiently on an 8-bit processor with limited working memory.

The second linear transformation is a byte permutation, denoted by π. The effect of π is a shift of the rows of a state. Every row is shifted a different amount. We have

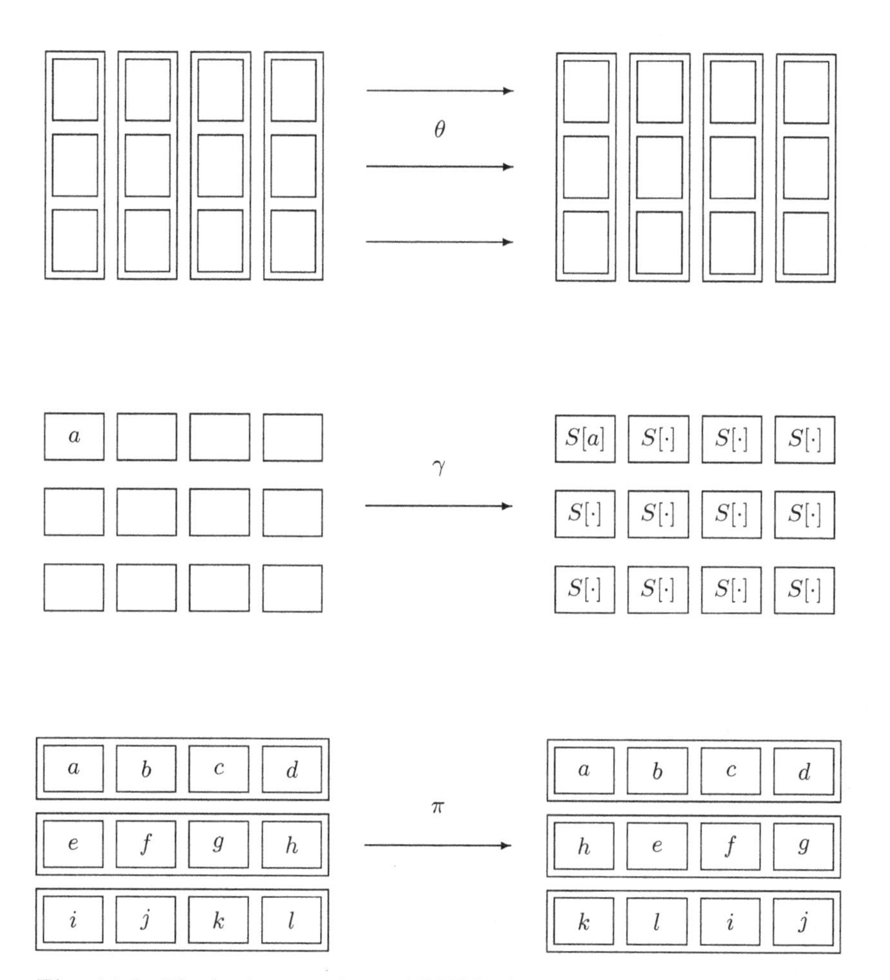

Fig. 11.2. The basic operations of BKSQ. θ is a mixing step with four parallel linear transformations. γ consists of 12 separate substitutions. π is a shift of the rows

$$\pi : \mathbf{b} = \pi(\mathbf{a}) \Leftrightarrow b_{i,j} = a_{i,j-i}. \tag{11.17}$$

The effect of π is that for every column of **a**, the three elements are moved to three different columns in $\pi(\mathbf{a})$.

The nonlinear transformation. The nonlinear transformation is a bricklayer permutation operating on bytes, denoted by γ. It operates on all bytes in the same way. We have

$$\gamma : \mathbf{b} = \gamma(\mathbf{a}) \Leftrightarrow b_{i,j} = \mathrm{S}_\gamma(a_{i,j}), \tag{11.18}$$

where S_γ is an invertible 8-bit substitution table or S-box. The inverse of γ consists of the application of the inverse substitution S_γ^{-1} to all bytes of a state. The S-box of BKSQ is exactly the same as the S-box of Rijndael.

The key addition. The key addition with key $\mathbf{k}^{(t)}$ is denoted by $\sigma[\mathbf{k}^{(t)}]$. It is defined analogously to the key addition of Square and Rijndael.

The cipher. The round transformation denoted by ρ is a sequence of three steps:

$$\rho = \pi \circ \gamma \circ \theta. \tag{11.19}$$

BKSQ is defined as R times the round operation, interleaved with $R + 1$ applications of the key addition and preceded by θ^{-1}:

$$\mathrm{BKSQ}[\mathbf{k}] = \sigma[\mathbf{k}^{(R)}] \circ \rho \circ \sigma[\mathbf{k}^{(R-1)}] \circ \rho \circ \cdots \circ \rho \circ \sigma[\mathbf{k}^{(0)}] \circ \theta^{-1}. \tag{11.20}$$

The number of rounds R depends on the key length that is used. For 96-bit keys, there are 10 rounds; for 144-bit keys, there are 14 rounds; and for 192-bit keys, the number of rounds is 18.

The key schedule. The derivation of the round keys $\mathbf{k}^{(t)}$ from the cipher key K is very similar to the key schedule of Rijndael. The round keys $\mathbf{k}^{(t)}$ are extracted from an expanded key array, denoted by W:

$$\mathbf{k}^{(t)} = \mathsf{W}[\cdot][4t] \parallel \mathsf{W}[\cdot][4t+1] \parallel \mathsf{W}[\cdot][4t+2] \parallel \mathsf{W}[\cdot][4t+3]. \tag{11.21}$$

As in Rijndael, the expansion of the cipher key $\mathbf{K}$ into the expanded key array W depends on the length of the cipher key. Let L denote the key length divided by 24. The array is constructed by repeated application of an invertible nonlinear transformation ψ: the first L columns are the columns of $\mathbf{K}$, the next L are given by $\psi(\mathbf{K})$, the following columns are given by $\psi(\psi(\mathbf{K}))$, etc. The transformation ψ operates on blocks of L columns and is defined in terms of the XOR operation, a byte-rotation rot that rotates the bytes of a column, and a nonlinear substitution γ' that operates in exactly the same way as γ, but takes as argument column vectors instead of arrays. For a detailed description of the key schedule, we refer to [48].

11.5 Conclusion

Rijndael is the result of a long design process with continuous improvements along the way. The earliest related design, SHARK, dates back to 1995. Most of the predecessors of Rijndael have been scrutinized intensively by cryptanalysts looking for security flaws, and by programmers interested in efficient implementations. The result of all this work has been taken into account in the design of Rijndael.

The design approach we used for Square and Rijndael has been adopted enthusiastically by a number of cipher designers all over the world. This demonstrates a worldwide belief that the strategy used is sound.

12. Correlation Analysis in $GF(2^n)$

This chapter is based on Appendix A of the first edition of this book and [52]. In the specification of Rijndael in Chap. 3, we have extensively used operations in a finite field, where the bytes of the state and key represent elements of $GF(2^8)$. Still, as for most block ciphers, Rijndael operates on plaintext blocks, ciphertext blocks and keys that are strings of bits. Apart from some exceptions such as interpolation attacks [73] and algebraically oriented analysis [61, 115], cryptanalysis of ciphers is also generally conducted at the bit level. In particular, linear cryptanalysis exploits high correlations between linear combinations of bits of the state in different stages of the encryption process; see Chap. 7. Differential cryptanalysis (see Chap. 8) exploits high propagation probabilities between bitwise differences in the state in different stages of the encryption.

In Section 12.4, we demonstrate how Rijndael can be specified completely with algebraic operations in $GF(2^8)$. How the elements of $GF(2^8)$ are represented in bytes can be seen as a detail of the specification. Addressing this representation issue in the specification is important for different implementations of Rijndael to be interoperable, but not more so than for instance the ordering of the bits within the bytes, or the way the bytes of the plaintext and ciphertext blocks are mapped onto the state bytes.

We can abstract away from the representation of the elements of $GF(2^8)$ and consider a block cipher that operates on strings of elements of $GF(2^8)$. We call this generalization RIJNDAEL-GF. Rijndael can be seen as an instance of RIJNDAEL-GF where the representation of the elements has been specified. In principle, this can be applied to most block ciphers. Each block cipher with block length and key length that are a multiple of n can in principle be generalized to operate on strings of elements of $GF(2^n)$. However, unlike for Rijndael, the specification of these generalized ciphers may become quite complicated.

Intuitively, it seems obvious that if Rijndael has a cryptographic weakness, this is inherited by RIJNDAEL-GF and any instance of it, whatever the representation of the elements of $GF(2^8)$. Still, in the correlation analysis as described in Chap. 7, we work at the bit level and must assume a specific representation to study the propagation properties. In this chapter, we

J. Daemen, V. Rijmen, *The Design of Rijndael*, Information Security and Cryptography,
https://doi.org/10.1007/978-3-662-60769-5_12

demonstrate how to conduct correlation analysis at the level of elements of $\mathrm{GF}(2^n)$, without having to deal with representation issues.

This chapter is devoted to functions over fields with characteristic two. However, building on the generalization of linear cryptanalysis published in [8] all properties and theorems can be generalized to finite fields with odd characteristic.

We start by describing correlation properties of functions over $\mathrm{GF}(2)^n$ and of functions over $\mathrm{GF}(2^n)$, with the focus on linear functions. This is further generalized to functions over $\mathrm{GF}(2^n)^\ell$. We then discuss representations and bases in $\mathrm{GF}(2)^n$ and show how propagation in functions over $\mathrm{GF}(2^n)$ maps to propagation in Boolean functions by the choice of a basis. Subsequently, we prove two theorems that relate representations of linear functions in $\mathrm{GF}(2)^n$ and functions in $\mathrm{GF}(2^n)$ that are linear over $\mathrm{GF}(2)$. Finally we specify RIJNDAEL-GF.

12.1 Description of Correlation in Functions over $\mathrm{GF}(2^n)$

In this section we study the correlation properties of the functions over $\mathrm{GF}(2^n)$:

$$f : \mathrm{GF}(2^n) \to \mathrm{GF}(2^n) : a \mapsto b = f(a).$$

For Boolean functions, correlation is defined between parities. For a function over $\mathrm{GF}(2^n)$, individual bits cannot be distinguished without adopting a representation, and hence speaking about parities does not make sense. A parity is a function that maps $\mathrm{GF}(2)^n$ to $\mathrm{GF}(2)$ and is linear over $\mathrm{GF}(2)$. In $\mathrm{GF}(2^n)$, we can find functions with the same properties. For that purpose, we use the *trace* function in a finite field (see Section 2.1.8).

It follows that the functions of the form

$$f(a) = \mathrm{Tr}(wa)$$

with $w \in \mathrm{GF}(2^n)$ are linear functions mapping $\mathrm{GF}(2^n)$ to $\mathrm{GF}(2)$. There are exactly 2^n such functions, one for each value of w. We call the function $\mathrm{Tr}(wa)$ a *trace parity*, and the corresponding value w a *trace mask*.

In the analysis of correlation properties of functions over $\mathrm{GF}(2^n)$, trace parities play the role that is played by the parities in the correlation analysis of Boolean functions, where $n = 1$. When a representation is chosen, these functions can be mapped one-to-one to parities (see Sect. 12.3.1).

By working with trace masks, it is possible to study correlation properties in functions over $\mathrm{GF}(2^n)$ without having to specify a basis. Hence, the obtained results are valid for all choices of basis. Once a basis is chosen, trace masks can be converted to the usual masks, which we will call *selection masks* in this chapter (see Theorem 12.3.1).

For a function f over $\mathrm{GF}(2^n)$, we denote the correlation between an input trace parity $\mathrm{Tr}(wa)$ and an output trace parity $\mathrm{Tr}(uf(a))$ by $\mathsf{C}^{(f)}_{u,w}$. We have

$$\begin{aligned}\mathsf{C}^{(f)}_{u,w} &= 2^{-n}\sum_a (-1)^{\mathrm{Tr}(wa)}(-1)^{\mathrm{Tr}(uf(a))}\\ &= 2^{-n}\sum_a (-1)^{\mathrm{Tr}(wa)+\mathrm{Tr}(uf(a))}\\ &= 2^{-n}\sum_a (-1)^{\mathrm{Tr}(wa+uf(a))}.\end{aligned}$$

The value of this correlation is determined by the number of values a that satisfy

$$\mathrm{Tr}(wa + uf(a)) = 0. \tag{12.1}$$

If this equation is satisfied by r such values, the correlation $\mathsf{C}^{(f)}_{u,w}$ is equal to $2^{1-n}r - 1$. If it has no solutions, the correlation is -1; if it is satisfied by all values a, the correlation is 1; and if it is satisfied by exactly half of the possible values a, the correlation is 0. By using the polynomial expression for f, (12.1) becomes a polynomial equation in a (see Section 2.1.8):

$$\mathrm{Tr}(wa + u\sum_i c_i a^i) = 0.$$

For some cases the number of solutions of these polynomials can be analytically determined, providing provable bounds for correlation properties. See for example the results on Kloosterman sums in [92] that provide bounds on the input-output correlation of the multiplicative inverse in $\mathrm{GF}(2^n)$.

Example 12.1.1. Let us consider the following operation:

$$b = f(a) = a + \mathrm{c},$$

where c is a constant. We can determine the correlation by finding the number of solutions of

$$\mathrm{Tr}(wa + u(a + \mathrm{c})) = 0.$$

This is equivalent to

$$\mathrm{Tr}((w + u)a + u\mathrm{c}) = 0.$$

If $w + u$ is different from 0, the trace is zero for exactly half of the values of a, and the correlation is 0. If $w = u$ this becomes

$$\mathrm{Tr}(uc) = 0.$$

This equation is true for all values of a if $\mathrm{Tr}(uc) = 0$, and has no solutions if $\mathrm{Tr}(uc) = 1$. It follows that the addition of a constant has no effect on the trace mask and that the sign of the correlation is equal to $(-1)^{\mathrm{Tr}(uc)}$.

12.1.1 Functions That Are Linear over GF(2^n)

The functions of GF(2^n) that are linear over GF(2^n) (see Sect. 2.1.2) are of the form

$$f(a) = l^{(0)}a,$$

where $l^{(0)}$ is an element of GF(2^n). Hence, there are exactly 2^n functions over GF(2^n) that are linear over GF(2^n).

To determine the correlation we can find the number of solutions of

$$\mathrm{Tr}(wa + ul^{(0)}a) = \mathrm{Tr}((w + ul^{(0)})a) = 0.$$

If the factor of a is different from 0, the correlation is 0. The correlation between $\mathrm{Tr}(wa)$ and $\mathrm{Tr}(uf(a))$ is equal to 1 iff

$$w = l^{(0)}u.$$

12.1.2 Functions That Are Linear over GF(2)

A function over GF(2^n) is linear over GF(2) if it satisfies the following:

$$\forall\, x, y \in \mathrm{GF}(2^n) : f(x + y) = f(x) + f(y)\ .$$

Observe that the functions that are linear over GF(2^n) are a subset of the functions that are linear over GF(2). For example, the function $f(x) = x^2$ is linear over GF(2), but not over GF(2^n):

$$\begin{aligned} f(x+y) &= (x+y)^2 = x^2 + xy + yx + y^2 = x^2 + y^2 \\ &= f(x) + f(y) \\ f(ax) &= a^2 f(x) \neq af(x) \text{ if } a \notin \mathrm{GF}(2). \end{aligned}$$

In general, the functions of GF(2^n) that are linear over GF(2) are the so-called linearized polynomials [95]:

$$f(a) = \sum_{t=0}^{n-1} l^{(t)} a^{2^t}\ , \text{ with } l^{(t)} \in \mathrm{GF}(2^n). \tag{12.2}$$

The relation between the trace mask at the input and the trace mask at the output is not trivial.

Theorem 12.1.1. *For a function* $b = \sum_{t=0}^{n-1} l^{(t)} a^{2^t}$ *an output trace parity* $\mathrm{Tr}(ub)$ *is correlated to input trace parity* $\mathrm{Tr}(wa)$ *with a correlation of 1 iff*

$$w = \sum_{t=0}^{n-1} (l^{(n-t \bmod n)} u)^{2^t}. \tag{12.3}$$

Proof. We will prove that $\mathrm{Tr}(wa) = \mathrm{Tr}(ub)$ and hence that $\mathrm{Tr}(wa + ub) = 0$ for all values of a if w is given by (12.3). All computations with variables t, s and r are performed modulo n, and all summations are from 0 to $n-1$.

$$\begin{aligned}
\mathrm{Tr}(wa) &= \mathrm{Tr}(ub) \\
\mathrm{Tr}\left(\sum_t (l^{(n-t)} u)^{2^t} a\right) &= \mathrm{Tr}\left(u \sum_t l^{(t)} a^{2^t}\right) \\
\sum_s \left(\sum_t l^{(n-t)^{2^t}} u^{2^t} a\right)^{2^s} &= \sum_s \left(\sum_t l^{(t)} u a^{2^t}\right)^{2^s} \\
\sum_s \sum_t l^{(n-t)^{2^{s+t}}} u^{2^{s+t}} a^{2^s} &= \sum_s \sum_t l^{(t)^{2^s}} u^{2^s} a^{2^{s+t}} \\
\sum_s \sum_t l^{(n-t)^{2^{s+t}}} u^{2^{s+t}} a^{2^s} &= \sum_{r=s+t} \sum_t l^{(t)^{2^{r-t}}} u^{2^{r-t}} a^{2^r} \\
\sum_s \sum_{r=n-t} l^{(r)^{2^{s-r}}} u^{2^{s-r}} a^{2^s} &= \sum_s \sum_t l^{(t)^{2^{s-t}}} u^{2^{s-t}} a^{2^s} \\
\sum_s \sum_t l^{(t)^{2^{s-t}}} u^{2^{s-t}} a^{2^s} &= \sum_s \sum_t l^{(t)^{2^{s-t}}} u^{2^{s-t}} a^{2^s}.
\end{aligned}$$

□

We illustrate this with the following example.

Example 12.1.2. We consider two transformations f and g over GF(2^3), defined by

$$\begin{aligned}
f(a) &= \alpha a \\
g(a) &= a^4 + (\alpha^2 + \alpha + 1)a^2.
\end{aligned}$$

For both functions, we want to derive a general expression that for any output trace mask u gives the input trace mask w it correlates with. We denote these expressions by f_{d} and g_{d}, respectively. Applying Theorem 12.1.1, we obtain for $f(a)$

$$l^{(0)} = \alpha,\ l^{(1)} = l^{(2)} = 0,$$

and hence

$$w = f_{\mathrm{d}}(u) = \alpha u. \tag{12.4}$$

Similarly, for $g(a)$ we have

$$l^{(0)} = 0,\ l^{(1)} = \alpha^2 + \alpha + 1,\ l^{(2)} = 1,$$

and hence

$$w = g_{\mathrm{d}} = u^2 + ((\alpha^2 + \alpha + 1)u)^4 = u^2 + (\alpha^2 + 1)u^4. \tag{12.5}$$

12.2 Description of Correlation in Functions over GF$(2^n)^\ell$

In this section we treat the correlation properties of functions that operate on arrays of ℓ elements of GF(2^n). We denote the arrays by

$$\mathbf{A} = [a_1\ a_2\ a_3\ \ldots\ a_\ell\ ]^{\mathrm{T}},$$

where the elements $a_i \in \mathrm{GF}(2^n)$. We have

$$Q : \mathrm{GF}(2^n)^\ell \to \mathrm{GF}(2^n)^\ell : \mathbf{A} \mapsto \mathbf{B} = F(\mathbf{A}).$$

The trace parities can be extended to vectors. We can define a trace mask vector as

$$\mathbf{W} = [w_1\ w_2\ w_3\ \ldots\ w_\ell]^{\mathrm{T}},$$

where the elements $w_i \in \mathrm{GF}(2^n)$. The trace parities for a vector are of the form

$$\sum \mathrm{Tr}(w_i a_i) = \mathrm{Tr}\left(\sum_i w_i a_i\right) = \mathrm{Tr}(\mathbf{W}^{\mathrm{T}}\mathbf{A}).$$

We can define a correlation between an input trace parity $\mathrm{Tr}(\mathbf{W}^{\mathrm{T}}\mathbf{A})$ and an output trace parity $\mathrm{Tr}(\mathbf{U}^{\mathrm{T}}Q(\mathbf{A}))$:

$$\begin{aligned}
\mathrm{C}^{(F)}_{\mathbf{U},\mathbf{W}} &= 2^{-n\ell}\sum_{\mathbf{A}}(-1)^{\mathrm{Tr}(\mathbf{W}^{\mathrm{T}}\mathbf{A})}(-1)^{\mathrm{Tr}(\mathbf{U}^{\mathrm{T}}Q(\mathbf{A}))} \\
&= 2^{-n\ell}\sum_{\mathbf{A}}(-1)^{\mathrm{Tr}(\mathbf{W}^{\mathrm{T}}\mathbf{A})+\mathrm{Tr}(\mathbf{U}^{\mathrm{T}}Q(\mathbf{A}))} \\
&= 2^{-n\ell}\sum_{\mathbf{A}}(-1)^{\mathrm{Tr}(\mathbf{W}^{\mathrm{T}}\mathbf{A}+\mathbf{U}^{\mathrm{T}}Q(\mathbf{A}))}.
\end{aligned}$$

12.2.1 Functions That Are Linear over GF(2^n)

If F is linear over GF(2^n), it can be denoted by a matrix multiplication. We have

$$\begin{bmatrix} b_1 \\ b_2 \\ b_3 \\ \vdots \\ b_\ell \end{bmatrix} = \begin{bmatrix} l_{1,1} & l_{1,2} & l_{1,3} & \cdots & l_{1,\ell} \\ l_{2,1} & l_{2,2} & l_{2,3} & \cdots & l_{2,\ell} \\ l_{3,1} & l_{3,2} & l_{3,3} & \cdots & l_{3,\ell} \\ \vdots & \vdots & \vdots & \ddots & \vdots \\ l_{\ell,1} & l_{\ell,2} & l_{\ell,3} & \cdots & l_{\ell,\ell} \end{bmatrix} \times \begin{bmatrix} a_1 \\ a_2 \\ a_3 \\ \vdots \\ a_\ell \end{bmatrix}.$$

Or for short $\mathbf{B} = \mathsf{L}\mathbf{A}$. The elements of the matrix are elements of GF(2^n).

For the correlation, we have

$$\begin{aligned} \mathrm{Tr}(\mathbf{W}^{\mathrm{T}}\mathbf{A} + \mathbf{U}^{\mathrm{T}}\mathsf{L}\mathbf{A}) &= \mathrm{Tr}(\mathbf{W}^{\mathrm{T}}\mathbf{A} + (\mathsf{L}^{\mathrm{T}}\mathbf{U})^{\mathrm{T}}\mathbf{A}) \\ &= \mathrm{Tr}((\mathbf{W} + \mathsf{L}^{\mathrm{T}}\mathbf{U})^{\mathrm{T}}\mathbf{A}). \end{aligned}$$

Hence, the correlation between $\mathrm{Tr}(\mathbf{W}^{\mathrm{T}}\mathbf{A})$ and $\mathrm{Tr}(\mathbf{U}^{\mathrm{T}}\mathbf{B})$ is equal to 1 if

$$\mathbf{W} = \mathsf{L}^{\mathrm{T}}\mathbf{U}. \tag{12.6}$$

12.2.2 Functions That Are Linear over GF(2)

Generalizing equation (12.2) to vectors of GF(2^n) yields

$$b_i = \sum_j \sum_t l_{i,j}^{(t)} a_j^{2^t} \quad 0 \le i < n.$$

If we introduce the following notation:

$$\mathbf{A}^{2^t} = \begin{bmatrix} a_1^{2^t} & a_2^{2^t} & a_3^{2^t} & \ldots & a_\ell^{2^t} \end{bmatrix},$$

this can be written as

$$\mathbf{B} = \sum_t \mathsf{L}^{(t)} \mathbf{A}^{2^t}.$$

For the relation between the input trace mask and the output trace mask, it can be proven that

$$\mathbf{W} = \sum_t (\mathsf{L}^{(n-t \bmod n)^{\mathrm{T}}} \mathbf{U})^{2^t}.$$

12.3 Boolean Functions and Functions in GF(2^n)

12.3.1 Relationship Between Trace Masks and Selection Masks

If we study correlations in GF$(2)^n$, then we have to use selection masks, and we need to specify a basis. We can avoid specification of a basis if we study instead the correlations in GF(2^n), and work with trace masks. Since there exists an isomorphism between GF$(2)^n$ and GF(2^n), we can expect that for every selection mask $\boldsymbol{w}$ there exists a trace mask w, and vice versa.

Since generally $\mathrm{Tr}(wa) \neq \phi_{\mathsf{e}}(w)^{\mathrm{T}} \boldsymbol{a}$, a selection mask $\boldsymbol{w} = \phi_{\mathsf{e}}(w)$, with ϕ defined in Sect. 2.1.9, usually does not correspond to the trace mask w. This is illustrated by the example below.

Example 12.3.1. We use basis e defined in Example 2.1.10. We take $w = \alpha$, hence $\boldsymbol{w}^{\mathrm{T}} = [011]$. Then it follows from Table 12.1 that $\mathrm{Tr}(wa) \neq \boldsymbol{w}^{\mathrm{T}} \boldsymbol{a}$.

Table 12.1. $\mathrm{Tr}(wa) \neq \boldsymbol{w}^{\mathrm{T}} \boldsymbol{a}$

a	$\boldsymbol{a}^{\mathrm{T}}$	$\mathrm{Tr}(\alpha a)$	$[011]^{\mathrm{T}} \boldsymbol{a}$
0	000	0	0
1	001	0	1
$\alpha + 1$	010	0	1
α	011	0	0
$\alpha^2 + \alpha + 1$	100	1	0
$\alpha^2 + \alpha$	101	1	1
α^2	110	1	1
$\alpha^2 + 1$	111	1	0

In the following theorem, we give and prove the correct relation between trace masks and selection masks.

Theorem 12.3.1. *Let $\boldsymbol{a} =_{\mathsf{e}} (a)$. Then the trace mask w corresponds to $\phi_{\mathsf{d}}(w)$ with d the dual basis of e.*

Proof. We prove that

$$\mathrm{Tr}(wa) = \boldsymbol{w}_{\mathsf{d}}{}^{\mathrm{T}} \boldsymbol{a},$$

and hence that the correlations in GF$(2)^n$ and GF(2^n) have the same value if the relation between the masks is satisfied. Applying (2.42) to w and a, we get

$$\mathrm{Tr}(wa) = \mathrm{Tr}\left(\left(\sum_i \mathrm{Tr}(e^{(i)} w) d^{(i)}\right)\left(\sum_j \mathrm{Tr}(d^{(j)} a) e^{(j)}\right)\right).$$

Since the output of the trace map lies in GF(2), and since the trace map is linear over GF(2), we can convert this to

$$\begin{aligned}\mathrm{Tr}(wa) &= \sum_i \mathrm{Tr}(e^{(i)}w) \sum_j \mathrm{Tr}(d^{(j)}a)\mathrm{Tr}(d^{(i)}e^{(j)}) \\ &= \sum_i \mathrm{Tr}(e^{(i)}w) \sum_j \mathrm{Tr}(d^{(j)}a)\delta(i \oplus j) \\ &= \sum_i \mathrm{Tr}(e^{(i)}w)\mathrm{Tr}(d^{(i)}a).\end{aligned}$$

Applying (2.41) twice completes the proof. □

12.3.2 Relationship Between Linear Functions in GF$(2)^n$ and GF(2^n)

A linear function of GF$(2)^n$ is completely specified by an $n \times n$ matrix M:

$$\boldsymbol{b} = \mathsf{M}\boldsymbol{a}.$$

A linear function of GF(2^n) is specified by the n coefficients $l^{(t)} \in \mathrm{GF}(2^n)$ in

$$b = \sum_{t=0}^{n-1} l^{(t)} a^{2^t}.$$

After choosing a basis e over GF(2^n), these two representations can be converted to one another.

Theorem 12.3.2. *Given the coefficients $l^{(t)}$ and a basis e, the elements of the matrix M are given by*

$$M_{ij} = \sum_{t=0}^{n-1} \mathrm{Tr}\left(l^{(t)} d^{(i)} {e^{(j)}}^{2^t}\right).$$

Proof. We will derive an expression of b_i as a linear combination of a_j in terms of the factors $l^{(t)}$. For a component b_i we have

$$\begin{aligned}b_i &= \mathrm{Tr}(bd^{(i)}) \\ &= \mathrm{Tr}\left(\sum_t l^{(t)} a^{2^t} d^{(i)}\right) \\ &= \sum_t \mathrm{Tr}(l^{(t)} a^{2^t} d^{(i)}).\end{aligned} \tag{12.7}$$

The powers of a can be expressed in terms of the components a_j:

$$
\begin{aligned}
a^{2^t} &= \left(\sum_j a_j e^{(j)}\right)^{2^t} \\
&= \sum_j a_j {e^{(j)}}^{2^t},
\end{aligned} \tag{12.8}
$$

where we use the fact that exponentiation by 2^t is linear over GF(2) to obtain (12.8). Substituting (12.8) in (12.7) yields

$$
\begin{aligned}
b_i &= \sum_t \mathrm{Tr}\left(l^{(t)} \sum_j a_j {e^{(j)}}^{2^t} d^{(i)}\right) \\
&= \sum_t \sum_j \mathrm{Tr}\left(l^{(t)} {e^{(j)}}^{2^t} d^{(i)} a_j\right) \\
&= \sum_j \left(\sum_t \mathrm{Tr}(l^{(t)} {e^{(j)}}^{2^t} d^{(i)})\right) a_j.
\end{aligned}
$$

It follows that

$$
M_{ij} = \sum_t \mathrm{Tr}\left(l^{(t)} {e^{(j)}}^{2^t} d^{(i)}\right),
$$

proving the theorem. □

Theorem 12.3.3. *Given matrix* M *and a basis* e*, the elements* $l^{(t)}$ *are given by*

$$
l^{(t)} = \sum_{i=1}^{n} \sum_{j=1}^{n} M_{ij} {d^{(j)}}^{2^t} e^{(i)}.
$$

Proof. We will express b as a function of powers of a in terms of the elements of the matrix M. We have

$$
b = \sum_i b_i e^{(i)}, \tag{12.9}
$$

and

$$
\begin{aligned}
b_i &= \sum_j M_{ij} a_j \\
&= \sum_j M_{ij} \mathrm{Tr}(a d^{(j)}) \\
&= \sum_j M_{ij} \sum_t a^{2^t} {d^{(j)}}^{2^t}.
\end{aligned} \tag{12.10}
$$

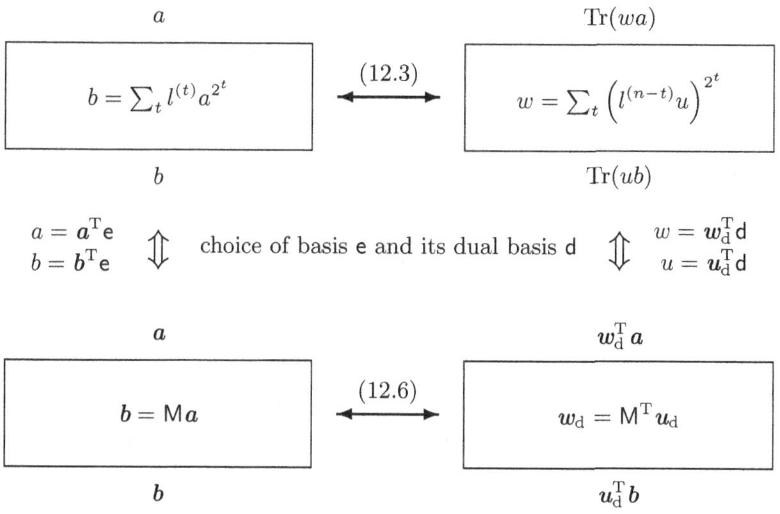

a | | $\mathrm{Tr}(wa)$
$b = \sum_t l^{(t)} a^{2^t}$ | $\xleftrightarrow{(12.3)}$ | $w = \sum_t \left(l^{(n-t)} u\right)^{2^t}$
b | | $\mathrm{Tr}(ub)$

$a = \boldsymbol{a}^{\mathrm{T}} \mathsf{e}$, $b = \boldsymbol{b}^{\mathrm{T}} \mathsf{e}$ $\Updownarrow$ choice of basis e and its dual basis d $\Updownarrow$ $w = \boldsymbol{w}_{\mathrm{d}}^{\mathrm{T}} \mathsf{d}$, $u = \boldsymbol{u}_{\mathrm{d}}^{\mathrm{T}} \mathsf{d}$

$\boldsymbol{a}$ | | $\boldsymbol{w}_{\mathrm{d}}^{\mathrm{T}} \boldsymbol{a}$
$\boldsymbol{b} = \mathsf{M} \boldsymbol{a}$ | $\xleftrightarrow{(12.6)}$ | $\boldsymbol{w}_{\mathrm{d}} = \mathsf{M}^{\mathrm{T}} \boldsymbol{u}_{\mathrm{d}}$
$\boldsymbol{b}$ | | $\boldsymbol{u}_{\mathrm{d}}^{\mathrm{T}} \boldsymbol{b}$

Fig. 12.1. The propagation of selection and trace masks through a function that is linear over GF(2)

Substituting (12.10) into (12.9) yields

$$\begin{aligned} b &= \sum_i \sum_j M_{ij} \sum_t a^{2^t} {d^{(j)}}^{2^t} e^{(i)} \\ &= \sum_t \left(\sum_i \sum_j M_{ij} {d^{(j)}}^{2^t} e^{(i)} \right) a^{2^t}. \end{aligned}$$

It follows that

$$l^{(t)} = \sum_i \sum_j M_{ij} {d^{(j)}}^{2^t} e^{(i)},$$

proving the theorem. □

Figure 12.1 illustrates the relations between the selection mask and trace mask at the input and output of linear functions in GF(2^n). Remember that we always express the *input* mask w as a function of the *output* mask u.

We illustrate this in the next example.

Example 12.3.2. We take the functions f and g of Example 12.1.2 and the bases e and d of Example 2.1.10. Table 12.2 shows the coordinates of the elements of GF(2^3), as well as the coordinates of the images of f and g with respect to e.

Table 12.2. Coordinates of the field elements, and the images of f and g with respect to the basis e

a	$\boldsymbol{a}$	$\boldsymbol{b} = \boldsymbol{f}(\boldsymbol{a})$	$\boldsymbol{b} = \boldsymbol{g}(\boldsymbol{a})$
0	000	000	000
1	001	011	101
$\alpha + 1$	010	101	001
α	011	110	100
$\alpha^2 + \alpha + 1$	100	111	100
$\alpha^2 + \alpha$	101	100	001
α^2	110	010	101
$\alpha^2 + 1$	111	001	000

Once the coordinates of the inputs and outputs of f and g have been determined, we can derive the matrices M and N that describe the functions $\boldsymbol{f}$ and $\boldsymbol{g}$ in the vector space:

$$\mathsf{M} = \begin{bmatrix} 1 & 1 & 0 \\ 1 & 0 & 1 \\ 1 & 1 & 1 \end{bmatrix}, \quad \mathsf{N} = \begin{bmatrix} 1 & 0 & 1 \\ 0 & 0 & 0 \\ 0 & 1 & 1 \end{bmatrix}.$$

The transformations to derive input selection masks from output selection masks are determined by M^{T} and N^{T}:

$$\boldsymbol{f}_{\mathrm{d}}(\boldsymbol{u}_{\mathrm{d}}) = \mathsf{M}^{\mathrm{T}} \boldsymbol{u}_{\mathrm{d}} \tag{12.11}$$

$$\boldsymbol{g}_{\mathrm{d}}(\boldsymbol{u}_{\mathrm{d}}) = \mathsf{N}^{\mathrm{T}} \boldsymbol{u}_{\mathrm{d}}. \tag{12.12}$$

Table 12.3 shows for all the elements of GF(2^3) the coordinates with respect to basis d in the first column, and the coordinates of the images of $\boldsymbol{f}_{\mathrm{d}}$ and $\boldsymbol{g}_{\mathrm{d}}$ calculated according to (12.11) and (12.12) in the second and third column. The fourth column gives the elements of GF(2^3), the fifth and the sixth column give the functions f and g according to (12.4)–(12.5). It can now be verified that the coordinates in the second, respectively the third column correspond to the field elements in the fifth, respectively the sixth column.

12.4 Rijndael-GF

We will now define RIJNDAEL-GF. This is a block cipher very much like Rijndael, but with keys, plaintext and ciphertexts that consist of sequences of elements of GF(2^8) rather than bytes. We will express constants in this specification by powers of α, where α is a root of the primitive polynomial $x^8 + x^4 + x^3 + x^2 + 1$ and hence a generator of the multiplicative group of GF(2^8).

We will first specify the RIJNDAEL-GF round transformation. It operates on a state in GF(2^8)$^{n_{\mathrm{t}}}$ where $n_{\mathrm{t}} \in \{16, 20, 24, 28, 32\}$.

Table 12.3. The functions f_{d} and g_{d}

u_{d}	$w_{\mathrm{d}} = f_{\mathrm{d}}(u_{\mathrm{d}})$	$w_{\mathrm{d}} = g_{\mathrm{d}}(u_{\mathrm{d}})$	u	$w = f_{\mathrm{d}}(u)$	$w = g_{\mathrm{d}}(u)$
000	000	000	0	0	0
001	111	011	α^2+1	1	$\alpha+1$
010	101	000	$\alpha^2+\alpha$	$\alpha^2+\alpha+1$	0
011	010	011	$\alpha+1$	$\alpha^2+\alpha$	$\alpha+1$
100	110	101	α	α^2	$\alpha^2+\alpha+1$
101	001	110	$\alpha^2+\alpha+1$	α^2+1	α^2
110	011	101	α^2	$\alpha+1$	$\alpha^2+\alpha+1$
111	100	110	1	α	α^2

The step `SubBytes-GF` operates on the individual elements of the state. It is composed of two sub-steps. The first step is taking the multiplicative inverse in $\mathrm{GF}(2^n)$:

$$g(a) = a^{-1}, \tag{12.13}$$

with 0 mapping to 0. The second sub-step consists of applying the following linearized polynomial:

$$f(a) = \alpha^2 a + \alpha^{199} a^2 + \alpha^{99} a^{2^2} + \alpha^{185} a^{2^3} + \alpha^{197} a^{2^4} + a^{2^5} + \alpha^{96} a^{2^6} + \alpha^{232} a^{2^7}, \tag{12.14}$$

followed by the addition of the constant α^{195}.

The step `ShiftRows-GF` is a transposition that does not modify the values of the elements in the state but merely changes their positions. It is the same as in Rijndael.

The mixing step `MixColumns-GF` operates independently on four-element columns and mixes them linearly by multiplication with the following matrix:

$$\begin{bmatrix} \alpha^{25} & \alpha & 1 & 1 \\ 1 & \alpha^{25} & \alpha & 1 \\ 1 & 1 & \alpha^{25} & \alpha \\ \alpha & 1 & 1 & \alpha^{25} \end{bmatrix}$$

Finally, the addition of a round key `AddRoundKey-GF` consists of the addition of a round key by a simple addition in $\mathrm{GF}(2^8)$.

The key expansion is the same as that in Rijndael, with the exception that the Rijndael S-boxes are replaced by the RIJNDAEL-GF S-box and the round constants defined as $\mathrm{RC}[i] = \alpha^{25(i-1)}$.

RIJNDAEL-GF, together with the choice of a representation of the elements of $\mathrm{GF}(2^8)$ as bytes constitutes a block cipher operating on bit strings. We can now show that RIJNDAEL-GF is equivalent to Rijndael. As a matter of fact, the choice of the following basis converts RIJNDAEL-GF into Rijndael:

$$\mathsf{e} = \left(1, \alpha^{25}, \alpha^{50}, \alpha^{75}, \alpha^{100}, \alpha^{125}, \alpha^{150}, \alpha^{175}\right).$$

We can compute the corresponding dual basis d by solving (2.40). This yields:

$$\mathsf{d} = \left(\alpha^{166}, \alpha^{187}, \alpha^{37}, \alpha^{26}, \alpha^{236}, \alpha^{191}, \alpha^{196}, \alpha^{48}\right).$$

In Rijndael the second sub-step of the S-box is specified as the multiplication with a binary matrix. This matrix can be reconstructed by applying Theorem 12.3.2 to (12.14) using these bases. The equivalence of the matrices of `MixColumns` and `MixColumns-GF` follows from the fact that $\phi_{\mathsf{e}}^{-1}(\mathtt{02}) = \alpha^{25}$ and $\phi_{\mathsf{e}}^{-1}(\mathtt{03}) = 1 + \alpha^{25} = \alpha$.

13. On the EDP and the ELP of Two and Four Rijndael Rounds

In Chaps. 7 and 8 we explain how the correlation and the difference propagation over a number of rounds of an iterative block cipher are composed of a number of linear trails and differential trails respectively. We show that in key-alternating ciphers the LP of linear trails and the weight of differential trails are both independent of the value of the key. Section 9.1 explains how to choose the number of rounds of a key-alternating cipher to offer resistance against linear and differential cryptanalysis. Although the existence of high correlations and difference propagation probabilities cannot be avoided, taking a number of rounds so that the contributions of the individual trails are below some limit makes the values of the patterns that exhibit large difference propagation probabilities or correlations very key-dependent. We count on this key-dependence to make the exploitation of these high correlations and difference propagation probabilities in cryptanalysis infeasible.

In our analysis in Sect. 9.1, we have neglected possible trail clustering: the fact that sets of trails tend to propagate along common intermediate patterns. If clustering of trails occurs, the small contributions of the individual trails may be compensated for by the fact that there are so many trails between an input pattern and an output pattern. The structure of Rijndael, and any cipher that operates on tuples rather than bits, can be suspected of trail clustering.

In this chapter we prove some properties of Boolean transformations with a maximum branch number. Subsequently, we give provable upper bounds for the EDP of differentials and ELP of linear hulls over two rounds and four rounds of ciphers with the Rijndael structure.

13.1 Properties of MDS Mappings

Consider a Boolean transformation F operating on vectors of n_t tuples. We have

$$\left[b_{(1)}\ b_{(2)}\ b_{(3)} \ldots\ b_{(n_t)}\right]^{\mathrm{T}} = \mathsf{F}(a_{(1)}, a_{(2)}, a_{(3)}, \ldots, a_{(n_t)}). \tag{13.1}$$

Figure 13.1 illustrates this with an example.

J. Daemen, V. Rijmen, *The Design of Rijndael*, Information Security and Cryptography,
https://doi.org/10.1007/978-3-662-60769-5_13

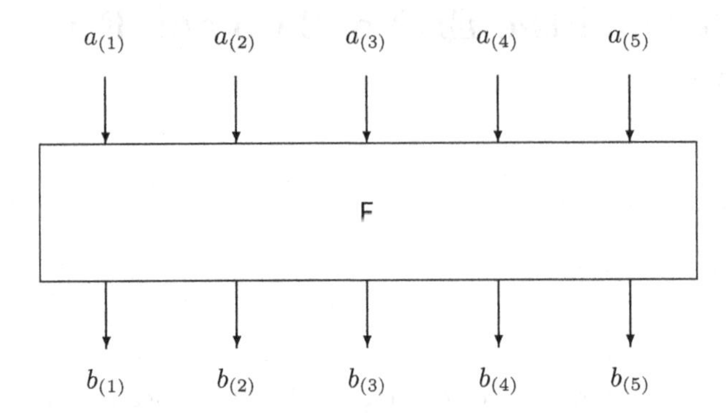

Fig. 13.1. Boolean transformation F operating on 5-byte vectors

Consider the following equation:

$$\mathsf{F}(x_{(1)}, x_{(2)}, x_{(3)}, \ldots, x_{(n_t)}) = \left[x_{(n_t+1)}\ x_{(n_t+2)}\ x_{(n_t+3)}\ \cdots\ x_{(2n_t)}\right]^{\mathrm{T}} \quad (13.2)$$

Clearly, (13.2) has exactly $2^{\mathrm{n_s} n_t}$ solutions, one for each choice of the vector $\left[x_{(1)}\ x_{(2)}\ x_{(3)}\ \cdots\ x_{(n_t)}\right]$.

We consider a partition Ξ of the set $\{1, 2, 3, \ldots, 2n_t\}$ that divides the set of indices into two equally sized subsets ξ and $\bar{\xi}$. We denote the vector with components x_i with $i \in \xi$ by $\mathbf{x}_\xi$. Given such a partition and a value for $\mathbf{x}_\xi$, we define the following set of equations:

$$\begin{cases} \mathsf{F}(y_{(1)}, y_{(2)}, \ldots, y_{(n_t)}) = \left[y_{(n_t+1)}\ y_{(n_t+2)}\ \cdots\ y_{(2n_t)}\right]^{\mathrm{T}} \\ \mathbf{y}_\xi = \mathbf{x}_\xi \end{cases} . \quad (13.3)$$

Theorem 13.1.1. *A Boolean transformation* F *has a maximum differential branch number, i.e.* $\mathcal{B}(\mathsf{F}) = n_t + 1$, *iff any set of equations of the form (13.3) has exactly one solution, whatever the choice of* ξ *(with* $\#\xi = n_t$*) and* $\mathbf{x}_\xi$*.*

Proof.

- ⇒ Assume that $\mathcal{B}(\mathsf{F}) = n_t + 1$, and that there is a choice of ξ and a value of $\mathbf{x}_\xi$ for which (13.3) has more than one solution. The solutions can only differ in at most n_t bytes, since the n_t components of $\mathbf{y}_\xi$ are fixed by $\mathbf{y}_\xi = \mathbf{x}_\xi$. However, if F has a differential branch number equal to $n_t + 1$, (13.2) cannot have two solutions that differ in less than $n_t + 1$ bytes. Hence, (13.3) has at most one solution.

  Now consider the $2^{n_t \mathrm{n_s}}$ solutions of (13.2). For some given choice of ξ, each of these solutions $\mathbf{a}$ is also a solution of exactly one set of equations of type (13.3), i.e. the one with $\mathbf{x}_\xi = \mathbf{a}_\xi$. As each set of equations (as in Eq. 13.3) has at most one solution and as the total number of sets of equations of type (13.3) for a given ξ is $2^{n_t \mathrm{n_s}}$, each of these sets has exactly one solution.

$\Leftarrow$ Assume the Boolean transformation F has a differential branch number that is smaller than $n_t + 1$. This implies that there must exist at least two solutions of (13.2) that differ in at most n_t bytes. We can now construct a set of equations (as in Eq. 13.3) that has two solutions as follows. We choose ξ to contain only byte positions in which the two solutions are the same, and $\mathbf{x}_\xi$ the vector containing the value of those bytes for the two solutions. This contradicts the premise and hence our initial hypothesis is proven to be false. □

Corollary 13.1.1. *For a Boolean transformation operating on n_t-byte vectors and with a maximum branch number, any set of n_t input and/or output bytes determines the remaining n_t output and/or input bytes completely.*

Hence, if we have a Boolean transformation F with a maximum branch number, any partition Ξ that divides the input and output bytes into two sets with an equal number of elements ξ and $\bar{\xi}$ also defines a Boolean transformation. We call this function F_ξ. This is illustrated with an example in Fig. 13.2. As for any value of ξ both F_ξ and $\mathsf{F}_{\bar{\xi}}$ are Boolean tranformations, it follows that all F_ξ are Boolean permutations. Note that with this convention, the permutation F corresponds to F_ξ with $\xi = \{1, 2, 3, \ldots, n_t\}$, and its inverse F^{-1} with F_ξ^{-1} with $\xi = \{n_t + 1, n_t + 2, n_t + 3, \ldots, 2n_t\}$.

In [144], S. Vaudenay defines the similar concept of *multipermutations*. An (r, n)-*multipermutation* is a function that maps a vector of r bytes to n bytes with a differential branch number that is larger than r. A Boolean transformation F with maximum differential branch number is hence an (n_t, n_t)-multipermutation. The name multipermutation is very appropriate for such a transformation, since it defines a permutation from any set of n_t input and/or output bytes to the complementary set.

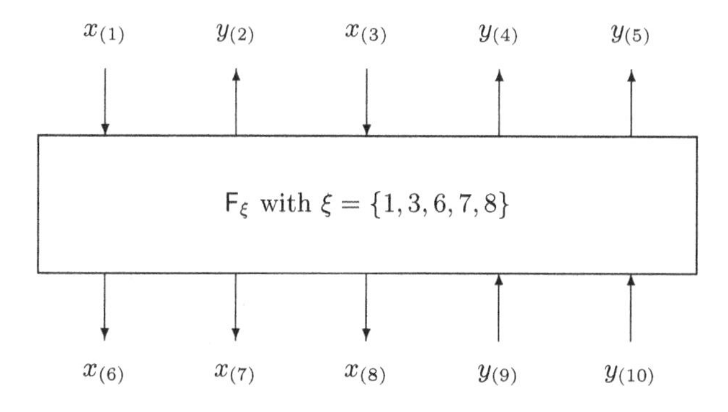

Fig. 13.2. Boolean function F

Theorem 13.1.2. *A Boolean transformation* F *has a maximum differential branch number iff it has a maximum linear branch number.*

Proof.

- $\Rightarrow$ Assume that F has a maximum differential branch number and not a maximum linear branch number. Consider now (13.2). If F does not have a maximum linear branch number, there is a mask w with a byte weight of less than $n_t + 1$ such that the parity $\mathbf{w}^T\mathbf{x}$ is correlated to 0. If we now consider F_ξ with only positions in ξ for which $w_i = 0$, this implies that a parity of output bits of F_ξ is correlated to 0, or in other words, is unbalanced. As F has a maximum differential branch number, any F_ξ must be a permutation and hence according to Theorem 7.5.1 all its output parities must be balanced. It follows that F cannot have a maximum differential branch number and a non-maximum linear branch number.
- $\Leftarrow$ Assume that F has a maximum linear branch number and not a maximum differential branch number. If F does not have a maximum differential branch number, (13.2) has at least two solutions that differ in at most n_t bytes. If we choose ξ such that the byte positions in which these two solutions differ are all in ξ, this means that the function F_ξ has two inputs with the same output and hence is no permutation. According to Theorem 7.5.1, this F_ξ must have output parities $w_{\bar{\xi}}{}^T x_{\bar{\xi}}$ that are not balanced. Hence, F must have a linear branch number that is maximally n_t. It follows that F cannot have a maximum linear branch number and a non-maximum differential branch number. □

13.2 Bounds for Two Rounds

For a cipher with a $\gamma\lambda$ round structure we can prove upper bounds for the EDP of differentials and the ELP of linear hulls (see Sect. 7.9.3) over two rounds.

Figure 13.3 depicts the sequence of steps in two rounds of a cipher with the $\gamma\lambda$ round structure. We study the probability of propagation of a difference in $\mathbf{a}^{(1)}$ to a difference in $\mathbf{a}^{(3)}$. The difference pattern in $\mathbf{a}^{(3)}$ completely determines the difference pattern in $\mathbf{b}^{(2)}$. Hence for this study we only have to consider the first round and the nonlinear step of the second round. For the correlation potentials, we study the correlation between parities of $\mathbf{a}^{(3)}$ and parities of $\mathbf{a}^{(1)}$. A parity of $\mathbf{a}^{(3)}$ is correlated to exactly one parity of $\mathbf{b}^{(2)}$ with a correlation of 1 or -1 depending on the value of a parity of round key $\mathbf{k}^{(2)}$. As we are not interested in the sign, again we can limit ourselves to studying the first round and the nonlinear step of the second round.

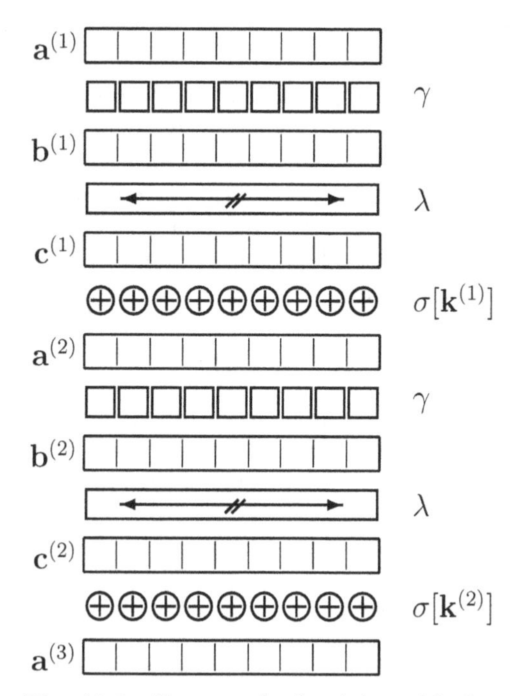

Fig. 13.3. Two rounds of a cipher with the $\gamma\lambda$ round structure

Figure 13.4 depicts the sequence of steps relevant in our analysis of difference and linear propagation over two rounds. In the remainder of this section we denote the difference patterns and masks in the state at the different intermediate stages by $\mathbf{a}, \mathbf{b}, \mathbf{d}$ and $\mathbf{e}$.

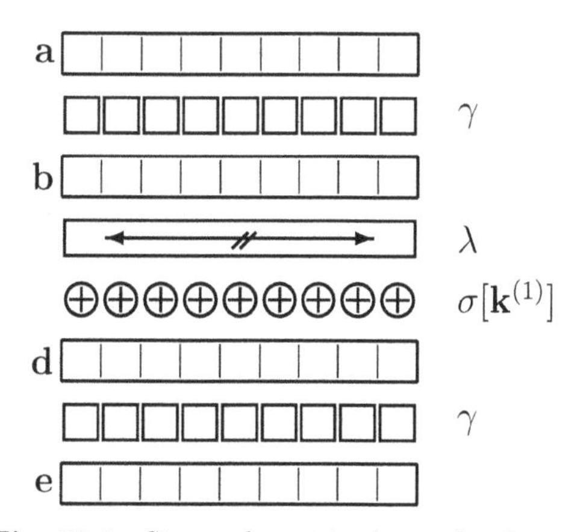

Fig. 13.4. Steps relevant in the study of two-round difference propagation

13.2.1 Difference Propagation

In this section we will denote the difference propagation probabilities of the γ S-box by $\mathrm{DP}^{(\mathrm{s})}(x, y)$. Moreover, we will assume that these difference propagation probabilities are below an upper bound, denoted MDP.

Consider now a differential trail from a difference pattern $\mathbf{a}$ to a difference pattern $\mathbf{e}$. We denote the set of positions with active bytes in $\mathbf{a}$ by α, and the set of positions with active bytes in $\mathbf{e}$ by δ. The number of active bytes in $\mathbf{a}$ is denoted by $\#\alpha$, and the number of active bytes in $\mathbf{e}$ is denoted by $\#\delta$. The active bytes of $\mathbf{a}$ propagate to active bytes in $\mathbf{b}$ through the γ S-boxes. The difference pattern $\mathbf{b}$ fully determines $\mathbf{d}$ by $\mathbf{d} = \lambda(\mathbf{b})$. The difference pattern $\mathbf{d}$ propagates to $\mathbf{e}$ through the γ S-boxes. The difference patterns $\mathbf{d}$ and $\mathbf{e}$ have active bytes in the same positions. As $\mathbf{b}$ completely determines the trail, together with $\mathbf{a}$ and $\mathbf{e}$, we denote this trail by $(\mathbf{a}, \mathbf{b}, \mathbf{e})$. The weight $\mathrm{w_r}(\mathbf{a}, \mathbf{b}, \mathbf{e})$ of this differential trail is the sum of the weights of the difference propagation over active S-boxes corresponding to the active bytes in $\mathbf{a}$ (or equivalently $\mathbf{b}$) and $\mathbf{d}$ (or equivalently $\mathbf{e}$). The sum of the number of active bytes in $\mathbf{b}$ and $\mathbf{d}$ is lower bounded by $\mathcal{B}(\lambda)$. Since $\mathbf{a}$ and $\mathbf{e}$ have active bytes at the same positions as $\mathbf{b}$ and $\mathbf{d}$, respectively, it follows that $\#\alpha + \#\delta$ is lower bounded by $\mathcal{B}(\lambda)$. An approximation for the probability of a differential trail is $2^{-\mathrm{w_r}(\mathbf{a},\mathbf{b},\mathbf{e})}$. This approximation should be interpreted with care, as it is obtained under the assumption that the restrictions are independent (see Sect. 8.4.2). We have

$$\mathrm{EDP}(\mathbf{a}, \mathbf{b}, \mathbf{e}) = \prod_{i \in \alpha} \mathrm{DP}^{(\mathrm{s})}(a_i, b_i) \prod_{j \in \delta} \mathrm{DP}^{(\mathrm{s})}(d_j, e_j). \tag{13.4}$$

and

$$\mathrm{EDP}(\mathbf{a}, \mathbf{e}) = \sum_{\mathbf{b}} \mathrm{EDP}(\mathbf{a}, \mathbf{b}, \mathbf{e}). \tag{13.5}$$

Example 13.2.1. Consider the propagation from a difference pattern $\mathbf{a}$ with a single active byte, in position 1. Equation (13.4) simplifies to

$$\mathrm{EDP}(\mathbf{a}, \mathbf{b}, \mathbf{e}) = \mathrm{DP}^{(\mathrm{s})}(a_1, b_1) \prod_{j \in \delta} \mathrm{DP}^{(\mathrm{s})}(d_i, e_i). \tag{13.6}$$

By using the upper bound for the difference propagation probability for the S-box, $\mathrm{DP}^{(\mathrm{s})}(x, y) \leq \mathrm{MDP}$, this can be reduced to

$$\mathrm{EDP}(\mathbf{a}, \mathbf{b}, \mathbf{e}) \leq \mathrm{DP}^{(\mathrm{s})}(a_1, b_1) \mathrm{MDP}^{\#\delta}. \tag{13.7}$$

Substitution into (13.5) yields

$$\mathrm{EDP}(\mathbf{a}, \mathbf{e}) \leq \sum_{b_1} \mathrm{DP}^{(\mathrm{s})}(a_1, b_1) \mathrm{MDP}^{\#\delta}. \tag{13.8}$$

By using the fact that $\sum_y \mathrm{DP}^{(\mathrm{s})}(x,y) = 1$, we obtain

$$\mathrm{EDP}(\mathbf{a},\mathbf{e}) \leq \mathrm{MDP}^{\#\delta}. \tag{13.9}$$

Since $\#\delta + 1$ is the sum of the number of active S-boxes of $\mathbf{b}$ and $\mathbf{d}$, it is lower bounded by $\mathcal{B}(\lambda)$. It follows that

$$\mathrm{EDP}(\mathbf{a},\mathbf{e}) \leq \mathrm{EDP}^{\mathcal{B}-1}. \tag{13.10}$$

We can now prove the following theorem.

Theorem 13.2.1. *If λ has a maximum branch number, the* EDP *of differentials over two rounds is upper bounded by* $\mathrm{MDP}^{n_\mathrm{t}}$.

Proof. Clearly $\#\alpha + \#\delta$ is lower bounded by $n_\mathrm{t}+1$. Let us now partition the byte positions of $\mathbf{b}$ and $\mathbf{d}$ into two equally sized sets ξ and $\bar{\xi}$, such that $\bar{\xi}$ has only active bytes. This is always possible as there must be at least $n_\mathrm{t}+1$ active bytes in $\mathbf{b}$ and $\mathbf{d}$ together. We have $\mathbf{d} = \lambda(\mathbf{b})$, and that λ has a maximum differential branch number. Hence according to Theorem 13.1.1, the values of the bytes of $\mathbf{b}$ and $\mathbf{d}$ that are in positions in ξ completely determine the values of the bytes with positions in $\bar{\xi}$.
We can convert (13.4) into

$$\mathrm{EDP}(\mathbf{a},\mathbf{b},\mathbf{e}) = \prod_{i\in\alpha\cap\xi} \mathrm{DP}^{(\mathrm{s})}(a_i,b_i) \prod_{j\in\delta\cap\xi} \mathrm{DP}^{(\mathrm{s})}(d_j,e_j) \times \tag{13.11}$$

$$\prod_{i\in\alpha\cap\bar{\xi}} \mathrm{DP}^{(\mathrm{s})}(a_i,b_i) \prod_{j\in\delta\cap\bar{\xi}} \mathrm{DP}^{(\mathrm{s})}(d_j,e_j). \tag{13.12}$$

Since all bytes in positions in $\bar{\xi}$ are active, the factor (13.12) is upper bounded by $\mathrm{MDP}^{n_\mathrm{t}}$. We obtain

$$\mathrm{EDP}(\mathbf{a},\mathbf{b},\mathbf{e}) \leq \mathrm{MDP}^{n_\mathrm{t}} \prod_{i\in\alpha\cap\xi} \mathrm{DP}^{(\mathrm{s})}(a_i,b_i) \prod_{j\in\delta\cap\xi} \mathrm{DP}^{(\mathrm{s})}(d_j,e_j). \tag{13.13}$$

The EDP of the differential $(\mathbf{a},\mathbf{e})$ can be found by summing over all possible trails. In this case, this implies summing over all possible values of the active bytes in $\mathbf{b}$ and $\mathbf{d}$ that have positions in ξ. We have

$$\mathrm{EDP}(\mathbf{a},\mathbf{e}) \leq \mathrm{MDP}^{n_\mathrm{t}} \prod_{i\in\alpha\cap\xi} \sum_{b_i} \mathrm{DP}^{(\mathrm{s})}(a_i,b_i) \prod_{j\in\delta\cap\xi} \sum_{d_j} \mathrm{DP}^{(\mathrm{s})}(d_j,e_j). \tag{13.14}$$

We can apply $\sum_y \mathrm{DP}^{(\mathrm{s})}(x,y) = 1$ to the factors $\sum_{b_i} \mathrm{DP}^{(\mathrm{s})}(a_i,b_i)$. Moreover, as the S-box is invertible, we also have $\sum_x \mathrm{DP}^{(\mathrm{s})}(x,y) = 1$. This can be applied to the factors $\sum_{d_j} \mathrm{DP}^{(\mathrm{s})}(d_j,e_j)$. We obtain

$$\mathrm{EDP}(\mathbf{a},\mathbf{e}) \leq \mathrm{MDP}^{n_\mathrm{t}},$$

proving the theorem. □

More generally, [126] proves that the EDP of differentials over two rounds is upper bounded by $\mathrm{MDP}^{(\mathcal{B}-1)}$.

13.2.2 Correlation

In this section we will denote the square input-output correlations of the γ S-box by $\mathrm{C}^2(x,y)$. We will assume that these correlation potentials are below an upper bound, denoted MLP. For an introduction to correlation potentials we refer to Chap. 7.

Consider now a linear trail from a mask $\mathbf{a}$ to a mask $\mathbf{e}$. We denote the set of positions with active bytes in $\mathbf{a}$ by α, and the set of positions with active bytes in $\mathbf{e}$ by δ. The active bytes of $\mathbf{a}$ propagate to active bytes in $\mathbf{b}$ through the γ S-boxes. Since λ is linear, the mask $\mathbf{b}$ fully determines $\mathbf{d}$. The mask $\mathbf{d}$ propagates to $\mathbf{e}$ through n_t S-boxes. The masks $\mathbf{d}$ and $\mathbf{e}$ have active bytes in the same positions. The correlation potential of this linear trail is the product of the correlation potentials of the active S-boxes corresponding to the active bytes in $\mathbf{a}$ (or equivalently $\mathbf{b}$) and $\mathbf{d}$ (or equivalently $\mathbf{e}$). The sum of the number of active bytes in $\mathbf{b}$ and $\mathbf{d}$ is lower bounded by $\mathcal{B}(\lambda)$. Since $\mathbf{a}$ and $\mathbf{e}$ have active bytes at the same positions as $\mathbf{b}$ and $\mathbf{d}$, respectively, it follows that $\#\alpha + \#\delta$ is lower bounded by $\mathcal{B}(\lambda)$. We have

$$\mathrm{LP}(\mathbf{a},\mathbf{b},\mathbf{e}) = \prod_{i\in\alpha} \mathrm{C}^2(a_i,b_i) \prod_{j\in\delta} \mathrm{C}^2(d_j,e_j), \tag{13.15}$$

and

$$\mathrm{ELP}(\mathbf{a},\mathbf{e}) = \sum_{\mathbf{b}} \mathrm{LP}(\mathbf{a},\mathbf{b},\mathbf{e}). \tag{13.16}$$

We can now prove the following theorem.

Theorem 13.2.2. *If λ has a maximum branch number, the* ELP *of linear hulls over two rounds is upper bounded by* MLP^{n_t}.

Proof. Clearly $\#\alpha + \#\delta$ is lower bounded by $n_t + 1$. Let us now partition the byte positions of $\mathbf{b}$ and $\mathbf{d}$ into two equally sized sets ξ and $\bar{\xi}$ such that ξ has only active bytes. This is always possible as there must be at least $n_t + 1$ active bytes in $\mathbf{b}$ and $\mathbf{d}$ together. As λ is a linear transformation with a maximum branch number, according to Theorem 13.1.1 the values of the bytes of $\mathbf{b}$ and $\mathbf{d}$ that are in positions in ξ completely determine the values of the bytes with positions in $\bar{\xi}$. We can convert (13.15) to

$$\mathrm{LP}(\mathbf{a},\mathbf{b},\mathbf{e}) = \prod_{i\in\alpha\cap\xi} \mathrm{C}^2(a_i,b_i) \prod_{j\in\delta\cap\xi} \mathrm{C}^2(d_j,e_j) \times \tag{13.17}$$

$$\prod_{i\in\alpha\cap\bar{\xi}} \mathrm{C}^2(a_i,b_i) \prod_{j\in\delta\cap\bar{\xi}} \mathrm{C}^2(d_j,e_j). \tag{13.18}$$

As all bytes in positions in $\bar{\xi}$ are active, the factor (13.18) is upper bounded by MLP^{n_t}. We obtain

$$\mathrm{LP}(\mathbf{a},\mathbf{b},\mathbf{e}) \leq \mathrm{MLP}^{n_t} \prod_{i \in \alpha \cap \xi} \mathrm{C}^2(a_i, b_i) \prod_{j \in \delta \cap \xi} \mathrm{C}^2(d_j, e_j). \tag{13.19}$$

The ELP of a hull $(\mathbf{a},\mathbf{e})$ can be found by summing over all possible trails. In this case, this implies summing over all possible values of the active bytes in $\mathbf{b}$ and $\mathbf{d}$ that have positions in ξ. We have

$$\mathrm{ELP}(\mathbf{a},\mathbf{e}) \leq \mathrm{MLP}^{n_t} \prod_{i \in \alpha \cap \xi} \sum_{b_i} \mathrm{C}^2(a_i, b_i) \prod_{j \in \delta \cap \xi} \sum_{d_j} \mathrm{C}^2(d_j, e_j). \tag{13.20}$$

Applying Parseval's theorem to the γ S-box yields $\sum_y \mathrm{C}^2(x,y) = 1$ and applying it to the inverse of the γ S-box yields $\sum_x \mathrm{C}^2(x,y) = 1$. Using this, we obtain

$$\mathrm{ELP}(\mathbf{a},\mathbf{e}) \leq \mathrm{MLP}^{n_t},$$

proving the theorem. □

13.3 Bounds for Four Rounds

For key-iterated ciphers with a $\gamma\pi\theta$ round structure, we can prove similar bounds for four rounds. In Theorem 9.5.1 we have shown that the analysis of such a cipher can be reduced to the analysis of a key-alternating cipher with two round transformations. In this section, we will study this key-alternating cipher structure.

As illustrated in Fig. 13.5, the relevant steps of four rounds can be grouped into a number of supersteps. The first step and last step consist of an application of γ, θ, key addition and again γ. This step operates independently on the columns of the state and can be considered as a γ step with big S-boxes. These big S-boxes are exactly the super boxes defined in Sect. 3.4.5. If θ has a maximum branch number at the level of the columns, the theorems of Sect. 13.2 provide upper bounds for the super box. The EDP is upper bounded by $\mathrm{MDP}^{\mathcal{B}-1}$ and the ELP by $\mathrm{MLP}^{\mathcal{B}-1}$.

In the four-round structure, the two steps in-between are a linear mixing step and a key addition. If the mixing step has a maximum branch number at the level of the columns, the theorems of Sect. 13.2 are also applicable at this level, giving upper bounds for four rounds. The EDP is upper bounded by ${\mathrm{MDP}'}^{\mathcal{B}'-1}$ and the ELP is upper bounded by ${\mathrm{MLP}'}^{\mathcal{B}'-1}$. In these expressions $\mathcal{B}'$ is the branch number of Θ, and MDP' and MLP' refer to the S-boxes of Γ. By substituting the values for the Γ S-boxes, we obtain upper limits of $\mathrm{MDP}^{(\mathcal{B}'-1)(\mathcal{B}-1)}$ and $\mathrm{MLP}^{(\mathcal{B}'-1)(\mathcal{B}-1)}$. In the case where the branch number of Θ and θ is the same, this is reduced to $\mathrm{MDP}^{(\mathcal{B}-1)^2}$ and $\mathrm{MLP}^{(\mathcal{B}-1)^2}$. This applies to Rijndael with a block length of 128 bits.

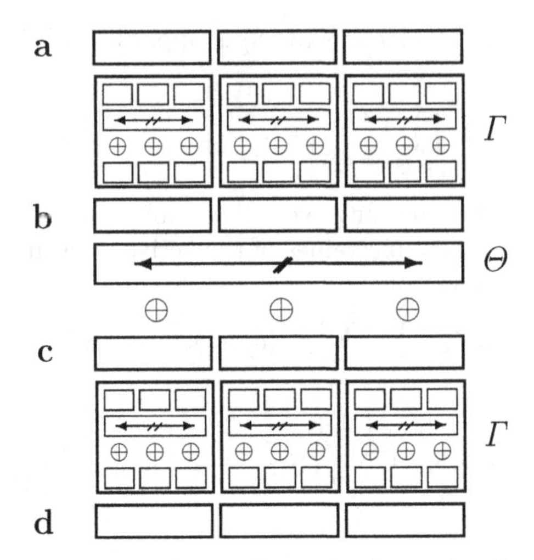

Fig. 13.5. Steps relevant in the study of upper bounds for four rounds

13.4 Conclusions

If we apply Theorems 13.2.1 and 13.2.2 to Rijndael, we find that the EDP of differentials and ELP of linear hulls over two rounds are upper bounded by 2^{-24}, and over four rounds by 2^{-96}. These upper bounds are however not tight. For the EDP it was proven in [80] that for two rounds the maximum is 13.25×2^{-32}, implying an upper bound on the EDP of four rounds of about 2^{-113}. For the ELP it was proven in the same paper that the maximum ELP for two rounds is $109,953,193 \times 2^{-54}$, implying an upper bound on the ELP of four rounds of about 2^{-109}.

In the next chapter we will use the specific properties of the Rijndael components to explain the value of the EDP over two rounds.

14. Two-Round Differential Trail Clustering

This chapter is based on material that appeared earlier in [50]. Bounds on the EDP of trails are proven in Sect. 9.5.2. Bounds on the EDP of differentials have been investigated in Chap. 13 and [79, 126, 127]. The EDP value of differentials is important in the resistance against differential cryptanalysis. In general, the EDP of differentials over multiple rounds of Rijndael is difficult to compute. In this chapter we study the probability of differentials and trails over two rounds of Rijndael with the objective to understand how the components of the Rijndael round transformation interact in this respect. We believe the analysis in this chapter can be used to obtain tighter bounds for the EDP over four rounds of Rijndael and generally a better understanding of its distribution. For this we concentrate on the Rijndael super box as introduced in Sect. 3.4.5.

In Sect. 14.1 we extend and correct the analysis of the differential properties of the multiplicative inverse in GF(2^n) given in [120]. In Sect. 14.2 we introduce the concept of bundles of trail cores, which forms the basis of our analysis. The bundles partition the trails into sets that share important characteristics. In Sect. 14.3 we express the conditions for a trail core in a bundle to contribute to the EDP of a differential. In Sect. 14.4 we use these characteristics to derive expressions for the number of trails in such a bundle and report on experiments confirming these expressions. In Sect. 14.5 we convert the results on number of trails per bundle to their EDP by taking into account the DP of S-box differentials, and we discuss the impact of the presence of the affine mapping in the S-box. In Sect. 14.6 we study the distributions of the EDP in differentials and finally in Sect. 14.7 we show how the differentials with maximum EDP over two-round Rijndael can be explained with our results.

14.1 The Multiplicative Inverse in GF(2^n)

In this section we discuss the differential properties of the single component in Rijndael that is nonlinear over GF(2): the multiplicative inverse in GF(2^n), extended with 0 being mapped to 0. In fact this is the operation of raising to the power $2^n - 2$. For readability we use the notation x^{-1} rather than

J. Daemen, V. Rijmen, *The Design of Rijndael*, Information Security and Cryptography,
https://doi.org/10.1007/978-3-662-60769-5_14

x^{2^n-2}. Differential properties of this map were previously studied in [120]. In the following, a and b denote arbitrary non-zero differences. We denote the probability of a differential (a, b) over the multiplicative inverse map as $\mathrm{DP}^{(\mathrm{i})}(a, b)$. This differential probability is non-zero if and only if the equation

$$(x + a)^{-1} + x^{-1} = b \tag{14.1}$$

has solutions. If $x = a$ or $x = 0$ is a solution of (14.1), we have $b = a^{-1}$ and both are solutions. Otherwise, we can transform (14.1) by multiplying with $b^{-1}x(x + a)$, yielding

$$x^2 + ax + ab^{-1} = 0.$$

If we substitute x by $a^{-1}y$, this becomes

$$y^2 + y + (ab)^{-1} = 0. \tag{14.2}$$

To investigate the condition for this equation to have solutions we have the following lemma.

Lemma 14.1.1 ([95, Theorem 2.25]). $\mathrm{Tr}(t) = 0$ *iff* $t = z^p - z$ *for some* $z \in GF(p^n)$.

Taking $p = 2$, we easily obtain the following.

Lemma 14.1.2. *For* $b \neq a^{-1}$*, equation (14.1) has two solutions if* $\mathrm{Tr}((ab)^{-1}) = 0$*, and zero solutions otherwise.*

Consider now the case $b = a^{-1}$. Let ν and ν^2 denote the elements of $\mathrm{GF}(2^n)$ of order 3. Then $\nu^2 + \nu = 1$ and $\mathrm{GF}(2^2) = \{0, 1, \nu, \nu^2\}$.

Lemma 14.1.3. *For even* n*, the solutions of*

$$(x + a)^{-1} + x^{-1} = a^{-1} \tag{14.3}$$

form the set $T_a = \{0, a, \nu a, \nu^2 a\}$.

Proof. $x = a$ and $x = 0$ are solutions of (14.3). Assume there are other solutions. We can write such a solution as a product of a with an element z different from 0 or 1. We have

$$(za + a)^{-1} + (za)^{-1} = a^{-1} . \tag{14.4}$$

Or, equivalently,

$$(z + 1)^{-1} + z^{-1} = 1 . \tag{14.5}$$

Multiplication with $z(z + 1)$ yields

$$z^2 + z + 1 = 0 \, . \tag{14.6}$$

According to Lemma 14.1.1, Equation (14.6) has two solutions iff $\mathrm{Tr}(1) = 0$ and none otherwise. $\mathrm{Tr}(1) = 0$ iff n is even. Since a solution of (14.6) satisfies $z^3 = 1$, its solutions are the two elements of $\mathrm{GF}(2^n)$ of order 3. □

From these lemmas follow several corollaries.

Corollary 14.1.1 ([120]). *For odd n,*

$$(x + a)^{-1} + x^{-1} = a^{-1}$$

has two solutions: 0 and a.

Corollary 14.1.2. *For even n, the possible output differences b for a given input difference a are those with $\mathrm{Tr}((ab)^{-1}) = 0$ except $b = 0$. For odd n, the possible output differences b for a given input difference a are those with $\mathrm{Tr}((ab)^{-1}) = 0$ except $b = 0$ and extended with $b = a^{-1}$.*

Together with the fact that (14.1) has four solutions only if $b = a^{-1}$, this leads to the following corollary.

Corollary 14.1.3. *For all non-zero $c \in GF(2^n)$ and for all positive integers t:*

$$\mathrm{DP}^{(\mathrm{i})}(a, b) = \mathrm{DP}^{(\mathrm{i})}(b, a) = \mathrm{DP}^{(\mathrm{i})}(ca, bc^{-1}) = \mathrm{DP}^{(\mathrm{i})}(a^{2^t}, b^{2^t}).$$

14.2 Bundles in the Rijndael Super Box

14.2.1 Differentials, Trails and Trail Cores

We study differentials and trails through the Rijndael super box, as depicted in Fig. 14.1. A trail through the Rijndael super box consists of a sequence of five differences: $\mathbf{a}$, $\mathbf{b}$, $\mathbf{c}$, $\mathbf{d}$ and $\mathbf{e}$.

We can partition the set of 4-byte vectors by considering *truncated* differences [84]. All vectors in a given equivalence class have zeroes in the same byte positions and non-zero values in the other byte positions. An equivalence class is characterized by an *activity pattern.* The activity pattern has a single bit for each byte position indicating whether its value must be 0 (passive) or not (active). The activity pattern of a differential $(\mathbf{a}, \mathbf{e})$ is the couple of the activity patterns of $\mathbf{a}$ and $\mathbf{e}$. We say that two differences are *compatible* if they have the same activity pattern. Due to the diffusion properties of $\mathrm{M_c}$, activity patterns of differentials must have a minimum of five active positions. In total there are 93 such activity patterns.

We denote by $\mathrm{DP}^{(\mathrm{s})}(\mathbf{a}, \mathbf{b})$ the differential probability of $(\mathbf{a}, \mathbf{b})$ over the transformation `SubBytes`. Since $\mathrm{S_{RD}}$ is invertible, $\mathrm{DP}^{(\mathrm{s})}(\mathbf{a}, \mathbf{b})$ can be non-zero only if $\mathbf{a}$ and $\mathbf{b}$ are compatible. Other necessary conditions to have $\mathrm{EDP} > 0$ are $\mathbf{c} = \mathbf{d}$, $\mathbf{d} = \mathrm{M_c}\mathbf{b}$, and $\mathbf{d}$ must be compatible with $\mathbf{e}$. Moreover, all components of $\mathbf{a}$ and $\mathbf{b}$ must satisfy $\mathrm{DP}^{(\mathrm{s})}(a_i, b_i) > 0$, with $\mathrm{DP}^{(\mathrm{s})}(a_i, b_i)$ denoting the differential probability of a differential over $\mathrm{S_{RD}}$. The same goes for $\mathbf{d}$ and $\mathbf{e}$.

In the rest of this chapter we use the term *trail* only when $\mathrm{EDP} > 0$ and we will omit $\mathbf{c}$ from the notation, specifying a trail as $(\mathbf{a}, \mathbf{b}, \mathbf{d}, \mathbf{e})$. Such a trail is fully determined by the differential $(\mathbf{a}, \mathbf{e})$ it is in and the intermediate difference $\mathbf{b}$. We call a sequence of intermediate differences $[\mathbf{b}, \mathbf{d}]$ with $\mathbf{d} = \mathrm{M_c}\mathbf{b}$ a *trail core*. A trail core $[\mathbf{b}, \mathbf{d}]$ contributes to a differential $(\mathbf{a}, \mathbf{e})$ if for all i we have $\mathrm{DP}^{(\mathrm{s})}(a_i, b_i) > 0$ and $\mathrm{DP}^{(\mathrm{s})}(d_i, e_i) > 0$. If that is the case, we say the trail core $[\mathbf{b}, \mathbf{d}]$ *extends* to a trail in differential $(\mathbf{a}, \mathbf{e})$. We denote the number of trails in a differential $(\mathbf{a}, \mathbf{e})$ by $\mathrm{N_t}(\mathbf{a}, \mathbf{e})$.

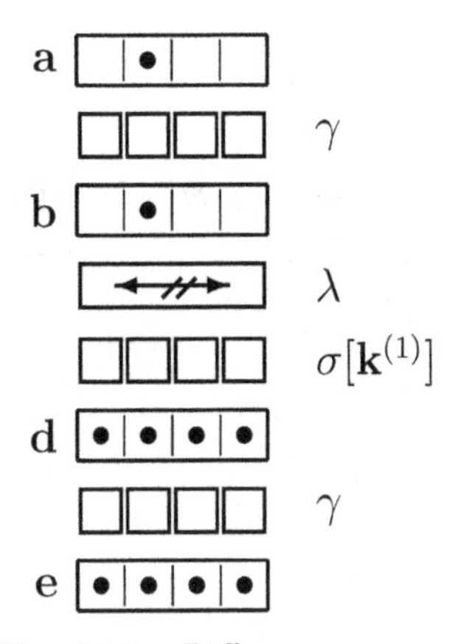

Fig. 14.1. Difference propagation over the Rijndael super box. Active bytes are indicated with bullets

In Chap. 12 we have shown that L is a linearized polynomial and hence it is not linear over $\mathrm{GF}(2^8)$. We denote its inverse by L_i. The additive group of the finite field $\mathrm{GF}(2^8)$ forms a vector space. We will sometimes tacitly switch from one representation to another.

For the EDP of a differential over the Rijndael super box, we have

$$\begin{aligned}\mathrm{EDP}(\mathbf{a},\mathbf{e}) &= \sum_{\mathbf{b}} \mathrm{EDP}(\mathbf{a},\mathbf{b},\mathrm{M_c}\mathbf{b},\mathbf{e}) \\ &= \sum_{\mathbf{b}} \mathrm{DP}^{(\mathrm{s})}(\mathbf{a},\mathbf{b})\mathrm{DP}^{(\mathrm{s})}(\mathrm{M_c}\mathbf{b},\mathbf{e}) \\ &= \sum_{\mathbf{b}} \left(\prod_i \mathrm{DP}^{(\mathrm{s})}(a_i,b_i) \prod_i \mathrm{DP}^{(\mathrm{s})}(d_i,e_i) \right) .\end{aligned}$$

In order to compute the EDP of a differential, we first determine the number of trails in the differential. To ease this task, we partition trail cores into so-called *bundles*, defined below. We start with an example.

Example 14.2.1. Consider a trail in a differential $(\mathbf{a},\mathbf{e})$ with $\mathbf{a} = [a_0, 0, 0, 0]^{\mathrm{T}}$. Then clearly we must have $\mathbf{b} = [b_0, 0, 0, 0]^{\mathrm{T}}$ and thanks to `MixColumns` we have $d_0 = 2b_0$, $d_1 = b_0$, $d_2 = b_0$ and $d_3 = 3b_0$, or equivalently $\mathbf{d} = b_0[2, 1, 1, 3]^{\mathrm{T}}$, where $b_0[2, 1, 1, 3]^{\mathrm{T}}$ denotes the scalar multiplication of the vector $[2, 1, 1, 3]^{\mathrm{T}}$ with the (non-zero) scalar b_0. For each non-zero value of b_0, there may be a trail in the differential, so there are at most 255.

This can be generalized to any Rijndael super box differential with five active S-boxes. If $Q = (\mathbf{a},\mathbf{b},\mathbf{d},\mathbf{e})$ and $Q' = (\mathbf{a},\mathbf{b}',\mathbf{d}',\mathbf{e})$ are two trails in the same differential with five active S-boxes, then there exists a γ such that for all positions i: $b_i = \gamma b'_i$ and $d_i = \gamma d'_i$.

14.2.2 Bundles

We define a *bundle* as follows.

Definition 14.2.1. *The* bundle $B([\mathbf{b},\mathbf{d}])$ *defined by a trail core* $[\mathbf{b},\mathbf{d}]$ *is a set of 255 trail cores:*

$$B([\mathbf{b},\mathbf{d}]) = \{\gamma[\mathbf{b},\mathbf{d}] | \gamma \in \mathrm{GF}(2^8) \textit{ and } \gamma \neq 0\} .$$

The linearity of `MixColumns` over $\mathrm{GF}(2^8)$ implies that $\mathrm{M_c}(\gamma\mathbf{b}) = \gamma(\mathrm{M_c}\mathbf{b})$ and hence multiplication of a trail core with a non-zero scalar γ results in another trail core. Moreover, scalar multiplication does not change the activity pattern of a trail core, hence if $[\mathbf{b},\mathbf{d}]$ is compatible with a differential $(\mathbf{a},\mathbf{e})$, then all trail cores in $B([\mathbf{b},\mathbf{d}])$ are. Hence, the bundles form a partition of the set of trail cores compatible with $(\mathbf{a},\mathbf{e})$.

We will represent a bundle with a canonical representative $[\mathbf{b},\mathbf{d}]$ and denote trail cores as $\gamma[\mathbf{b},\mathbf{d}]$ with $[\mathbf{b},\mathbf{d}]$ a canonical representative of its bundle.

We can count the number of trails in differential $(\mathbf{a},\mathbf{e})$ by counting the number of trails in each bundle and adding the results. In the following, we will explain how the number of trails in a bundle can be counted.

Table 14.1. Activity patterns with five active S-boxes and the corresponding canonical trail cores $[\mathbf{b}, \mathbf{d}]$ (in hexadecimal notation)

Activity Pattern	b	d
(1000; 1111)	$[1,0,0,0]^T$	$[2,1,1,3]^T$
(1100; 1011)	$[2,1,0,0]^T$	$[7,0,3,7]^T$
(0110; 1110)	$[0,1,1,0]^T$	$[2,1,3,0]^T$
(0011; 1011)	$[0,0,1,3]^T$	$[2,0,7,7]^T$
(1001; 1110)	$[2,0,0,3]^T$	$[7,1,7,0]^T$
(1010; 1101)	$[2,0,1,0]^T$	$[5,1,0,7]^T$
(0101; 0111)	$[0,1,0,3]^T$	$[0,1,4,7]^T$
(1101; 1010)	$[5,1,0,7]^T$	$[E,0,D,0]^T$
(0111; 0101)	$[0,1,4,7]^T$	$[0,9,0,B]^T$
(1110; 1001)	$[7,1,3,0]^T$	$[E,0,0,B]^T$
(1011; 0011)	$[2,0,3,7]^T$	$[0,0,D,B]^T$
(1110; 0110)	$[2,1,7,0]^T$	$[0,9,D,0]^T$
(1011; 1100)	$[7,0,7,7]^T$	$[E,9,0,0]^T$
(1111; 1000)	$[E,9,D,B]^T$	$[1,0,0,0]^T$

14.2.3 Number of Bundles with a Given Number of Active Bytes

As explained in our example, a differential with five active S-boxes only has a single bundle of trail cores. Table 14.1 lists the activity patterns with five active S-boxes and the corresponding canonical trail cores $[\mathbf{b}, \mathbf{d}]$ we have chosen. In total there are 56 patterns. They can be derived by rotation of the 14 patterns listed.

For the bundles with six active positions, we can construct canonical representatives by taking (almost) all possible combinations of two canonical trail cores with five active positions. We illustrate this with an example.

Example 14.2.2. For activity pattern (1110; 1110) we combine the canonical trail cores for (1010; 1110) and (0110; 1110) as given by Table 14.1. This gives $\mathbf{b} = [1,0,3,0]^T + z[0,1,1,0]^T = [1,z,3+z,0]^T$ and $\mathbf{d} = [1,4,7,0]^T + z[2,1,3,0]^T = [1+2z, 4+z, 7+3z, 0]^T$.

This results in 255 different bundles, one for each non-zero value of z. However, for $[\mathbf{b}, \mathbf{d}]$ to have activity pattern (1110; 1110) the value of z must be different from 3, 1/2, 4 and 7/3, where x/y denotes $x.y^{-1}$ in GF(2^8). Hence, for a given activity pattern with six active S-boxes, there are $255 - 4 = 251$ bundles of trail cores.

In general, the number of bundles with a given activity pattern is fully determined by the Hamming weight of that pattern. The total number of trail cores is $2^{32} - 1$. Each bundle groups 255 trail cores, so the total number of bundles is

$$\frac{2^{32} - 1}{2^8 - 1} = 2^{24} + 2^{16} + 2^8 + 1 .$$

The bundles with a given activity pattern A with Hamming weight $\nu \geq 5$ are those with activity patterns B that are zero in the positions where A is zero, minus the bundles with those activity patterns B that differ from A. The former is $255^{\nu-5}$ and the latter correspond to activity patterns with smaller Hamming weight. This allows the number of bundles to be computed recursively. To build the canonical trail cores for a given activity pattern with Hamming weight ν takes the linear combination of $\nu - 4$ canonical trail cores with weight 5. Each of these has $\nu - 4$ positions that cannot become 0. There remain four positions that can become zero and hence these must be excluded. If we denote the number of bundles for an activity pattern with x active S-boxes by $\mathrm{BN}(x)$, we have

$$\begin{aligned}
\mathrm{BN}(5) &= 1 \\
\mathrm{BN}(6) &= 255 - \binom{4}{1}\mathrm{BN}(5) = 251 \\
\mathrm{BN}(7) &= 255^2 - \binom{4}{1}\mathrm{BN}(6) - \binom{4}{2}\mathrm{BN}(5) = 64,015 \\
\mathrm{BN}(8) &= 256^3 - \binom{4}{1}\mathrm{BN}(7) - \binom{4}{2}\mathrm{BN}(6) - \binom{4}{3}\mathrm{BN}(5) = 16,323,805
\end{aligned}$$

The number of trails with ν active S-boxes is

$$\binom{8}{\nu} 255\mathrm{BN}(\nu)127^{\nu} .$$

This results in a total of about 2.8×10^{26} trails.

14.3 Conditions for a Trail Core to Extend to a Trail

A trail core $[\mathbf{b}, \mathbf{d}]$ extends to a trail in differential $(\mathbf{a}, \mathbf{e})$ if both $(\mathbf{a}, \mathbf{b})$ and $(\mathbf{d}, \mathbf{e})$ are differentials over `SubBytes` with $\mathrm{DP}^{(\mathrm{s})} > 0$. We will now study the conditions this imposes on the trail cores within a bundle.

14.3.1 The Naive Super Box

We will first study a simplified version of the super box, where we remove the affine mapping from the `SubBytes` S-box. We call this the *naive super box.* Assume we have a differential $(\mathbf{a}, \mathbf{e})$ and a bundle $B([\mathbf{b}, \mathbf{d}])$ compatible with it. We can formulate conditions for a trail core $\gamma[\mathbf{b}, \mathbf{d}]$ in this bundle to extend to a trail in the differential.

We denote by $\mathrm{DP}^{(i)}(\mathbf{a}, \mathbf{b})$ the differential probability of $(\mathbf{a}, \mathbf{b})$ over four parallel applications of the multiplicative inverse in $\mathrm{GF}(2^8)$. We have from Corollary 14.1.2

$$\mathrm{DP}^{(i)}(\mathbf{a}, \gamma\mathbf{b}) > 0 \Leftrightarrow \begin{cases} \mathrm{Tr}((a_i b_i \gamma)^{-1}) = 0 \\ a_i \neq 0 \text{ iff } b_i \neq 0 \end{cases}, \ 0 \leq i < 4, \tag{14.7}$$

and

$$\mathrm{DP}^{(i)}(\mathbf{d}, \gamma\mathbf{e}) > 0 \Leftrightarrow \begin{cases} \mathrm{Tr}((d_i e_i \gamma)^{-1}) = 0 \\ a_i \neq 0 \text{ iff } b_i \neq 0 \end{cases}, \ 0 \leq i < 4. \tag{14.8}$$

In these equations a_i, b_i, d_i and e_i are constants and γ is a variable.

Let x be γ^{-1} and let V be the set of constants $(a_i b_i)^{-1}$ where a_i (and b_i) are non-zero and $(d_i e_i)^{-1}$ where d_i (and e_i) are non-zero. The equations now become

$$\mathrm{DP}(\mathbf{a}, \gamma\mathbf{b}, \gamma\mathbf{d}, \mathbf{e}) > 0 \Leftrightarrow \mathrm{Tr}(\mathrm{V}_j x) = 0 \text{ for all } j. \tag{14.9}$$

Thanks to the linearity of the trace map over GF(2), the solution space of $\mathrm{Tr}(\mathrm{V}_0 x) = 0$ is a vector space of dimension 7 over GF(2). The intersection of $\mathrm{Tr}(\mathrm{V}_0 x) = 0$ and $\mathrm{Tr}(\mathrm{V}_1 x) = 0$ is a vector space of dimension 6 or 7. If the dimension is 7, this implies $\mathrm{V}_0 = \mathrm{V}_1$. In general, the dimension of the intersection of a system of equations $\mathrm{Tr}(\mathrm{V}_j x) = 0$ is equal to $8 - \alpha$ with α the dimension of the vector space generated by the elements V_j. For example, the solution space of $\mathrm{Tr}(\mathrm{V}_0 x) = \mathrm{Tr}(\mathrm{V}_1 x) = \mathrm{Tr}(\mathrm{V}_2 x) = 0$ with $\mathrm{V}_0 \neq \mathrm{V}_1 \neq \mathrm{V}_2 \neq \mathrm{V}_0$ has dimension 6 if $\mathrm{V}_2 = \mathrm{V}_0 + \mathrm{V}_1$ and dimension 5 otherwise. The number of non-zero solutions equals $2^{8-\alpha} - 1$.

14.3.2 Sharp Conditions and Blurred Conditions

If we consider differentials over `SubBytes` then we have to take into account the effect of the linear transformation L_i in the Rijndael S-box. In order to determine the number of trail cores $[\mathbf{b}, \mathbf{d}]$ that may lead to a given output difference $\mathbf{e}$, it suffices to replace V by

$$\mathrm{V}_a = \{(d_0 L_i(e_0))^{-1}, (d_1 L_i(e_1))^{-1}, (d_2 L_i(e_2))^{-1}, (d_3 L_i(e_3))^{-1}\} \,. \tag{14.10}$$

We call the corresponding conditions on γ the *sharp* conditions on trails.

When determining the number of trail cores $[\mathbf{b}, \mathbf{d}]$ that may result from a given input difference $\mathbf{a}$, (14.7) becomes

$$\mathrm{Tr}((a_i L_i(\gamma b_i))^{-1}) = 0 \,, \ 0 \leq i < 4 \,.$$

Equations of this type cannot easily be reworked and are hard to analyze. Therefore we call these conditions the *blurred* conditions.

14.4 Number of Trail Cores in a Bundle Extending to a Trail

The number of trail cores in a bundle $B([\mathbf{b},\mathbf{d}])$ that extend to trails in a given differential $(\mathbf{a},\mathbf{e})$ is the number of γ values that satisfy the sharp conditions due to $(\gamma\mathbf{d},\mathbf{e})$ over `SubBytes` and the blurred conditions due to $(\mathbf{a},\gamma\mathbf{b})$ over `SubBytes`. In this section we first derive formulas to estimate the number of trails in $B([\mathbf{b},\mathbf{d}])$ for the special case of a differential with one active S-box in the first round. Then we will derive formulas for the general case.

14.4.1 Bundles with One Active S-Box in the First Round

Consider a differential $(\mathbf{a},\mathbf{e})$ with activity pattern (1000; 1111). There is a single bundle $B([\mathbf{b},\mathbf{d}])$ with $\mathbf{b}=[1,0,0,0]^{\mathrm{T}}$ and $\mathbf{d}=[2,1,1,3]^{\mathrm{T}}$. The sharp conditions become

$$\begin{aligned}
\mathrm{Tr}((2L_i(e_0))^{-1}\gamma^{-1}) &= 0\\
\mathrm{Tr}((L_i(e_1))^{-1}\gamma^{-1}) &= 0\\
\mathrm{Tr}((L_i(e_2))^{-1}\gamma^{-1}) &= 0\\
\mathrm{Tr}((3L_i(e_3))^{-1}\gamma^{-1}) &= 0\,.
\end{aligned}$$

If $\mathbf{e}=[L(z/2),L(z),L(z),L(z/3)]^{\mathrm{T}}$ with z a non-zero value, then $\mathrm{V}_a=\{z^{-1}\}$. This implies $\alpha=1$ and hence there are 127 trail cores satisfying the sharp conditions.

The presence of the blurred condition does not allow us to determine the number of trails from just the values of a_0 and $\mathbf{e}$. However, for a given $\mathbf{e}$, we can say something about the distribution of the number of trails over the values of a_0. To obtain that distribution, we model the effect of the blurred condition as a sampling process.

The space sampled is the set of 255 trail cores $B([\mathbf{b},\mathbf{d}])$. Of these 255 trail cores, 127 satisfy the blurred condition and the remaining 128 do not. We call the former the *good ones* and the latter the *bad ones*. The joint sharp conditions take a sample with size $2^{8-\alpha}-1$. This gives rise to a hypergeometric distribution $H(\mathrm{N_t};n,m,N)$ [148] with the following parameters:

- Number of ways for a good selection $n=127$.
- Number of ways for a bad selection $m=255-127=128$.
- Sample size N: $2^{8-\alpha}-1$.

In [148] we can read that this distribution has mean

$$E\left[\mathrm{N_t}\right]=\frac{n}{m+n}N=\frac{127}{255}(2^{8-\alpha}-1),$$

and variance

$$\sigma^2(\mathrm{N_t}) = \frac{mnN(m+n-N)}{(m+n)^2(m+n-1)} = \frac{128 \times 127(2^{8-\alpha}-1)(256-2^{8-\alpha})}{255^2 254} .$$

14.4.2 Any Bundle

A differential $(\mathbf{a}, \mathbf{e})$ imposes on γ a number of sharp conditions, determined by $\mathbf{e}$ and $\mathbf{d}$, and a number of blurred conditions, determined by $\mathbf{a}$ and $\mathbf{b}$. Following (14.10), the sharp conditions state that γ^{-1} has to be orthogonal to

$$\mathrm{V}_a = \{v_0, v_1, v_2, v_3\},$$

with $v_i^{-1} = d_i L_i(e_i)$. The parameter α is defined as the dimension of V_a. Hence γ^{-1} is in a vector space of dimension $8-\alpha$ ranging from 4 to 7.

The number of blurred conditions is denoted by β, and given by the number of different non-zero elements in the following set of couples:

$$\{(a_0, b_0), (a_1, b_1), (a_2, b_2), (a_3, b_3)\}.$$

For the vast majority of differentials, β equals the number of active S-boxes in $\mathbf{a}$. β is smaller only when two a_i values are the same and the corresponding b_i in the bundle are also equal. Hence a reduction of β occurs much less often than a reduction of α. Both α and β range from 1 to 4.

We can now generalize the approach of Sect. 14.4.1 to determine the distribution of $\mathrm{N_t}$ over all bundles with given values of α and β.

Lemma 14.4.1. *If the blurred conditions are independent, the number of trails in the bundle $B([\mathbf{b}, \mathbf{d}])$ is a stochastic variable with expected value and variance given by*

$$E[\mathrm{N_t}] = \left(\frac{127}{255}\right)^{\beta} (2^{8-\alpha} - 1) , \tag{14.11}$$

$$\sigma^2(\mathrm{N_t}) = E[\mathrm{N_t}] \times \left[1 - \left(\frac{127}{255}\right)^{\beta} + (2^{8-\alpha} - 2)\left(\left(\frac{63}{127}\right)^{\beta} - \left(\frac{127}{255}\right)^{\beta}\right)\right] . \tag{14.12}$$

Proof. We generalize the sampling model introduced in Sect. 14.4.1. The space sampled is now the set of β-component vectors where each of the components can take any non-zero value in GF(2^8). There are 255^{β} such vectors. In a good selection the first component satisfies the first condition, the second component satisfies the second condition and so on. There are 127^{β} such vectors. This gives rise to a hypergeometric distribution $H(\mathrm{N_t}; n, m, N)$ with the following parameters:

Table 14.2. Mean of the number of trails for a differential given α and β

α, β	1	2	3	4
1	63.25	31.50	15.69	7.81
2	31.38	15.63	7.78	3.88
3	15.44	7.69	3.83	1.91
4	7.47	3.72	1.85	0.92

Table 14.3. Variance of the number of trails for a differential given α and β

α, β	1	2	3	4
1	16.00	15.89	10.86	6.38
2	11.91	9.85	6.11	3.40
3	6.83	5.33	3.19	1.73
4	3.54	2.70	1.59	0.85

- Number of ways for a good selection $n = 127^{\beta}$.
- Number of ways for a bad selection $m = 255^{\beta} - 127^{\beta}$.
- Sample size N: $2^{8-\alpha} - 1$.

Filling in these values of n, m and N in the expressions for the mean yields (14.11) and in the expressions for the variance yields (14.12). □

The numerical values computed with these formulas are given in Table 14.2 and Table 14.3. We have conducted a large number of experiments that confirm the mean and variance predicted by (14.11) and (14.12) for any combination of α and β.

14.4.3 Experimental Verification

We have experimentally verified the distributions of the number of trails per differential for all 16 combinations of α and β. For the combination of (α, β) equal to $(1, 1)$, $(2, 1)$, $(3, 1)$, $(4, 1)$ and $(1, 2)$ we were able to do this exhaustively, covering all possible cases. As a side result we found for these values of (α, β) the minimum and maximum values for the number of trails per differential, which are listed in Table 14.4.

For the other values of (α, β), the number of combinations becomes too large to compute exhaustively. Still, our sampling experiments confirm the shape predicted by formulas (14.11) and (14.12). As α and β grow, the mean and variance of the distributions shrink. Clearly, the majority of differentials with five active S-boxes and $\alpha = 1$ and $\beta = 1$ have more trails than any differential with five active S-boxes where $\alpha + \beta$ has a higher value. Figure 14.2 depicts the four distributions for $\beta = 1$ on a logarithmic scale. The distributions appear as slightly skewed parabolas, the typical shape of hypergeometric distributions.

Table 14.4. Minimum and maximum number of trails in differentials with five active S-boxes given (α, β)

(α, β)	minimum	maximum
$(1,1)$	48	82
$(2,1)$	14	48
$(3,1)$	3	29
$(4,1)$	0	15
$(1,2)$	10	56

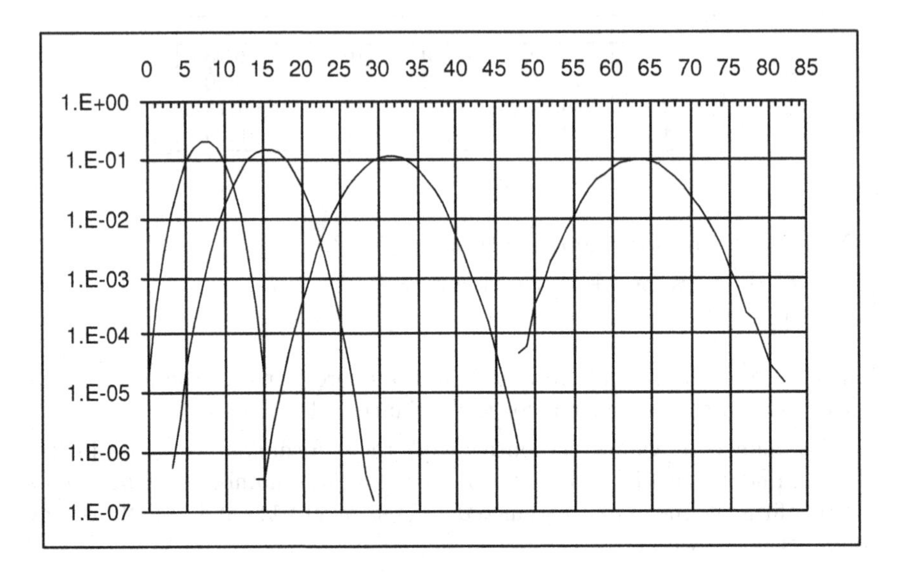

Fig. 14.2. Distributions of the number of trails per differential for $\beta = 1$ and for α ranging from 4 (leftmost) to 1 (rightmost)

14.5 EDP of a Bundle

The distributions for the number of trail cores in a bundle that extend to trails in a differential can be converted to distributions of the EDP contribution of a bundle to that differential by taking into account the EDP of the trails. The EDP of a trail is the product of the DP values of its active S-box differentials. If we apply Sect. 14.1 to the Rijndael S-box, we see that for an S-box differential with given input (output) difference, there are 126 output (input) differences with DP $= 2^{-7}$ and a single output (input) difference with DP $= 2^{-6} = 2 \times 2^{-7}$. We call the latter a *double differential.* It follows that the EDP of a trail is $2^m 2^{-7s}$ with s the number of active S-boxes and m the number of double differentials. One could say that the presence of m double differentials multiplies the EDP of the trail by a factor 2^m.

A trail core $\gamma[\mathbf{b}, \mathbf{d}]$ has a double differential in the ith S-box of the first round if and only if

$$\gamma b_i = L({a_i}^{-1}) \Leftrightarrow \gamma = (b_i)^{-1} L({a_i}^{-1}). \tag{14.13}$$

The condition for a double differential in the second round is:

$$\gamma d_j = L_i(e_j)^{-1} \Leftrightarrow \gamma = (d_j L_i(e_j))^{-1}. \tag{14.14}$$

Hence each double differential occurs in exactly one trail core of the bundle. Two observations can be made here.

14.5.1 Multiple Solutions

The EDP of a trail core $\gamma[\mathbf{b}, \mathbf{d}]$ increases with the number of equations in (14.13) and (14.14) that it satisfies, provided it extends to a trail. Consider for example a differential with five active S-boxes. There are seven different cases, with the following two as extremes:

'Poker': the double differentials are all in the same trail core,

'No Pair': the double differentials occur in five different trail cores.

The other five cases are 'One Pair', 'Two Pairs', 'Three of a Kind', 'Full House' and 'Four of a Kind'. The occurrence of these cases is related to the values of α and β. The number of different solutions for (14.14) equals the number of different elements in V_a. If α is 1 or 4, this number is equal to α. If α is 2 or 3 and the number of active S-boxes in e is higher than α, the number of solutions can also be $\alpha + 1$. The number of solutions for (14.14) usually equals β, but it can also be smaller. It follows that the double differentials tend to cluster together if α and β are small.

In any case, for a given input difference $\mathbf{a}$ there can be at most one output difference $\mathbf{e}$ with all double differentials in the same trail.

14.5.2 Occurrence in Trails

The solutions of (14.13) and (14.14) still have to satisfy the remaining sharp conditions and blurred conditions in order for the trail core to extend to a trail. Clearly, the expected number of trail cores satisfying the remaining conditions decreases when there are more conditions, i.e. when α and β increase. A 'Poker' trail core, i.e. one in which the differentials over all active S-boxes are double, always extends to a trail.

14.5.3 How L_i Makes a Difference

If we remove L_i from the S-box, the set of blurred conditions is replaced by a second set of sharp conditions. The number of trails in a bundle is then given by $2^{8-\alpha} - 1$, with $1 \leq \alpha < 8$. The maximum EDP occurs for differentials with five active S-boxes and $\alpha = 1$. There are 56×255 such differentials in the super box. For these, the double differentials are in the same trail and hence the EDP is equal to $2^5 \times 2^{-35} + 126 \times 2^{-35} = 19.75 \times 2^{-32}$, where for Rijndael this is 13.25×2^{-32} [80].

14.6 EDP of a Differential

Differentials with five active S-boxes contain only a single bundle, hence they are covered by the previous sections. For differentials with more active S-boxes, there are more bundles. Given a differential $(\mathbf{a}, \mathbf{e})$, we can compute for each of its bundles the value of (α, β). With α and β we can estimate the number of trails in a bundle by the mean and variance of its distribution. The mean number of trails in a differential is the sum of the mean number of trails in these bundles. For the variance of the number of trails, the sum of the variances in the bundles gives a good idea.

The values of the differences $\mathbf{a}$ and $\mathbf{e}$ determine the distribution of α and β over the different bundles in the differential $(\mathbf{a}, \mathbf{e})$. As the number of active S-boxes grows, the analysis becomes more and more involved. Therefore we start with an example.

14.6.1 Differentials with Activity Pattern (1110; 1110)

There are in total 251 bundles with activity pattern (1110; 1110). The distribution of α over the 251 bundles in $(\mathbf{a}, \mathbf{e})$ is completely determined by $\mathbf{e}$, or more specifically, by the couple $(L_i(e_1)/L_i(e_0), L_i(e_2)/L_i(e_0))$. Table 14.5 lists the seven distributions that are possible and gives for each of them the number of output differences $\mathbf{e}$ for which they occur.

The distribution of β depends on the values of a_0, a_1 and a_2. If they are three different values, then β is equal to 3 for all bundles. For this case, Table 14.5 gives the theoretical mean and standard deviation of the number of trails (assuming independence between the bundles). If two of the values a_0, a_1 and a_2 are equal, then β will be 2 for at most one bundle and 3 for all other bundles. If they are all three equal, then either β will be 2 for at most three bundles, or β will be 1 for at most one bundle and 3 for all other bundles.

In principle, the distributions for α and β combine to form a two-dimensional distribution. In the worst case, the small values of β occur in

Table 14.5. Distribution of α for differentials with activity pattern (1110; 1110)

α distribution			# couples $(L_i(e_1)/L_i(e_0),$ $L_i(e_2)/L_i(e_0))$	mean	standard deviation	
$\alpha = 3$	$\alpha = 2$	$\alpha = 1$			theory	experim.
250	1	0	21	965.2	28.42	25.65
249	2	0	1501	969.1	28.47	25.14
248	3	0	31170	973.1	28.53	25.15
247	4	0	2175	977.0	28.58	25.16
246	5	0	29907	981.0	28.63	25.23
250	0	1	3	973.1	28.42	23.28
249	1	1	248	977.0	28.47	25.01

bundles with a small value of α. All in all, there are only few bundles where β is smaller than 3, hence we can approximate by working with $\beta = 3$ for all bundles.

We have experimentally verified this theory by computing the number of trails for a large set of differentials with six, seven and eight active S-boxes. The measured mean values coincide with the theoretically predicted values. The measured standard deviations, also listed in Table 14.5, are systematically smaller than the theoretical ones, implying that the numbers of trails in the bundles of a differential are not independent.

14.6.2 A Bound on the Multiplicity

In Sect. 14.2.2 we have shown that the bundles with activity pattern (1110; 1110) can be enumerated by $\mathbf{b} = [1, z, 3 + z, 0]^{\mathrm{T}}$ and $\mathbf{d} = [1 + 2z, 4 + z, 7 + 3z, 0]^{\mathrm{T}}$ with z different from 0, 3, 1/2, 4 and 7/3.

Lemma 14.6.1. *If two double differentials occur in the same trail core of one bundle with activity pattern* (1110; 1110), *then they occur in different trail cores for the* 250 *other bundles with the same activity pattern.*

Proof. Assume we have a bundle where the double differentials in the first and the second S-box of the second round occur in the same trail core. Then we have from (14.14):

$$((1 + 2z)L_i(e_1))^{-1} = ((4 + 7z)L_i(e_2))^{-1} .$$

This equation is linear in z and has at most one solution. Hence the double differentials cannot be in the same trail core for any other bundle. The same holds for any other pair of active S-box positions. □

The expected contribution of the double differentials to the EDP of a differential is maximum when there is a bundle in which they are all six in the same trail. This trail contributes 64×2^{-42} to the EDP of the differential.

Lemma 14.6.1 implies that in the remaining 250 bundles, there can be no trails with more than one double differential. Hence each of these bundles will contribute at most $(\mathrm{N_t} + \min(6, \mathrm{N_t}))2^{-42}$ to the EDP of the differential. On the average the presence of the double differentials makes the contribution of these trails only rise from $\mathrm{N_t}2^{-42}$ for the hypothetical case where no double differentials exist to $(132/127)\mathrm{N_t}2^{-42}$.

We conclude that for this type of differential, the distribution of the EDP values is much more centered around its mean value than is the case for differentials with five active S-boxes. This is mainly due to the fact that the distribution of the EDP of the differential is the convolution of the distributions of many bundles. Moreover, Lemma 14.6.1 implies that the different bundles compensate for one another.

The same phenomenon can be observed for the other types of differentials with six active S-boxes. For differentials with seven or eight active S-boxes the average number of trails is even much higher and the EDP values much smaller. Furthermore, the individual trails all have very small EDP values. This all means that the EDP values of differentials with six or more active S-boxes have a very narrow distribution.

14.7 Differentials with the Maximum EDP Value

The maximum EDP value obtained in [80] occurs for exactly 12 differentials over the Rijndael super box. Due to the rotational symmetry of the Rijndael super box, they come in three sets, where the differentials in a set are just rotated versions of each other. It is no surprise that they are differentials with five active S-boxes, where the deviations from the average value 2^{-32} are largest. Moreover, they have $\alpha = 1$ and $\beta = 1$ for which the expected number of trails is the highest over all differentials with five active S-boxes, as is clear from Figure 14.2 in Sect. 14.4.3. The differentials are the following:

$$\left([x, \mathtt{0}, \mathtt{0}, \mathtt{0}]^{\mathrm{T}}, [L(y/2), L(y), L(y), L(y/3)]^{\mathrm{T}}\right),$$
$$\left([x, x, \mathtt{0}, \mathtt{0}]^{\mathrm{T}}, [L(y), L(y/3), \mathtt{0}, L(y/2)]^{\mathrm{T}}\right),$$
$$\left([x, x, x, \mathtt{0}]^{\mathrm{T}}, [\mathtt{0}, \mathtt{0}, L(y/2), L(y/3)]^{\mathrm{T}}\right),$$

with $x = \mathtt{75}$ and $y = \mathtt{41}$. For these differentials, the number of trails is 75: 74 trails with EDP 2^{-35} and one with EDP 2^{-30}, resulting in EDP value $2^{-30} + 74 \times 2^{-35} = 13.25 \times 2^{-32}$. Clearly all five double differentials are in the same trail. Note that there are differentials with five active S-boxes that have 82 trails (see Sect. 14.4.3) but these have a lower EDP value due to the fact that the double differentials are not in the same trail.

To prove the correctness of the maximum EDP value, [80] uses so-called 5-lists, a concept similar to, but different from, the bundles defined in this chapter. Both bundles and 5-lists group sets of 255 *b*-differences. Bundles with five active S-boxes correspond to the 5-lists of type 1. In bundles with more than five active S-boxes the ratios between the components of the trail cores are fixed, while in 5-lists of type 2, a number of components of the trail cores are fixed. Their goal is also different: the concept of 5-lists helps in efficiently finding bounds, while bundles help us to gain insight into the distribution of trails in differentials.

14.8 Conclusions

The Rijndael super box can be compared with an idealized keyed 32-bit map constructed as a family of 2^{32} randomly selected permutations (one permutation for each value of the key). In this idealized model, the distribution of the EDP over all differentials $(\mathbf{a}, \mathbf{b})$ with both $\mathbf{a}$ and $\mathbf{b}$ different from zero has a normal distribution with expected value 2^{-32} and standard deviation $2^{-47.5}$.

The Rijndael super box differentials deviate from the idealized model: differentials with four or fewer active S-boxes have EDP $= 0$, and differentials with five active S-boxes can have EDP values as large as 13.25×2^{-32} [80]. Our results on differentials with six active S-boxes indicate that for differentials with six or more active S-boxes the distribution of the EDP is very narrowly centered around 2^{-32}. Further analysis can lead to strict bounds.

Clearly, the linear transformation L_i in the Rijndael S-box does not influence the $\mathrm{DP}^{(\mathrm{s})}$ histograms of S-box differentials and the bounds on the EDP of trails. Our results explain how the presence of L_i influences the EDP of two-round differentials.

Bounds on the EDP of two-round differentials can be used to derive bounds on the EDP of four-round differentials [79]. Our results allow us to describe the full distribution of the EDP of two-round differentials. We expect that this information can be used to derive sharper bounds on the EDP of four-round differentials.

15. Plateau Trails

This chapter is based on our work in [50] and [51]. It has been reported before that the fixed-key probability of trails depends on the value of the key [14, 38, 6]. We define *plateau trails*, where the dependency on the value of the key is very structured. The fixed-key probability of these trails is either zero or 2^h, with h a value that depends only on the trail and not on the key. We show that for a large class of ciphers, all two-round trails and a fraction of the more-round trails are plateau trails. This fraction is very close to 100% for Rijndael. Our results show that the distribution of the key-dependent probability $\mathrm{DP}[\mathbf{k}]$ is not narrow and hence the widely made assumption that it can be approximated by the EDP is not justified.

Applying our results to Rijndael, we see that for almost all values of the key there are two-round trails with a fixed-key probability equal to $32/2^{32}$, while the EDP of two-round trails is at most $4/2^{32}$ (cf. Sect. 9.5) and the EDP of two-round *differentials* [93] is at most $13.25/2^{32}$ (cf. Chap. 14 and [80, 126, 127]).

15.1 Motivation

When we examine the resistance of Rijndael and related ciphers against differential cryptanalysis, we cannot ignore the difference between the fixed-key probability (DP) of trails and their average probability (EDP). For instance, for trails over four or more rounds, the EDP values are already below 2^{-150}, which is much smaller than the smallest possible non-zero DP-value $(2^{1-n_b} = 2^{-127})$. In this chapter we describe completely the distribution of DP of all trails over two rounds of Rijndael. We also give results for four and more rounds. The following example illustrates in a simple way the effects we want to examine for Rijndael.

Example 15.1.1. Consider the keyed map $E[\mathbf{k}]$, defined as

$$E[\mathbf{k}](\mathbf{x}) = \rho^{-1}(\mathbf{k} + \rho(\mathbf{x})), \tag{15.1}$$

where ρ is an arbitrary invertible transformation [59]. Since $E[\mathbf{0}]$ is the identity transformation, for all differences $\mathbf{a}$ the differential $(\mathbf{a}, \mathbf{a})$ over the map

J. Daemen, V. Rijmen, *The Design of Rijndael*, Information Security and Cryptography,
https://doi.org/10.1007/978-3-662-60769-5_15

E has fixed-key probability $\mathrm{DP}[\mathbf{0}](\mathbf{a}, \mathbf{a}) = 1$. This property holds whatever value $\mathrm{EDP}(\mathbf{a}, \mathbf{a})$ takes.

This example is contrived. In practice we do not expect $\mathrm{DP}[\mathbf{k}]$ to deviate this strongly from EDP. However, we observe effects that go in this direction. We found that for several ciphers, including Rijndael, $\mathrm{DP}[\mathbf{k}]$ has a distribution with a surprisingly rich structure.

15.2 Two-Round Plateau Trails

For a large class of ciphers, two-round trails have a $\mathrm{DP}[\mathbf{k}]$ that can take only two values.

15.2.1 Planar Differentials and Maps

As customary, we consider ordered pairs of inputs [26], but we denote them using curled braces '{}' in order to avoid confusion with differentials. Let $F_{(\mathbf{a},\mathbf{b})}$ denote the set containing the inputs x for which the pair $\{\mathbf{x}, \mathbf{x} + \mathbf{a}\}$ follows the differential $(\mathbf{a}, \mathbf{b})$ over an unkeyed map. Let $G_{(\mathbf{a},\mathbf{b})}$ denote the set containing the corresponding outputs. Similarly, let $F_{\mathsf{Q}}[\mathbf{k}]$ denote the set containing the inputs $\mathbf{x}$ for which the pair $\{\mathbf{x}, \mathbf{x}+\mathbf{a}\}$ is a pair that follows the trail Q over a keyed map. Let $G_{\mathsf{Q}}[\mathbf{k}]$ denote the set containing the corresponding outputs. We introduce the concept of planar differentials.

Definition 15.2.1. *A differential* $(\mathbf{a}, \mathbf{b})$ *is* planar *if* $F_{(\mathbf{a},\mathbf{b})}$ *and* $G_{(\mathbf{a},\mathbf{b})}$ *form affine subspaces:*

$$\begin{aligned} F_{(\mathbf{a},\mathbf{b})} &= \mathbf{u} + U_{(\mathbf{a},\mathbf{b})} \\ G_{(\mathbf{a},\mathbf{b})} &= \mathbf{v} + V_{(\mathbf{a},\mathbf{b})} \end{aligned}$$

with $U_{(\mathbf{a},\mathbf{b})}$ *and* $V_{(\mathbf{a},\mathbf{b})}$ *vector spaces,* $\mathbf{u}$ *any element in* $F_{(\mathbf{a},\mathbf{b})}$ *and* $\mathbf{v}$ *any element in* $G_{(\mathbf{a},\mathbf{b})}$.

If $F_{(\mathbf{a},\mathbf{b})}$ contains an element $\mathbf{x}$, then it also contains $\mathbf{x} + \mathbf{a}$. Hence if $F_{(\mathbf{a},\mathbf{b})}$ is not empty, then $\mathbf{a} \in U_{(\mathbf{a},\mathbf{b})}$. The number of elements in $F_{(\mathbf{a},\mathbf{b})}$ is $2^{\dim(U_{(\mathbf{a},\mathbf{b})})}$, so $\dim(U_{(\mathbf{a},\mathbf{b})}) = n_{\mathrm{b}} - \mathrm{w_r}(\mathbf{a}, \mathbf{b})$. Similarly, we have $\mathbf{b} \in V_{(\mathbf{a},\mathbf{b})}$ and $\dim(V_{(\mathbf{a},\mathbf{b})}) = n_{\mathrm{b}} - \mathrm{w_r}(\mathbf{a}, \mathbf{b})$. We can now prove the following lemmas.

Lemma 15.2.1. *A differential* $(\mathbf{a}, \mathbf{b})$ *that is followed by exactly two pairs is planar.*

Proof. Denote the pairs by $\{\mathbf{p}, \mathbf{p}+\mathbf{a}\}$, $\{\mathbf{p}+\mathbf{a}, \mathbf{p}\}$. The elements $\mathbf{p}$ and $\mathbf{p}+\mathbf{a}$ form an affine subspace of dimension 1 with offset $\mathbf{u} = \mathbf{p}$ and the basis of $U_{(\mathbf{a},\mathbf{b})}$ equal to $(\mathbf{a})$. A similar argument is valid for the elements of the pairs at the output. □

Lemma 15.2.2. *A differential* $(\mathbf{a}, \mathbf{b})$ *that is followed by exactly four pairs is planar.*

Proof. Denote the inputs of the pairs by $\mathbf{p}, \mathbf{p}+\mathbf{a}, \mathbf{q}$ and $\mathbf{q}+\mathbf{a}$. These four elements lie in an affine subspace of dimension 2 with offset $\mathbf{u} = \mathbf{p}$ and the basis of $U_{(\mathbf{a},\mathbf{b})}$ equal to $(\mathbf{a}, \mathbf{p}+\mathbf{q})$. A similar argument is valid for the elements of the pairs at the output. □

Lemma 15.2.3. *Any differential with* $\mathrm{DP} = 1$ *is a planar differential.*

Proof. $F_{(\mathbf{a},\mathbf{b})}$ and $G_{(\mathbf{a},\mathbf{b})}$ form the complete input space and output space respectively. □

Examples of differentials with $\mathrm{DP} = 1$ are the trivial differential $(0, 0)$ and differentials over linear maps. If $\mathrm{DP}(\mathbf{a}, \mathbf{b}) = 2^{t-n_\mathrm{b}}$, with $t \notin \{1, 2, n_\mathrm{b}\}$, the differential may or may not be planar.

Definition 15.2.2. *A map is planar if all differentials over it are planar.*

Any map for which all non-trivial differentials have $\mathrm{DP}(\mathbf{a}, \mathbf{b}) \leq 2^{2-n_\mathrm{b}}$ is planar. Such maps are called differentially 4-uniform [120]. Now we give two lemmas on planar differentials over composed maps. The first lemma applies for instance to a substitution step in a block cipher, consisting of the parallel application of some S-boxes.

Lemma 15.2.4. *Let* $\mathbf{y} = \alpha(\mathbf{x})$ *be a map consisting of a set of parallel maps* $\mathbf{y}_i = \alpha_i(\mathbf{x}_i)$ *with* $\mathbf{x} = (\mathbf{x}_0, \mathbf{x}_1, \ldots, \mathbf{x}_t)$ *and* $\mathbf{y} = (\mathbf{y}_0, \mathbf{y}_1, \ldots, \mathbf{y}_t)$*. A differential* $(\mathbf{a}, \mathbf{b})$ *for which the differentials* $(\mathbf{a}_i, \mathbf{b}_i)$ *are planar, is planar.*

We have

$$U_{(\mathbf{a},\mathbf{b})} = U_{(\mathbf{a}_0,\mathbf{b}_0)} \times U_{(\mathbf{a}_1,\mathbf{b}_1)} \times \cdots \times U_{(\mathbf{a}_t,\mathbf{b}_t)}$$
$$V_{(\mathbf{a},\mathbf{b})} = V_{(\mathbf{a}_0,\mathbf{b}_0)} \times V_{(\mathbf{a}_1,\mathbf{b}_1)} \times \cdots \times V_{(\mathbf{a}_t,\mathbf{b}_t)}$$

with $\times$ denoting the direct product. The following lemma applies to a sequence of maps.

Lemma 15.2.5. *If* $(\mathbf{a}, \mathbf{b})$ *is a planar differential of* α*, then for any pair of affine maps* L_1 *and* L_2 *with* L_1 *invertible, the differential* $(L_1(\mathbf{a}), L_2(\mathbf{b}))$ *is planar over* $L_2 \circ \alpha \circ L_1^{-1}$.

Examples of ciphers in which single-round differentials are planar are Rijndael, but also 3-Way [43], SHARK (Sect. 11.2), Square (Sect. 11.3), Camellia [7], Serpent [4] and Noekeon [46]. Some other popular maps that are planar are the bitwise majority function $f(\mathbf{x}, \mathbf{y}, \mathbf{z}) = \mathbf{xy} + \mathbf{xz} + \mathbf{yz}$ and the bitwise 'if' function $g(\mathbf{x}, \mathbf{y}, \mathbf{z}) = \mathbf{xy} + (\neg\mathbf{x})\mathbf{z}$.

15.2.2 Plateau Trails

Similar to the concept of plateaued functions [150], for which the Walsh spectrum takes only two values (in absolute value), we introduce here *plateau trails* as trails for which the $\mathrm{DP}[\mathbf{k}]$ takes only two values (where one value is always zero). The *height* of a plateau trail determines how high the non-zero $\mathrm{DP}[\mathbf{k}]$ value of the plateau trail is.

Definition 15.2.3. *A trail* Q *is a* plateau trail *with height* height(Q) *if and only if the following holds:*

1. *For a fraction* $2^{n_\mathrm{b}-(\mathrm{w_r}(\mathrm{Q})+\mathrm{height}(\mathrm{Q}))}$ *of the keys,* $\mathrm{DP}[\mathbf{k}](\mathrm{Q}) = 2^{\mathrm{height}(\mathrm{Q})-n_\mathrm{b}}$, *and*
2. *For all other keys,* $\mathrm{DP}[\mathbf{k}](\mathrm{Q}) = 0$.

The height of a plateau trail can be bounded as follows. Firstly, $\mathrm{height}(\mathrm{Q}) \leq n_\mathrm{b}$. Secondly, height(Q) is maximal when all but one key have DP equal to zero. Denoting the number of keys by 2^{n_κ}, we obtain that in this case the EDP equals 2^{-n_κ} times the non-zero DP value. Taking the logarithm, we obtain $-\mathrm{w_r}(\mathrm{Q}) = -n_\kappa + \mathrm{height}(\mathrm{Q}) - n_\mathrm{b}$. Hence, we have in all cases $\mathrm{height}(\mathrm{Q}) \leq n_\kappa + n_\mathrm{b} - \mathrm{w_r}(\mathrm{Q})$. We can now prove the following result on an n_b-bit map consisting of two steps and an addition with an n_b-bit key in between (hence $n_\kappa = n_\mathrm{b}$).

Theorem 15.2.1 (Two-Round Plateau Trail Theorem). *A trail* $\mathrm{Q} = (\mathbf{a}, \mathbf{b}, \mathbf{c})$ *over a map consisting of two steps with a key addition in between, and in which the differentials* $(\mathbf{a}, \mathbf{b})$ *and* $(\mathbf{b}, \mathbf{c})$ *are planar is a plateau trail with* $\mathrm{height}(\mathrm{Q}) = \dim(V_{(\mathbf{a},\mathbf{b})} \cap U_{(\mathbf{b},\mathbf{c})})$.

Proof. The proof is based on geometrical arguments. For pairs that follow the trail the values at the output of the first step are in $G_{(\mathbf{a},\mathbf{b})}$. The values at the input of the second step are in $F_{(\mathbf{b},\mathbf{c})}$, or equivalently, the values at the output of the first step are in $\mathbf{k} + F_{(\mathbf{b},\mathbf{c})}$. It follows that the values at the output of the first step are in

$$H = G_{(\mathbf{a},\mathbf{b})} \cap (\mathbf{k} + F_{(\mathbf{b},\mathbf{c})}) \ .$$

Since both differentials are planar, there exist offsets $\mathbf{u}, \mathbf{v}$ such that

$$H = (\mathbf{v} + V_{(\mathbf{a},\mathbf{b})}) \cap (\mathbf{k} + \mathbf{u} + U_{(\mathbf{b},\mathbf{c})}) \ ,$$

with $V_{(\mathbf{a},\mathbf{b})}$ and $U_{(\mathbf{b},\mathbf{c})}$ vector spaces. We start by deriving the condition that H is non-empty. First we translate the affine subspaces in the equation by the vector $\mathbf{v}$:

$$\mathbf{v} + H = V_{(\mathbf{a},\mathbf{b})} \cap (\mathbf{k} + \mathbf{u} + \mathbf{v} + U_{(\mathbf{b},\mathbf{c})}) \ .$$

$\mathbf{v}+H$ is a translated version of H and has the same number of elements. Now $\mathbf{v}+H$ is non-empty iff there is a vector $\mathbf{x} \in V_{(\mathbf{a},\mathbf{b})}$ and a vector $\mathbf{y} \in U_{(\mathbf{b},\mathbf{c})}$ such that $\mathbf{x} = \mathbf{k}+\mathbf{u}+\mathbf{v}+\mathbf{y}$ or formally, iff

$$\exists \mathbf{x} \in V_{(\mathbf{a},\mathbf{b})}, \mathbf{y} \in U_{(\mathbf{b},\mathbf{c})} : \mathbf{k}+\mathbf{u}+\mathbf{v} = \mathbf{x}+\mathbf{y} .$$

This is equivalent to saying that $(\mathbf{k}+\mathbf{u}+\mathbf{v}) \in (V_{(\mathbf{a},\mathbf{b})} + U_{(\mathbf{b},\mathbf{c})})$. If we denote $\mathbf{u}+\mathbf{v}+(V_{(\mathbf{a},\mathbf{b})} + U_{(\mathbf{b},\mathbf{c})})$ by K_{Q}, this corresponds to saying that H is non-empty iff $\mathbf{k} \in K_{\mathrm{Q}}$. Consider now the case that H is non-empty and let $\mathbf{w}$ be an element of H. Clearly, $\mathbf{w}$ is an element of both $G_{(\mathbf{a},\mathbf{b})}$ and $\mathbf{k}+F_{(\mathbf{b},\mathbf{c})}$. It follows that $\mathbf{w}+\mathbf{k}+F_{(\mathbf{b},\mathbf{c})} = U_{(\mathbf{b},\mathbf{c})}$ and hence $G_{(\mathbf{a},\mathbf{b})} = \mathbf{w}+V_{(\mathbf{a},\mathbf{b})}$ and $\mathbf{k}+F_{(\mathbf{b},\mathbf{c})} = \mathbf{w}+U_{(\mathbf{b},\mathbf{c})}$. We have

$$H = (\mathbf{w}+V_{(\mathbf{a},\mathbf{b})}) \cap (\mathbf{w}+U_{(\mathbf{b},\mathbf{c})}) .$$

Translation by $\mathbf{w}$ yields

$$\mathbf{w}+H = V_{(\mathbf{a},\mathbf{b})} \cap U_{(\mathbf{b},\mathbf{c})} .$$

Let $W_{\mathrm{Q}} = V_{(\mathbf{a},\mathbf{b})} \cap U_{(\mathbf{b},\mathbf{c})}$. The number of pairs in H is $2^{\dim(W_{\mathrm{Q}})}$ if $\mathbf{k} \in K_{\mathrm{Q}}$ and zero otherwise. The number of elements in K_{Q} is determined by the dimension of $V_{(\mathbf{a},\mathbf{b})} + U_{(\mathbf{b},\mathbf{c})}$. We use the subspace dimension theorem: $\dim(U)+\dim(V) = \dim(U+V)+\dim(U \cap V)$. This gives

$$\begin{aligned}\dim(V_{(\mathbf{a},\mathbf{b})} + U_{(\mathbf{b},\mathbf{c})}) &= \dim(V_{(\mathbf{a},\mathbf{b})}) + \dim(U_{(\mathbf{b},\mathbf{c})}) - \dim(V_{(\mathbf{a},\mathbf{b})} \cap U_{(\mathbf{b},\mathbf{c})}) \\ &= (\dim(V_{(\mathbf{a},\mathbf{b})}) + \dim(U_{(\mathbf{b},\mathbf{c})})) - \dim(V_{(\mathbf{a},\mathbf{b})} \cap U_{(\mathbf{b},\mathbf{c})}) \\ &= (2n_{\mathrm{b}} - \mathrm{w_r}(\mathrm{Q})) - \dim(W_{\mathrm{Q}})\end{aligned}$$

If we now denote $\mathrm{height}(\mathrm{Q}) = \dim(W_{\mathrm{Q}})$, we have $\mathrm{DP}(\mathrm{Q}) = 2^{\mathrm{height}(\mathrm{Q})-n_{\mathrm{b}}}$ for $2^{2n_{\mathrm{b}}-\mathrm{w_r}(\mathrm{Q})-\mathrm{height}(\mathrm{Q})}$ keys on the total number of $2^{n_{\mathrm{b}}}$ keys, and zero for all other keys. □

This theorem is valid for all ciphers in which single-round differentials are planar and round keys are applied with XOR. This includes all ciphers mentioned in Sect. 15.2.1.

Like any other trail, a plateau trail has $\mathrm{EDP}(\mathrm{Q}) = \mathrm{DP}(\mathbf{a},\mathbf{b})\mathrm{DP}(\mathbf{b},\mathbf{c}) = 2^{-\mathrm{w_r}(\mathrm{Q})}$. Only if $\mathrm{height}(\mathrm{Q}) = n_{\mathrm{b}} - \mathrm{w_r}(\mathrm{Q})$, it holds that $\mathrm{DP}[\mathbf{k}](\mathrm{Q}) = \mathrm{EDP}(\mathrm{Q})$ for all keys. This can only be the case for trails with $\mathrm{w_r}(\mathrm{Q}) < n_{\mathrm{b}}$.

15.2.3 Plateau Trails in Super Boxes

Let Q be a plateau trail in a super box where the linear map has the maximal branch number, i.e. $n_{\mathrm{t}}+1$. Then we can prove the following upper bound on height(Q).

Theorem 15.2.2. *Let* Q *be a plateau trail over a super box. Let the sets* γ_j *denote all possible selections of* n_t *S-boxes from the super box. Then*

$$\mathrm{height}(\mathrm{Q}) \leq n_\mathrm{b} - \max_j \left(\sum_{i \in \gamma_j} \mathrm{w_r}(\mathbf{x}_i, \mathbf{y}_i) \right) ,$$

where $(\mathbf{x}_i, \mathbf{y}_i)$ *denotes a differential over an S-box.*

Proof. Since $\mathrm{height}(\mathrm{Q}) = \dim(V_{(\mathbf{a},\mathbf{b})} \cap U_{(\mathbf{d},\mathbf{e})})$, we have $\mathrm{height}(\mathrm{Q}) \leq \dim(V_{(\mathbf{a},\mathbf{b})})$. Taking for γ_1 the selection of the n_t S-boxes of the first step, we have from Definition 15.2.1:

$$\dim(V_{(\mathbf{a},\mathbf{b})}) = n_\mathrm{b} - \mathrm{w_r}(\mathbf{a}, \mathbf{b}) = n_\mathrm{b} - \sum_{i=1}^{n_t} \mathrm{w_r}(\mathbf{a}_i, \mathbf{b}_i) \ . \tag{15.2}$$

Secondly, observe that the vectors $(\mathbf{b}, \mathbf{d})^\mathrm{T} = (\mathbf{b}, \mathrm{M}(\mathbf{b}))^\mathrm{T}$ are code vectors of a linear code over $\mathrm{GF}(2^{\mathrm{n_s}})$ with length $2n_t$ and dimension n_t. Since the branch number of M equals $n_t + 1$, the minimal distance between code vectors is $n_t + 1$, hence the code is MDS. Any n_t symbols of the codewords may be taken as message symbols [100]. Hence, we can always pick n_t out of the symbols, consider them as the message symbols (input) and compute the check symbols (output) from them. This leads to the definition of alternative vector spaces V' that bound height(Q) as before. The new definition of input and output leads to a new definition of input difference $\mathbf{a}'$, output difference $\mathbf{e}'$ and intermediate differences $\mathbf{b}', \mathbf{d}'$. This in turn leads to a definition of a new vector space $V' = V_{\mathbf{a}',\mathbf{b}'}$, which bounds height(Q) in the same way as in (15.2). □

Hence, given a trail over a super box, one chooses the n_t S-box differentials with the highest weight and adds them. The bound for the height is n_b minus this weight.

Theorem 15.2.3. *Consider a super box with* $n_t = 4$. *Then we have the following bounds on the height of trails where all active S-boxes have weight* $\mathrm{n_s} - 1$*:*

$$\begin{aligned} \textit{5 active boxes:}\ & \mathrm{height}(\mathrm{Q}) \leq 3 \\ \textit{6 active boxes:}\ & \mathrm{height}(\mathrm{Q}) \leq 2 \\ \textit{7 or 8 active boxes:}\ & \mathrm{height}(\mathrm{Q}) = 1. \end{aligned}$$

The theorem can be proven by going through all possibilities and counting. There is also a link to the existence of codes: a trail with i active boxes and height h exists only if there is a binary linear code with length i, distance $i - 3$ and dimension h that contains the vector $(1, 1, 1, \ldots, 1)$.

15.3 Plateau Trails over More Than Two Rounds

In this section we derive conditions for trails over more than two rounds to be plateau trails. For ciphers with a Rijndael-like block structure (Rijndael, Square, SHARK, ...), the results of this section cover the majority of the trails. For ciphers without the block structure (3-Way, Serpent, Noekeon, ...), only a small fraction of the trails is covered.

We can extend the *planar* property of differentials to trails.

Definition 15.3.1. *A trail* Q *is* input-planar *(respectively output-planar) if for all values of the key it holds that* $F_{\mathsf{Q}}[\mathbf{k}]$ *(respectively* $G_{\mathsf{Q}}[\mathbf{k}]$*) is either* $\emptyset$ *or an affine subspace.*

Lemma 15.3.1. *Plateau trails with height 1 or 2 are both input-planar and output-planar.*

Proof. Let Q be a plateau trail. Then $F_{\mathsf{Q}}[\mathbf{k}] = G_{\mathsf{Q}}[\mathbf{k}] = \emptyset$ or $\#F_{\mathsf{Q}}[\mathbf{k}] = \#G_{\mathsf{Q}}[\mathbf{k}] = 2^{\text{height}(\mathsf{Q})}$. If height(Q) is 1 or 2, then the proofs of Lemma 15.2.1 and Lemma 15.2.2 can be extended to the case of trails. □

For a cipher with S-boxes that are differentially 4-uniform or differentially 2-uniform we have the following result.

Theorem 15.3.1 (Planar Trail Extension Theorem). *Let* $\mathsf{Q} = (\mathsf{Q}_1, \mathbf{q}_n)$ *be a trail composed of an output-planar trail* $\mathsf{Q}_1 = (\mathbf{q}_0, \ldots, \mathbf{q}_{n-1})$*, followed by a (one-step) differential* $(\mathbf{q}_{n-1}, \mathbf{q}_n)$*. If all S-boxes in* $(\mathbf{q}_{n-1}, \mathbf{q}_n)$ *are active, then* Q *is output-planar.*

Proof. In this proof, we drop $[\mathbf{k}]$ from the notation, which we illustrate in Fig. 15.1. If we look at the output of Q_1, the elements of the pairs that follow Q_1 form the affine subspace V_{Q_1}. Since the differential $(\mathbf{q}_{n-1}, \mathbf{q}_n)$ is planar, the elements of the pairs that follow Q form an affine subspace $H = V_{\mathsf{Q}_1} \cap U_{(\mathbf{q}_{n-1}, \mathbf{q}_n)}$. We denote by H_i, $0 \leq i < n_t$, the projection of H onto the coordinate i: H_i contains the inputs of one S-box in the last step, for the pairs that follow Q. Since H_i is a projection of an affine subspace, it is also an affine subspace. We denote by O_i the corresponding outputs of the S-box. The set O_i is nothing else than the projection of the set G_{Q} onto the coordinate i.

Since the S-boxes are differentially 4-uniform (or 2-uniform), H_i and O_i contain one, two or four elements and O_i is also an affine subspace. Using Lemma 15.2.4, we conclude that G_{Q} is also an affine subspace. □

Note that this theorem also holds for a trail composed of a planar differential in which all S-boxes are active followed by an input-planar trail. Planar plateau trails can be composed to plateau trails. Theorem 15.2.1 can easily be extended.

$$\begin{array}{ccccc} \mathsf{Q}_1 & & & (\mathsf{q}_{n-1}, \mathsf{q}_n) & \\ F_{\mathsf{Q}_1} \longrightarrow G_{\mathsf{Q}_1} & F_{(\mathsf{q}_{n-1},\mathsf{q}_n)} & \longrightarrow & G_{(\mathsf{q}_{n-1},\mathsf{q}_n)} \\ & H = G_{\mathsf{Q}_1} \cap F_{(\mathsf{q}_{n-1},\mathsf{q}_n)} & \longrightarrow & G_{\mathsf{Q}} \subseteq G_{(\mathsf{q}_{n-1},\mathsf{q}_n)} \\ & \text{proj.} \downarrow & & \downarrow \text{proj.} \\ & H_i & \longrightarrow & O_i \end{array}$$

Fig. 15.1. Notation used in the proof of Theorem 15.3.1

Theorem 15.3.2 (Planar Trail Composition Theorem).
Let $\mathsf{Q} = (\mathsf{q}_0, \mathsf{q}_1, \ldots, \mathsf{q}_{i-1}, \mathsf{q}_i, \ldots, \mathsf{q}_n)$ *be a trail over a map consisting of* n *steps with a key addition as* i*th step. If the trails* $\mathsf{Q}_1 = (\mathsf{q}_1, \mathsf{q}_2, \ldots, \mathsf{q}_{i-1})$ *and* $Q_2 = (\mathsf{q}_i, \ldots, \mathsf{q}_n)$ *are plateau trails with* Q_1 *output-planar and* Q_2 *input-planar, then* Q *is a plateau trail with* $\text{height}(\mathsf{Q}) = \dim(V_{\mathsf{Q}_1} \cap U_{\mathsf{Q}_2})$.

The proof is similar to the proof of Theorem 15.2.1. From this theorem follows this corollary.

Corollary 15.3.1. *Let* $\mathsf{Q} = (\mathsf{Q}_1, \mathsf{q}_n)$ *be a trail composed of a trail* $\mathsf{Q}_1 = (\mathsf{q}_0, \ldots, \mathsf{q}_{n-1})$*, followed by a (one-round) planar differential* $(\mathsf{q}_{n-1}, \mathsf{q}_n)$*. If* Q_1 *is a plateau trail with height 1 or 2, then* Q *is a plateau trail.*

This is a special case of Theorem 15.3.2: according to Lemma 15.3.1, Q_1 is output-planar and the differential can be seen as a (one-round) input-planar plateau trail.

This extension of a plateau trail by a single round can be performed iteratively: an r-round plateau trail with height 1 or 2 can be extended by an arbitrary number of rounds, as long as the appended differentials are planar.

Plateau trails with height larger than 2 are in general neither input-planar nor output-planar. For instance, assume that we have a plateau trail Q consisting of an output-planar trail Q_1 followed by an input-planar trail Q_2 with $\mathsf{Q} = (\mathsf{Q}_1, \mathsf{Q}_2)$. Then it follows from the proof of Theorem 15.2.1 that the elements of the pairs that follow Q form an affine subspace H at the junction of Q_1 and Q_2. The set H is transformed into the set $G_{\mathsf{Q}}[\mathbf{k}]$ at the output. $G_{\mathsf{Q}}[\mathbf{k}]$ is a subset of $G_{\mathsf{Q}_2}[\mathbf{k}]$. $G_{\mathsf{Q}_2}[\mathbf{k}]$ is an affine subspace, but $G_{\mathsf{Q}}[\mathbf{k}]$ in general is not an affine subspace.

15.4 Further Observations

15.4.1 Effect of the Key Schedule

When we consider more than two rounds, the round keys are typically not independent. They are related by means of the key schedule. The key schedule doesn't change whether a trail is a plateau trail. The only visible effect of the key schedule is on the size of the set K_{Q}, which contains the keys for which $\mathrm{DP}[\mathbf{k}](\mathsf{Q}) > 0$ (see the proof of Theorem 15.2.1). The key schedule determines which values are possible for the expanded key. If relatively many of the expanded keys are in K_{Q}, then the average of $\mathrm{DP}[\mathbf{k}](\mathsf{Q})$ will be larger than $\mathrm{EDP}(\mathsf{Q})$.

15.4.2 Impact on the DP of Differentials

The dependence of the DP of trails on the key value means that the DP of differentials also depends on the key. Assume we have a differential for which all trails are plateau trails. If we denote the trails that contribute to a differential $(\mathbf{a}, \mathbf{b})$ by Q_i we have

$$\mathrm{DP}[\mathbf{k}](\mathbf{a}, \mathbf{b}) = 2^{-n_b} \sum_{i|\mathbf{k} \in K_{\mathsf{Q}_i}} 2^{\mathrm{height}(\mathsf{Q}_i)} . \qquad (15.3)$$

Hence this value varies per key $\mathbf{k}$ depending on the number of affine subspaces K_{Q_i} it is in.

15.5 Two-Round Trails in Rijndael

We now apply the results of the previous sections to Rijndael. We also compute the heights of all two-round trails for Rijndael and for a simplified variant. Since the sequence of two rounds of Square is equivalent to the sequence of two rounds of Rijndael, the distribution of the heights of trails is the same in both cases.

15.5.1 Trails in the Rijndael Super Box

A differential trail through the Rijndael super box (cf. Sect. 3.4.5 and Sect. 14.2) consists of a sequence of five differences: $\mathbf{a}$, $\mathbf{b}$, $\mathbf{c}$, $\mathbf{d}$ and $\mathbf{e}$. In a trail through the Rijndael super box, we always have $\mathbf{c} = \mathbf{d}$ (so we omit $\mathbf{c}$) and $\mathbf{d} = \mathrm{M_c}\mathbf{b}$. We denote these trails by $(\mathbf{a}, \mathbf{b}, \mathbf{d}, \mathbf{e})$.

The Rijndael super box satisfies the criteria of Theorem 15.2.1 and hence all trails Q in the Rijndael super box are plateau trails. $\mathrm{DP}[\mathbf{k}](\mathsf{Q})$ can be described by defining $W = V_{(\mathbf{a},\mathbf{d})} \cap U_{(\mathbf{d},\mathbf{e})}$ and $V_{(\mathbf{a},\mathbf{d})} = \mathrm{M_c}(V_{(\mathbf{a},\mathbf{b})})$, where $\mathrm{M_c}(V) = \{\mathrm{M_c}\mathbf{v} | \mathbf{v} \in V\}$.

Applying Theorem 15.2.2 to Rijndael results in the following bounds. It holds always that $\mathrm{height}(\mathsf{Q}) \leq 8$. If all active S-boxes have weight 7, then $\mathrm{height}(\mathsf{Q}) \leq 4$. Only if at most three S-boxes have weight 7, $\mathrm{height}(\mathsf{Q})$ can be larger than 4. Theorem 15.2.3 further decreases the bounds when all S-boxes have weight 7.

We have determined the weight and height of all trails over the Rijndael super box. An overview of the results is given in Table 15.1. Because of the large number of trails, the entries in the table were not computed by checking the height of each trail individually. We used the following observations to speed up the computations. Let $\mathsf{Q} = (\mathbf{a}, \mathbf{b}, \mathbf{d}, \mathbf{e})$ be a trail over the super box.

Lemma 15.5.1. *For all non-zero* $\mathbf{a}, \mathbf{b}$: $\mathbf{a} \in U_{(\mathbf{a},\mathbf{b})}$ *and* $\mathbf{b} \in V_{(\mathbf{a},\mathbf{b})}$.

Lemma 15.5.2. *For differentials with weight 7 over a single S-box* $U_{(\mathbf{a},\mathbf{b})} = \{\mathbf{0}, \mathbf{a}\}$ *and* $V_{(\mathbf{a},\mathbf{b})} = \{\mathbf{0}, \mathbf{b}\}$.

Hence $U_{(\mathbf{a},\mathbf{b})}$ is independent of the output difference $\mathbf{b}$ and $V_{(\mathbf{a},\mathbf{b})}$ of the input difference $\mathbf{a}$.

Lemma 15.5.3. *Let* $\mathsf{Q}' = (\mathbf{a}', \mathbf{b}, \mathbf{d}, \mathbf{e}')$ *be a trail in which all S-box differentials have weight 6. Then for all* $\mathsf{Q} = (\mathbf{a}, \mathbf{b}, \mathbf{d}, \mathbf{e})$,

$$W_{\mathsf{Q}} \subseteq W_{\mathsf{Q}'} \ .$$

Proof. Remember that $W_{\mathsf{Q}} = \mathrm{M_c}(V_{(\mathbf{a},\mathbf{b})}) \cap U_{(\mathbf{d},\mathbf{e})}$. From Lemma 15.5.1 and Lemma 15.5.2, we have that $V_{(\mathbf{a},\mathbf{b})} \subseteq V_{(\mathbf{a}',\mathbf{b})}$ and $U_{(\mathbf{d},\mathbf{e})} \subseteq U_{(\mathbf{d}',\mathbf{e})}$. □

Consequently, it is only needed to check the height of each trail with $(\mathbf{a}, \mathbf{e})$ chosen such that all active S-boxes have weight 6, and then to evaluate the effect of increasing the weight of the active S-boxes by one. For this last step, only one out of the 126 possible differences $\mathbf{a}_i$, respectively $\mathbf{e}_i$, needs to be tried for each active S-box.

15.5.2 Observations

We see in Table 15.1 that the trail weight ranges from 30 to 56 and the height from 1 to 5. It follows from the data in the table that the ratio

$$\mathrm{DP}[\mathbf{k}](\mathsf{Q})/\mathrm{EDP}(\mathsf{Q}) = 2^{\mathrm{height}(\mathsf{Q})-32+\mathrm{w_r}(\mathsf{Q})} \tag{15.4}$$

ranges from 1 to 2^{25}. We call trails for which the ratio is 1 *flat trails* because for these the equality $\mathrm{DP}[\mathbf{k}](\mathsf{Q}) = \mathrm{EDP}(\mathsf{Q})$ holds for all keys. Table 15.1 shows that there are in total $2^{20.9}$ flat trails: those with weight 30 and height 2, and those with weight 31 and height 1.

The trails for which the ratio is 2^{25} are the trails with weight 56 and height 1. Only for a fraction $2^{-32.9}$ of the trails this ratio is smaller than 2^{25}. Since the sequence of two Rijndael rounds can be described as the parallel application of four super boxes, it follows that for most trails over two rounds of Rijndael there are keys with $\mathrm{DP}[\mathbf{k}](\mathsf{Q}) = 2^{100}\mathrm{EDP}(\mathsf{Q})$. The trails that are

Table 15.1. Number of trails (binary logarithm) per number of active S-boxes, weight and height for the Rijndael super box

No. active S-boxes	trail weight	height 1	2	3	4	5
5	30	—	12.6	12.6	10.6	6.2
	31	20.9	22.1	21.2	18.1	11.0
	32	29.8	30.0	28.2	23.4	—
	33	37.1	36.9	33.7	26.4	—
	34	43.2	42.9	36.2	—	—
	35	48.0	47.5	—	—	—
6	36	20.7	15.6	8.3	3.8	—
	37	30.3	24.2	16.3	11.6	—
	38	38.6	31.5	23.1	17.5	—
	39	46.1	38.1	28.9	—	—
	40	52.6	44.0	33.4	—	—
	41	58.3	49.3	—	—	—
	42	62.7	53.4	—	—	—
7	42	27.0	15.7	5.3	—	—
	43	36.8	24.3	13.1	—	—
	44	45.3	31.7	19.5	—	—
	45	53.1	38.0	24.9	—	—
	46	60.0	43.5	—	—	—
	47	66.3	48.0	—	—	—
	48	71.7	50.9	—	—	—
	49	75.9	—	—	—	—
8	48	32.0	14.7	1.0	—	—
	49	41.9	23.7	9.0	—	—
	50	50.7	31.4	15.0	—	—
	51	58.7	38.3	—	—	—
	52	66.0	44.5	—	—	—
	53	72.7	49.9	—	—	—
	54	78.7	54.1	—	—	—
	55	83.7	—	—	—	—
	56	87.9	—	—	—	—
total		87.9	55.0	36.6	26.6	11.0

the most interesting for standard differential attacks are the trails with the lowest weight. They are in the top rows of the table. We see that exactly these trails have the highest heights, hence the most variation between DP values for different keys. Trails with height 5 have a DP equal to $32/2^{32}$, which is almost three times higher than the maximal EDP of a differential ($13.25/2^{32}$ [80, 126, 127]).

There are 72 trails of height 5 and weight 30. These trails have non-zero DP[**k**] for a fraction $2^{32-30-5} = 2^{-3}$ of all keys. For a given key this results in an expected value of nine such trails with DP[**k**] $= 2^5/2^{32}$. Similarly, there are 2^{11} trails of height 5 and weight 31 resulting in an expected value of 2^7 such trails with DP[**k**] $= 2^5/2^{32}$. This totals to an expected number of 137 trails with DP[**k**] $= 2^5/2^{32}$ per key for the Rijndael super box. For two Rijndael rounds, this is 548.

The table shows also that it is easy to find trails Q_1, Q_2 with $\mathrm{EDP}(Q_1) < \mathrm{EDP}(Q_2)$ and $\mathrm{height}(Q_1) > \mathrm{height}(Q_2)$.

15.5.3 Influence of L

If we remove the linear transformation L and the constant **q** from the S-box, we obtain a super box with a simpler algebraic structure than the Rijndael super box (the *naive super box*). We computed the heights of all trails over the naive super box. The results are summarized in Table 15.2. The comparison with the results in Table 15.1 shows us something about the effect of adding L. (It can be shown that the choice of **q** has no impact here.) For instance, we see that for the naive super box, there are no trails with height 5. Trails where all active S-boxes have weight 6 always have an even-numbered height, and those with exactly one S-box with weight 7 always have an odd-numbered height.

15.6 Trails over Four or More Rounds in Rijndael

Four-round Rijndael can be described with a super box-like structure, where again $n_t = 4$ but now the elements are 32-bit words [130, 42]. The Rijndael super boxes we defined before serve now as (key-dependent) S-boxes. A trail through such a super box-like structure consists of five to eight smaller trails, each over a Rijndael super box. If the trails over the Rijndael super boxes of the first step are output-planar and the trails over the Rijndael super boxes of the second step are input-planar, then according to Theorem 15.3.2 the four-round trail is a plateau trail. These conditions are fulfilled if the trails over the Rijndael super boxes

– have height 1 or 2 (by Lemma 15.3.1), or

Table 15.2. Number of trails (binary logarithm) per number of active S-boxes, weight and height for the naive super box

No. active S-boxes	trail weight	height 1	2	3	4
5	30	—	12.6	—	13.0
	31	21.9	—	22.3	—
	32	29.9	29.6	28.8	—
	33	37.1	37.0	33.5	—
	34	43.2	42.9	—	—
	35	48.0	47.5	—	—
6	36	—	20.8	—	12.8
	37	30.3	—	22.4	—
	38	38.7	30.1	29.0	—
	39	46.1	37.7	34.2	—
	40	52.6	44.0	—	—
	41	58.3	49.3	—	—
	42	62.7	53.4	—	—
7	42	—	27.0	—	11.0
	43	36.8	—	20.8	—
	44	45.3	28.9	27.3	—
	45	53.1	36.4	31.9	—
	46	60.0	42.4	—	—
	47	66.3	46.9	—	—
	48	71.7	—	—	—
	49	75.9	—	—	—
8	48	—	32.0	—	13.0
	49	41.9	—	23.0	—
	50	50.7	31.4	29.6	—
	51	58.7	39.0	34.7	—
	52	66.0	45.3	—	—
	53	72.7	50.6	—	—
	54	78.7	54.4	—	—
	55	83.7	—	—	—
	56	87.9	—	—	—
total		87.9	55.1	36.0	14.6

– have four active S-boxes at the output, respectively at the input (by Theorem 15.3.1).

This implies that most of the four-round trails over Rijndael are plateau trails. We have not determined the distribution of the heights, but Theorem 15.2.2 and a straightforward generalization of Theorem 15.2.3 apply. Since the overwhelming majority of the trails over the Rijndael super box have height 1, we expect that also the vast majority of the trails over four rounds of Rijndael will have height 1. Corollary 15.3.1 would then imply that the vast majority of trails over more than four Rijndael rounds are plateau trails with height 1.

15.7 DP of Differentials in Rijndael

The exact distribution of $\mathrm{DP}[\mathbf{k}](\mathbf{a}, \mathbf{b})$ depends on the relative positions of the affine subspaces K_{Q_i} and the height of the trails. Flat trails add a constant term and do not contribute to the variability. The larger the height of a trail, the more it contributes to the variability. In the Rijndael super box there is at most one flat trail per differential with five active S-boxes and none for differentials with more than five active S-boxes.

Rijndael has no flat plateau trails over four rounds or more, and the vast majority of trails have height equal to 1. Under the assumption that the affine subspaces K_{Q_i} are independent, the distribution of the $\mathrm{DP}[\mathbf{k}]$ of any four-round differential is the convolution of a huge number of distributions with a high peak in 0 and a very small peak in 2^{-127}. We conjecture that this gives rise to a Poisson distribution.

15.8 Related Differentials

In this section we define related differences and related differentials. We show that the existence of related differentials influences the height of the trails through super boxes, and this is *independent of the choice of S-boxes.* While in the previous sections of this chapter, $\mathbf{x}$ denoted a binary vector, i.e. with coordinates in GF(2), in the remainder of this chapter $\mathbf{x}$ denotes a vector with coordinates in $\mathrm{GF}(2^8)$.

15.8.1 Definitions

Definition 15.8.1. *Two vectors* $\mathbf{x}$, $\mathbf{x}^\diamond$ *each containing* n_t *elements of* $\mathrm{n_s}$ *bits are* related *if and only if*

$$x_j x_j^\diamond (x_j + x_j^\diamond) = 0, \ \textit{for}\ j = 0, 1, 2, \ldots, n_\mathrm{t} - 1. \tag{15.5}$$

The all-zero vector is trivially related to all vectors, and we exclude it from now on. If $\mathbf{x}$, $\mathbf{x}^\diamond$ are two related differences, then the differences $\mathbf{x}, \mathbf{x}+\mathbf{x}^\diamond$ are also related. The following condition is equivalent to (15.5):

$$x_j = 0 \text{ or } x_j^\diamond = t_j x_j, t_j \in \{0,1\},\ j = 0,1,\ldots,n_\mathrm{t}-1. \tag{15.6}$$

Two related differences define a special type of second-order differential [84]. Any second-order differential defines quartets $\{\mathbf{p}, \mathbf{p}+\mathbf{x}, \mathbf{p}+\mathbf{x}^\diamond, \mathbf{p}+\mathbf{x}+\mathbf{x}^\diamond\}$. If the differences $\mathbf{x}$ and $\mathbf{x}^\diamond$ are related, then it follows that the sets

$$\{p_j, p_j + x_j, p_j + x_j^\diamond, p_j + x_j + x_j^\diamond\},\quad j = 0,1,\ldots,n_\mathrm{t}-1$$

contain only two different elements. This is illustrated in Fig. 15.2. Related differences can be combined into related differentials.

Definition 15.8.2. *Two differentials* $(\mathbf{b},\mathbf{c})$, $(\mathbf{b}^\diamond,\mathbf{c}^\diamond)$ *for a linear map* M *are* related differentials *if and only if* $\mathbf{c} = \mathrm{M}(\mathbf{b})$, $\mathbf{c}^\diamond = \mathrm{M}(\mathbf{b}^\diamond)$, *the differences* $\mathbf{b}, \mathbf{b}^\diamond$ *are related differences and the differences* $\mathbf{c}, \mathbf{c}^\diamond$ *are related differences.*

The following differentials are related differentials over the map $\mathbf{c} = \mathrm{M_c}\mathbf{b}$:

$$\begin{aligned} \mathbf{b} &= [\mathtt{0,1,4,7}]^\mathrm{T}, & \mathbf{c} &= \mathrm{M_c}\mathbf{b} &&= [\mathtt{0,9,0,B}]^\mathrm{T}\\ \mathbf{b}^\diamond &= [\mathtt{5,1,0,7}]^\mathrm{T}, & \mathbf{c}^\diamond &= \mathrm{M_c}\mathbf{b}^\diamond &&= [\mathtt{D,0,E,0}]^\mathrm{T}\\ \mathbf{b}+\mathbf{b}^\diamond &= [\mathtt{5,0,4,0}]^\mathrm{T}, & \mathbf{c}+\mathbf{c}^\diamond &= \mathrm{M_c}(\mathbf{b}+\mathbf{b}^\diamond) &&= [\mathtt{D,9,E,B}]^\mathrm{T}. \end{aligned} \tag{15.7}$$

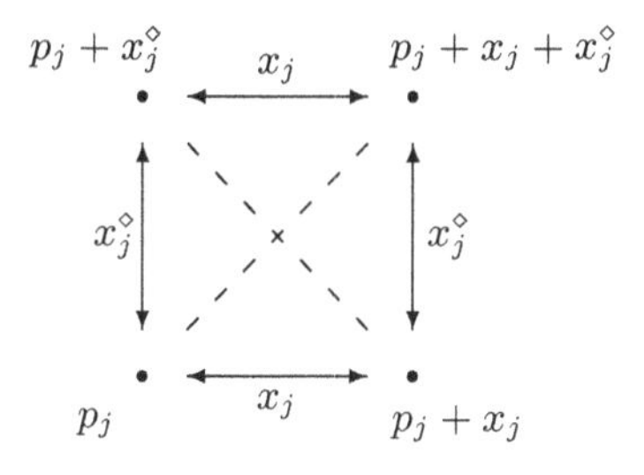

Fig. 15.2. A second-order differential and the associated quartet can be represented by a square. If one of the differences x_j, $x_j^\diamond$ or $x_j + x_j^\diamond$ equals zero, then the square collapses to a line

15.8.2 Related Differentials and Plateau Trails

Theorem 15.8.1. *Let* $\mathsf{Q} = (\mathbf{a},\mathbf{b},\mathbf{d},\mathbf{e})$ *be a trail through a super box with* $\mathrm{EDP}(\mathsf{Q}) > 0$. *If* $(\mathbf{b},\mathbf{d})$ *is in a set of related differentials over the mixing map, then* Q *is a plateau trail with* $\mathrm{height}(\mathsf{Q}) \geq 2$.

Proof: Theorem 15.2.1 states that all trails through a super box are plateau trails. Hence we only need to show that for each pair $P_x = \{\mathbf{x}, \mathbf{x} + \mathbf{a}\}$ that follows Q for a certain round key value $\mathbf{k}$, there exists a pair

$$P_y = \{\mathbf{x} + \mathbf{a}^\diamond, \mathbf{x} + \mathbf{a}^\diamond + \mathbf{a}\} \neq \{\mathbf{x} + \mathbf{a}, \mathbf{x}\}$$

that also follows Q for the same value of the round key. Let $(\mathbf{b}^\diamond, \mathbf{d}^\diamond)$ denote a differential that is related to $(\mathbf{b}, \mathbf{d})$. Define the difference $\mathbf{a}^\diamond$ to be the solution of

$$\mathrm{S}(\mathbf{x} + \mathbf{a}^\diamond) + \mathrm{S}(\mathbf{x}) = \mathbf{b}^\diamond \Leftrightarrow \mathbf{a}^\diamond = \mathrm{S}^{-1}(\mathbf{b}^\diamond + \mathrm{S}(\mathbf{x})) + \mathbf{x}. \tag{15.8}$$

Then the pair P_y follows Q because of the following:

- If S were a linear map, then (15.8) would imply that $\mathrm{S}(\mathbf{x}+\mathbf{a}+\mathbf{a}^\diamond)+\mathrm{S}(\mathbf{x}+\mathbf{a}) = \mathbf{b}^\diamond$ and we would have

$$\mathrm{S}(\mathbf{x}+\mathbf{a}^\diamond)+\mathrm{S}(\mathbf{x}+\mathbf{a}^\diamond+\mathbf{a}) = (\mathrm{S}(\mathbf{x})+\mathbf{b}^\diamond)+(\mathrm{S}(\mathbf{x}+\mathbf{a})+\mathbf{b}^\diamond) = \mathrm{S}(\mathbf{x})+\mathrm{S}(\mathbf{x}+\mathbf{a}) = \mathbf{b}. \tag{15.9}$$

  Since S is not linear, the equality doesn't hold for all $\mathbf{a}^\diamond$. However, because $\mathbf{b}$ and $\mathbf{b}^\diamond$ are related differences, for each j at least one of the three differences b_j, $b_j^\diamond$ and $b_j + b_j^\diamond$ equals zero. Since S uses invertible S-boxes, it follows that for each j also at least one of the three differences a_j, $a_j^\diamond$ and $a_j + a_j^\diamond$ equals zero. Hence the sets

$$\{x_j, x_j + a_j, x_j + a_j^\diamond, x_j + a_j + a_j^\diamond,\}, \quad j = 0, 1, 2, \ldots, n_t - 1$$

  contain only two different elements. Consequently, (15.9) holds and P_y follows Q through the first substitution step.
- P_y follows Q through the mixing step:

$$\mathrm{M}(\mathrm{S}(\mathbf{x}+\mathbf{a}^\diamond))+\mathrm{M}(\mathrm{S}(\mathbf{x}+\mathbf{a}^\diamond+\mathbf{a})) = \mathrm{M}(\mathrm{S}(\mathbf{x}+\mathbf{a}^\diamond)+\mathrm{S}(\mathbf{x}+\mathbf{a}^\diamond+\mathbf{a})) = \mathrm{M}(\mathbf{b}) = \mathbf{d}.$$

- Let $\mathbf{v}$ denote the vector $\mathrm{M}(\mathrm{S}(\mathbf{x})) + \mathbf{k}$, which is the output of the round key addition for the input $\mathbf{x}$. Since $\mathbf{x}$ is in the pair P_x, which follows Q, we know that $\mathbf{v} \in F_{(\mathbf{d},\mathbf{e})}$. From the previous steps, it follows also that $\mathrm{S}(\mathbf{x} + \mathbf{a}^\diamond) + \mathrm{S}(\mathbf{x}) = \mathbf{b}^\diamond$ and

$$\mathrm{M}(\mathrm{S}(P_y)) + \mathbf{k} = \{\mathbf{v} + \mathbf{d}^\diamond, \mathbf{v} + \mathbf{d}^\diamond + \mathbf{d}\}.$$

  Now we use the fact that $\mathbf{d}$ and $\mathbf{d}^\diamond$ are related differences and by following reasoning similar to the first substitution step, we conclude that P_y follows Q through the last substitution step and hence it follows Q through the super box.

15.9 Determining the Related Differentials

In this section, we derive an algorithm that for a given linear mixing map produces all the sets of related differentials. To simplify the description, we will assume that the linear map consists of multiplication by an $n_t \times n_t$ matrix M_c and that the associated linear code is an MDS code. We start with an example.

15.9.1 First Example

Assume we have a differential $(\mathbf{b}, \mathbf{d})$ with $\mathbf{d} = M_c\mathbf{b}$ through the mixing map of the Rijndael super box with activity pattern (0111; 0101). We want to determine whether there exist differences $\mathbf{b}^\diamond, \mathbf{d}^\diamond = M_c\mathbf{b}^\diamond$ satisfying (15.5).

Firstly, we know that the linear code associated with `MixColumns` has minimal distance 5, and hence if $b_0 = d_0 = d_2 = 0$, then all other b_j, d_j are different from zero. Equation (15.5) doesn't put any constraints on $b_0^\diamond, d_0^\diamond$, and $d_2^\diamond$. From $d^\diamond = M_c b^\diamond$ we can derive one equation from which these three elements are eliminated:

$$\mathtt{3}d_1^\diamond + d_3^\diamond = \mathtt{7}b_1^\diamond + \mathtt{4}b_2^\diamond + b_3^\diamond.$$

Using (15.6) we obtain

$$\mathtt{3}t_5 d_1 + t_7 d_3 = \mathtt{7}t_1 b_1 + \mathtt{4}t_2 b_2 + t_3 b_3,\ t_j \in \{0, 1\}. \tag{15.10}$$

Secondly, we know from Sect. 14.2.2 that there is only one bundle with activity pattern (0111; 0101). Hence

$$(\mathbf{b}, \mathbf{d})^{\mathrm{T}} = ([\mathtt{0}, \gamma, \mathtt{4}\gamma, \mathtt{7}\gamma], [\mathtt{0}, \mathtt{9}\gamma, \mathtt{0}, \mathtt{B}\gamma]) \text{ with } \gamma \in \mathrm{GF}(2^8) \setminus \{\mathtt{0}\}. \tag{15.11}$$

Combining (15.10) and (15.11) gives

$$\mathtt{7}\gamma t_1 + \mathtt{10}\gamma t_2 + \mathtt{7}\gamma t_3 + \mathtt{1B}\gamma t_5 + \mathtt{B}\gamma t_7 = 0.$$

For any value of γ, we obtain a system of linear equations over GF(2). The five unknowns are t_1, t_2, t_3, t_5, t_7. The number of independent solutions depends on the dimension of the vector space spanned by

$$\{\mathtt{7}\gamma, \mathtt{10}\gamma, \mathtt{7}\gamma, \mathtt{1B}\gamma, \mathtt{B}\gamma\}.$$

Note that the dimension here is determined over GF(2): linear dependencies between vectors must have binary coefficients. This dimension is always at most 4, since we know that setting all $t_j = 1$ gives the solution $(\mathbf{b}, \mathbf{d})$. In this case, the dimension equals 3, which means that there is one other solution $(\mathbf{b}^\diamond, \mathbf{d}^\diamond)$, hence the weight of a trail with $(\mathbf{b}, \mathbf{d})$ as input, respectively output difference for `MixColumns` has height at least two.

15.9.2 For Any Given Differential

Let $(\mathbf{b}, \mathbf{d})$ be the given differential for which we want to find a related differential over a given linear map. Denote by z the number of non-zero elements in $(\mathbf{b}, \mathbf{d})$, minus n_t. This implies that the number of zero elements in $(\mathbf{b}, \mathbf{d})$ equals $n_t - z$. Denote by H the check matrix of the linear code associated with the linear map. We know that $(\mathbf{b}, \mathbf{d})$ and the related differential(s) $(\mathbf{b}^\diamond, \mathbf{d}^\diamond)$ must correspond to vectors of the associated code:

$$\mathsf{H}(\mathbf{b}, \mathbf{d}) = \mathsf{H}(\mathbf{b}^\diamond, \mathbf{d}^\diamond) = 0.$$

This defines a first set of n_t constraints, one in each row of H. Secondly, for the indices j where b_j or d_j are different from zero, we get the conditions (15.6) on t_j. Every time we have a b_j or d_j equal to zero, we have no condition on the corresponding $b_j^\diamond$ or $d_j^\diamond$. Therefore we eliminate these unknowns from the set of conditions.

We denote by H_p the check matrix of a new linear code, where the code vectors can take any value in the positions where the activity pattern of $(\mathbf{b}, \mathbf{d})$ is zero, and where the conditions on the values in the other positions are the same as in the code associated with the linear map. We denote the elements of H_p by $h_{i,j}$ and write

$$\sum_j h_{i,j} b_j^\diamond + \sum_j h_{i,j+n_t} d_j^\diamond = 0, \; i = 0, \ldots, z-1.$$

In these equations, we fill out (15.6) and obtain

$$\sum_{j=0, b_j \neq 0}^{n_t - 1} h_{i,j} b_j t_j + \sum_{j=0, d_j \neq 0}^{n_t - 1} h_{i,j+n_t} d_j t_{j+n_t} = 0, \; i = 0, \ldots, z-1.$$

The solutions t_j are the codewords of the binary code with check matrix

$$\mathsf{D}_p = \begin{bmatrix} h_{0,0} b_0 & h_{0,1} b_1 & \ldots & h_{0,2n_t-1} d_{n_t-1} \\ h_{1,0} b_0 & h_{1,1} b_1 & \ldots & h_{1,2n_t-1} d_{n_t-1} \\ \ldots & \ldots & \ldots & \ldots \\ h_{z-1,0} b_0 & h_{z-1,1} b_1 & \ldots & h_{z-1,2n_t-1} d_{n_t-1} \end{bmatrix}, \tag{15.12}$$

except for the codeword $(1 \; 1 \; \ldots \; 1)$, which corresponds to the original difference $(\mathbf{b}, \mathbf{d})$. The number of independent solutions for this set of equations depends on the rank of D_p: if $\text{rank}(\mathsf{D}_p) < z + m - 1$, then related differentials exist. Note that D_p is a matrix containing elements of $\text{GF}(2^{n_s})$, but we determine the rank over GF(2): linear dependencies must have binary coefficients.

15.9.3 For All Differentials with the Same Activity Pattern

The previous method can be done in parallel for all differences with a given activity pattern p. Denote by G the generator matrix of the associated linear code. Each differential can be written as the transpose of a linear combination of rows of G. The differentials with activity pattern p (plus the zero vector) form a subspace of the linear code. The subspace contains only the code vectors that are zero in the positions where the activity pattern is zero. We denote the generator matrix for this subspace by G_p.

We now define z parameters ϵ_k, write $(\mathbf{b};\mathbf{d})^{\mathrm{T}} = (\epsilon_0, \ldots, \epsilon_{z-1})\mathsf{G}_p = \epsilon\mathsf{G}_p$ and apply the method described in Sect. 15.9.2 to determine related differentials. The elements of matrix D_p of (15.12) now depend on the vector ϵ. Denoting the elements of G_p by $g_{i,j}$ and the elements of $\mathsf{D}_p(\epsilon)$ by $d_{i,j}$, we obtain

$$d_{i,j} = h_{ij} \sum_{k=0}^{z-1} \epsilon_k g_{kj}.$$

The number of dependent columns in $\mathsf{D}_p(\epsilon)$ may depend on ϵ. Any nonzero vector ϵ for which there are more than two codewords in the binary code with $\mathsf{D}_p(\epsilon)$ as check matrix defines a difference $(\mathbf{b}, \mathbf{d})$ for which there exists a related differential. The related differential is again determined by the codeword that is different from $(1, 1, \ldots, 1)$. Figure 15.3 summarizes the algorithm.

15.9.4 Second Example

We now illustrate the algorithm described in Fig. 15.3. Consider the activity pattern $p = (1010; 1111)$ for the mixing map of the Rijndael super box. We have $z = 2$ and

$$\mathsf{H} = \begin{bmatrix} 2\,3\,1\,1\,1\,0\,0\,0 \\ 1\,2\,3\,1\,0\,1\,0\,0 \\ 1\,1\,2\,3\,0\,0\,1\,0 \\ 3\,1\,1\,2\,0\,0\,0\,1 \end{bmatrix}, \quad \mathsf{SH} = \begin{bmatrix} 7\,0\,7\,1\,2\,3\,0\,0 \\ 3\,1\,2\,0\,1\,1\,0\,0 \\ \mathrm{B}\,0\,9\,0\,7\,4\,1\,0 \\ \mathrm{E}\,0\,\mathrm{D}\,0\,5\,7\,0\,1 \end{bmatrix},$$

$$\mathsf{H}_p = \begin{bmatrix} \mathrm{B}\,0\,9\,0\,7\,4\,1\,0 \\ \mathrm{E}\,0\,\mathrm{D}\,0\,5\,7\,0\,1 \end{bmatrix}.$$

Further,

$$\mathsf{G} = \begin{bmatrix} 1\,0\,0\,0\,2\,1\,1\,3 \\ 0\,1\,0\,0\,3\,2\,1\,1 \\ 0\,0\,1\,0\,1\,3\,2\,1 \\ 0\,0\,0\,1\,1\,1\,3\,2 \end{bmatrix} = \mathsf{TG}, \ \mathsf{G}_p = \begin{bmatrix} 1\,0\,0\,0\,2\,1\,1\,3 \\ 0\,0\,1\,0\,1\,3\,2\,1 \end{bmatrix}.$$

Input: $n_t \times n_t$ matrix M_c defining the linear map, activity pattern p.
Output: Related differentials $(\mathbf{b}, \mathbf{d})$ (with activity pattern p) and $(\mathbf{b}^\diamond, \mathbf{d}^\diamond)$.
Algorithm:

1. Compute $\mathsf{G} = [\mathsf{I}\ M_c]$ and $\mathsf{H} = [{M_c}^T\ \mathsf{I}]$. Let $z = w_h(p) - n_t$.
2. Perform elementary row operations on H to compute an equivalent matrix SH where z rows have zeroes in the $n_t - z$ columns corresponding to the zero bits in p. Denote the submatrix of SH consisting of these rows by H_p, with elements $h_{i,j}$.
3. Similarly, perform elementary row operations on G to compute an equivalent matrix TG where z rows have zeroes in the $n_t - z$ columns corresponding to the zero bits in p. Denote the submatrix of TG consisting of these rows by G_p, with elements $g_{i,j}$.
4. Define the matrix $\mathsf{D}(\epsilon)$ as follows:
$$d_{i,j} = h_{i,j} \sum_{k=0}^{z-1} \epsilon_k g_{kj},$$
where $\epsilon = (\epsilon_0, \epsilon_1, \ldots, \epsilon_{z-1})$ is a vector of parameters.
5. Compute the values of ϵ for which the rank of $\mathsf{D}_p(\epsilon)$ is below $z + m - 1$.
6. For each of the outputs of the previous step compute $(\mathbf{b}; \mathbf{c})^T = \epsilon \mathsf{G}_p$. Compute the binary codeword $\mathbf{t} = (t_0, \ldots, t_{2n_t - 1})$ from $\mathsf{D}_p \mathbf{t} = 0$ and $\mathbf{t} \neq (1, \ldots, 1)$. For the positions where $b_j, c_j \neq 0$, compute $b_j^\diamond = t_j b_j$, $d_j^\diamond = t_{j+n_t} d_j$. Determine the remaining $b_j^\diamond, d_j^\diamond$ such that $\mathsf{H}(b_j^\diamond, d_j^\diamond) = 0$.
7. Output $(\mathbf{b}, \mathbf{d})$ and $(\mathbf{b}^\diamond, \mathbf{d}^\diamond)$.

Fig. 15.3. Algorithm to compute related differentials where one of the differentials has a given activity pattern p. If the algorithm terminates without finding related differentials, then there exist none for this activity pattern

This gives

$$\mathsf{D}_p = \begin{bmatrix} \mathtt{B}\epsilon_0 & \mathtt{0} & \mathtt{9}\epsilon_1 & \mathtt{0} & (2\epsilon_0+\epsilon_1)\mathtt{7} & (\epsilon_0+3u_1)\mathtt{4} & \epsilon_0+2\epsilon_1 & \mathtt{0} \\ \mathtt{E}\epsilon_1 & \mathtt{0} & \mathtt{D}\epsilon_1 & \mathtt{0} & (2\epsilon_0+\epsilon_1)\mathtt{5} & (\epsilon_0+3\epsilon_1)\mathtt{7} & \mathtt{0} & 3\epsilon_0+\epsilon_1 \end{bmatrix}.$$

For all values of the parameters ϵ_0, ϵ_1, the eight columns sum to zero. The sum of the third, the sixth and the eighth column equals

$$\begin{bmatrix} \mathtt{4}\epsilon_0 + (\mathtt{9}+\mathtt{3}\cdot\mathtt{4})\epsilon_1 \\ (\mathtt{7}+\mathtt{3})\epsilon_0 + (\mathtt{D}+\mathtt{3}\cdot\mathtt{7}+\mathtt{1})\epsilon_1 \end{bmatrix} = \begin{bmatrix} \mathtt{4}\epsilon_0 + \mathtt{5}\epsilon_1 \\ \mathtt{4}\epsilon_0 + \mathtt{5}\epsilon_1 \end{bmatrix}.$$

Hence for $\mathtt{4}\epsilon_0 = \mathtt{5}\epsilon_1$, these three columns are dependent. A non-trivial solution is $t_2 = t_5 = t_7 = \mathtt{1}$, $t_0 = t_4 = t_6 = \mathtt{0}$. This gives

$$\begin{aligned} (\mathbf{b}^\diamond, \mathbf{d}^\diamond)^{\mathrm{T}} &= ([\mathtt{0}, b_1^\diamond, b_2, b_3^\diamond], [\mathtt{0}, d_1, \mathtt{0}, d_3]) \\ (\mathbf{b}, \mathbf{d})^{\mathrm{T}} + (\mathbf{b}^\diamond, \mathbf{d}^\diamond)^{\mathrm{T}} &= ([b_0, b_1^\diamond, \mathtt{0}, b_3^\diamond], [d_0, \mathtt{0}, d_2, \mathtt{0}]). \end{aligned}$$

Filling out $\mathtt{4}\epsilon_0 = \mathtt{5}\epsilon_1$, we see that we obtain again the vectors of the previous example.

15.9.5 A Combinatorial Bound

If we want to check the existence of related differentials for a given map, then in principle we need to repeat the algorithm of Fig. 15.3 for all possible activity patterns. We present here an observation that reduces the number of activity patterns that need to be considered if the linear code associated with the map is an MDS code.

The three differentials $(\mathbf{b}, \mathbf{d})$, $(\mathbf{b}^\diamond, \mathbf{d}^\diamond)$ and $(\mathbf{b}+\mathbf{b}^\diamond, \mathbf{d}+\mathbf{d}^\diamond)$ correspond to vectors of an MDS code with minimal distance $n_{\mathrm{t}}+1$. Hence they can have at most $n_{\mathrm{t}}-1$ components equal to zero. On the other hand, from (15.5) we see that we need to distribute at least $2n_{\mathrm{t}}$ zeroes over these three differentials. A simple counting argument results in the following bound.

Lemma 15.9.1. *If* $(\mathbf{b}, \mathbf{d})$, $(\mathbf{b}^\diamond, \mathbf{d}^\diamond)$ *are related differentials over a linear map with an associated code that is an MDS code with length* $2n_{\mathrm{t}}$ *and distance* $n_{\mathrm{t}}+1$, *then*

$$\min\{\mathrm{w_t}(\mathbf{b}, \mathbf{d}), \mathrm{w_t}(\mathbf{b}^\diamond, \mathbf{d}^\diamond), \mathrm{w_t}(\mathbf{b}+\mathbf{b}^\diamond, \mathbf{d}+\mathbf{d}^\diamond)\} \le n_{\mathrm{t}} + \lfloor n_{\mathrm{t}}/3 \rfloor.$$

This means that if related differentials exist, they will be revealed when we check all the differentials with weights up to $n_{\mathrm{t}} + \lfloor n_{\mathrm{t}}/3 \rfloor$. For instance, if $n_{\mathrm{t}} = 4$, then two of the three differentials need to have zeroes in at least three positions. Consequently, if we check all activity patterns of weight 5 for the existence of related differentials, then we have determined all the related differentials. Table 15.3 lists for $n_{\mathrm{t}} = 4, 5, 6, 7, 8, 9$ the possible distributions of $2n_{\mathrm{t}}$ zeroes and the differentials that need to be checked for related differentials.

Table 15.3. Possible distributions of $2n_t$ zeroes over three vectors, where each vector counts at most $n_t - 1$ zeroes. The last column gives the weights of the activity patterns that need to be checked in order to determine all sets of related differentials

n_t	possible distributions of zeroes	weights of activity patterns to be checked
4	(3,3,2)	5
5	(4,4,2), (4,3,3)	6
6	(5,5,2), (5,4,3), (4,4,4)	7, 8
7	(6,6,2), (6,5,3), (6,4,4), (5,5,4)	8, 9
8	(7,7,2), (7,6,3), (7,5,4), (6,6,4), (6,5,5)	9, 10
9	...	10, 11, 12

15.10 Implications for Rijndael-Like Super Boxes

15.10.1 Related Differentials over Circulant Matrices

The fact that `MixColumns` has related differentials is no coincidence. This can be understood easily if we write out the equivalent of Table 14.1 for a general 4×4 circulant matrix. Denote the matrix and its inverse by

$$\mathrm{M_c} = \begin{bmatrix} a & b & c & d \\ d & a & b & c \\ c & d & a & b \\ b & c & d & a \end{bmatrix}, \qquad \mathrm{M_c}^{-1} = \begin{bmatrix} e & f & g & h \\ h & e & f & g \\ g & h & e & f \\ f & g & h & e \end{bmatrix}. \tag{15.13}$$

Table 15.4 gives the bundles for a mixing map using this matrix. Looking at the second and third row, respectively 12th and 13th row, we notice that they define related differentials. All rotations of these bundles also define related differentials, as do all scalar multiples.

15.10.2 Related Differentials in `MixColumns`

Besides the related differentials described in the previous section, `MixColumns` has eight more. Table 15.5 lists the four pairs of bundles from which all related differentials can be derived by means of rotation and/or multiplication by a scalar. We know from Sect. 14.2.2 that a differential with weight 5 is determined uniquely by its activity pattern. This implies that 3/7 of the differentials with $\mathrm{w_t}(\mathbf{b}, \mathbf{d}) = 5$ form part of a set of related differentials.

Four rounds of Rijndael are, up to a linear transformation, equivalent to a large super box structure where the S-boxes are exactly the super boxes we described before. The mixing transformation of this large super box structure is equivalent to the sequence `Shiftrows` followed by `MixColumns` followed by `ShiftRows`. Also this map has related differentials, and their activity patterns are the same as the activity patterns of the differentials in Table 15.5. Hence, also a large set of trails over four rounds of Rijndael has height at least 2.

Table 15.4. Bundles for a mixing map based on the circulant matrix defined in (15.13)

$\mathbf{b}^{\mathrm{T}}$				$\mathbf{d}^{\mathrm{T}}$			
1	0	0	0	a	b	c	d
a	0	c	0	a^2+c^2	$ab+cd$	0	$ad+bc$
0	b	0	d	0	$ab+cd$	b^2+d^2	$ad+bc$
a	0	0	d	a^2+bd	$ab+cd$	d^2+ac	0
a	b	0	0	a^2+bd	0	b^2+ac	$ad+bc$
0	b	c	0	c^2+bd	$ab+cd$	b^2+ac	0
0	0	c	d	c^2+bd	0	d^2+ac	$ad+bc$
e^2+fh	$ef+gh$	h^2+eg	0	e	0	0	h
e^2+fh	0	f^2+eg	$eh+fg$	e	f	0	0
g^2+fh	$ef+gh$	f^2+eg	0	0	f	g	0
g^2+fh	0	h^2+eg	$eh+fg$	0	0	g	h
e^2+g^2	$ef+gh$	0	$eh+fg$	e	0	g	0
0	$ef+gh$	f^2+h^2	$eh+fg$	0	f	0	h
e	f	g	h	1	0	0	0

Table 15.5. The sets of related differentials over `MixColumns`

$\mathbf{b}^{\mathrm{T}}$	$\mathbf{d}^{\mathrm{T}}$	$\mathbf{b}^{\diamond\mathrm{T}}$	$\mathbf{d}^{\diamond\mathrm{T}}$	$(\mathbf{b}+\mathbf{b}^{\diamond})^{\mathrm{T}}$	$(\mathbf{d}+\mathbf{d}^{\diamond})^{\mathrm{T}}$
[0, 1, 4, 7]	[0, 9, 0, B]	[5, 1, 0, 7]	[E, 0, D, 0]	[5, 0, 4, 0]	[E, 9, D, B]
[0, 1, 0, 3]	[0, 1, 4, 7]	[2, 0, 1, 0]	[5, 1, 0, 7]	[2, 1, 1, 3]	[5, 0, 4, 0]
[7, 0, 7, 7]	[E, 9, 0, 0]	[7, 7, 7, 0]	[0, 0, E, 9]	[0, 7, 0, 7]	[E, 9, E, 9]
[0, 3, 2, 0]	[7, 0, 7, 1]	[2, 0, 0, 3]	[7, 1, 7, 0]	[2, 3, 2, 3]	[0, 1, 0, 1]

15.10.3 Avoiding Related Differentials

There exist 4×4 matrices over $\mathrm{GF}(2^8)$ without related differentials, even matrices with special structure, for instance matrices with the Hadamard structure, as in Anubis [9]. Denote a 4×4 Hadamard matrix and its inverse as follows:

$$\mathrm{M_c} = \begin{bmatrix} a & b & c & d \\ b & a & d & c \\ c & d & a & b \\ d & c & b & a \end{bmatrix}, \qquad \mathrm{M_c}^{-1} = \begin{bmatrix} e & f & g & h \\ f & e & h & g \\ g & h & e & f \\ h & g & f & e \end{bmatrix}. \tag{15.14}$$

Table 15.6 gives the bundles for this matrix. It can be seen that, in general, there are no related differentials with five active positions. From Lemma 15.9.1 we know that this means there are in general no related differentials. Anubis uses

$$\mathrm{M_{Anubis}} = \begin{bmatrix} 1\,2\,4\,6 \\ 2\,1\,6\,4 \\ 4\,6\,1\,2 \\ 6\,4\,2\,1 \end{bmatrix}.$$

This matrix has related differentials $([0,0,4,6]^{\mathrm{T}}, [4,0,8,\mathrm{E}]^{\mathrm{T}})$ and $([8,\mathrm{E},4,0]^{\mathrm{T}}, [4,6,0,0]^{\mathrm{T}})$ (and their sum $([8,\mathrm{E},0,6]^{\mathrm{T}}, [0,6,8,\mathrm{E}]^{\mathrm{T}})$). However, if the four 6s are replaced by 9s, then there are no related differentials.

Table 15.6. Bundles for a mixing map based on the Hadamard matrix defined in (15.14)

$\mathbf{b}^{\mathrm{T}}$				$\mathbf{d}^{\mathrm{T}}$			
1	0	0	0	a	b	c	d
a	b	0	0	a^2+b^2	0	$ac+bd$	$ad+bc$
0	0	c	d	c^2+d^2	0	$ac+bd$	$ad+bc$
a	0	c	0	a^2+c^2	$ab+cd$	0	$ad+bc$
0	b	0	d	b^2+d^2	$ab+cd$	0	$ad+bc$
a	0	0	d	a^2+d^2	$ab+cd$	$ac+bd$	0
0	b	c	0	b^2+c^2	$ab+cd$	$ac+bd$	0
f^2+g^2	$ef+gh$	$eg+fh$	0	0	f	g	0
e^2+h^2	$ef+gh$	$eg+fh$	0	e	0	0	h
f^2+h^2	$ef+gh$	0	$eh+fg$	0	f	0	h
e^2+g^2	$ef+gh$	0	$eh+fg$	e	0	g	0
g^2+h^2	0	$eg+fh$	$eh+fg$	0	0	g	h
e^2+f^2	0	$eg+fh$	$eh+fg$	e	f	0	0
e	f	g	h	1	0	0	0

15.11 Conclusions and Further Work

We believe that an analysis of resistance against differential cryptanalysis needs to take into account more than the average behavior of a key-dependent map. Currently, no attacks on block ciphers are known that exploit non-uniformities in the distribution of the $\mathrm{DP}[\mathbf{k}]$ values. However, in the case of iterated mappings without a key, for instance hash functions, they are relevant. For example, they influence the complexity of rebound attacks (Sect. 10.7, [94]).

We showed that the $\mathrm{DP}[\mathbf{k}]$ of certain trails is distributed in a very structured way. For trails over the Rijndael super box, we showed that $\mathrm{DP}[\mathbf{k}] \in 0, 2^{h-32}$, with h an integer value between 1 and 5. We think that the results are somewhat surprising and deserve to be investigated in further detail. We have illustrated our analysis only on Rijndael, but several other ciphers using differentially 4-uniform S-boxes will show a similar behavior.

The presence of related differentials in `MixColumns` is one reason why such high values for the height occur for so many trails. We studied how related differentials can be discovered for any given linear map.

It would be interesting to find out what the maximum DP[$\mathbf{k}$] is for trails over more than two rounds. If the impact on the DP[$\mathbf{k}$] of differentials over more than two rounds can be investigated, then this could lead to new insights about the security margin of Rijndael and other ciphers.

A. Substitution Tables

In this appendix, we list some tables that represent various mappings used in Rijndael.

A.1 $\mathrm{S_{RD}}$

This section includes several representations of $\mathrm{S_{RD}}$ and related mappings. More explanation about the alternative representations for the mappings used in the definition of $\mathrm{S_{RD}}$ can be found in Sect. 3.4.1. Tabular representations of $\mathrm{S_{RD}}$ and ${\mathrm{S_{RD}}}^{-1}$ are given in Tables A.1 and A.2.

Table A.1. Tabular representation of $\mathrm{S_{RD}}(xy)$

		y															
		0	1	2	3	4	5	6	7	8	9	A	B	C	D	E	F
	0	63	7C	77	7B	F2	6B	6F	C5	30	01	67	2B	FE	D7	AB	76
	1	CA	82	C9	7D	FA	59	47	F0	AD	D4	A2	AF	9C	A4	72	C0
	2	B7	FD	93	26	36	3F	F7	CC	34	A5	E5	F1	71	D8	31	15
	3	04	C7	23	C3	18	96	05	9A	07	12	80	E2	EB	27	B2	75
	4	09	83	2C	1A	1B	6E	5A	A0	52	3B	D6	B3	29	E3	2F	84
	5	53	D1	00	ED	20	FC	B1	5B	6A	CB	BE	39	4A	4C	58	CF
	6	D0	EF	AA	FB	43	4D	33	85	45	F9	02	7F	50	3C	9F	A8
	7	51	A3	40	8F	92	9D	38	F5	BC	B6	DA	21	10	FF	F3	D2
x	8	CD	0C	13	EC	5F	97	44	17	C4	A7	7E	3D	64	5D	19	73
	9	60	81	4F	DC	22	2A	90	88	46	EE	B8	14	DE	5E	0B	DB
	A	E0	32	3A	0A	49	06	24	5C	C2	D3	AC	62	91	95	E4	79
	B	E7	C8	37	6D	8D	D5	4E	A9	6C	56	F4	EA	65	7A	AE	08
	C	BA	78	25	2E	1C	A6	B4	C6	E8	DD	74	1F	4B	BD	8B	8A
	D	70	3E	B5	66	48	03	F6	0E	61	35	57	B9	86	C1	1D	9E
	E	E1	F8	98	11	69	D9	8E	94	9B	1E	87	E9	CE	55	28	DF
	F	8C	A1	89	0D	BF	E6	42	68	41	99	2D	0F	B0	54	BB	16

For hardware implementations, it might be useful to use the following decomposition of $\mathrm{S_{RD}}$:

$$\mathrm{S_{RD}}[a] = \mathrm{Aff_8}(\mathrm{Inv_8}(a)), \tag{A.1}$$

where $\mathrm{Inv_8}(a)$ is the mapping

$$a \rightarrow a^{-1} \text{ in } \mathrm{GF}(2^8) \tag{A.2}$$

J. Daemen, V. Rijmen, *The Design of Rijndael*, Information Security and Cryptography,
https://doi.org/10.1007/978-3-662-60769-5

Table A.2. Tabular representation of $\mathrm{S_{RD}}^{-1}(xy)$

										y							
		0	1	2	3	4	5	6	7	8	9	A	B	C	D	E	F
	0	52	09	6a	d5	30	36	a5	38	bf	40	a3	9e	81	f3	d7	fb
	1	7c	e3	39	82	9b	2f	ff	87	34	8e	43	44	c4	de	e9	cb
	2	54	7b	94	32	a6	c2	23	3d	ee	4c	95	0b	42	fa	c3	4e
	3	08	2e	a1	66	28	d9	24	b2	76	5b	a2	49	6d	8b	d1	25
	4	72	f8	f6	64	86	68	98	16	d4	a4	5c	cc	5d	65	b6	92
	5	6c	70	48	50	fd	ed	b9	da	5e	15	46	57	a7	8d	9d	84
	6	90	d8	ab	00	8c	bc	d3	0a	f7	e4	58	05	b8	b3	45	06
	7	d0	2c	1e	8f	ca	3f	0f	02	c1	af	bd	03	01	13	8a	6b
x	8	3a	91	11	41	4f	67	dc	ea	97	f2	cf	ce	f0	b4	e6	73
	9	96	ac	74	22	e7	ad	35	85	e2	f9	37	e8	1c	75	df	6e
	A	47	f1	1a	71	1d	29	c5	89	6f	b7	62	0e	aa	18	be	1b
	B	fc	56	3e	4b	c6	d2	79	20	9a	db	c0	fe	78	cd	5a	f4
	C	1f	dd	a8	33	88	07	c7	31	b1	12	10	59	27	80	ec	5f
	D	60	51	7f	a9	19	b5	4a	0d	2d	e5	7a	9f	93	c9	9c	ef
	E	a0	e0	3b	4d	ae	2a	f5	b0	c8	eb	bb	3c	83	53	99	61
	F	17	2b	04	7e	ba	77	d6	26	e1	69	14	63	55	21	0c	7d

extended with 0 → 0, and $\mathrm{Aff}_8(a)$ is an affine mapping. Since $\mathrm{Inv}_8(a)$ is self-inverse, we have

$$\mathrm{S_{RD}}^{-1}[a] = \mathrm{Inv}_8^{-1}(\mathrm{Aff}_8^{-1}(a)) = \mathrm{Inv}_8(\mathrm{Aff}_8^{-1}(a)). \tag{A.3}$$

The tabular representations of Aff_8, Aff_8^{-1} and Inv_8 are given in Tables A.3–A.5.

Algebraic representations of $\mathrm{S_{RD}}$ have also received a lot of attention in the literature, especially in the cryptanalytic literature. Mappings over a finite domain can *always* be represented by polynomial functions with a finite number of terms. As a consequence, mappings from $\mathrm{GF}(2^8)$ to $\mathrm{GF}(2^8)$ can always be represented by a polynomial function over $\mathrm{GF}(2^8)$. In Section 12.4 we derived expression (12.14), which we repeat here:

$$\begin{aligned}\mathrm{S_{RD}}[x] = {} & \alpha^2 x^{254} + \alpha^{199} x^{253} + \alpha^{99} x^{251} + \alpha^{185} x^{247} \\ & + \alpha^{197} x^{239} + a^{223} + \alpha^{96} x^{191} + \alpha^{232} x^{127} + \alpha^{195}\end{aligned}$$

with α a root of $x^8 + x^4 + x^3 + x^2 + 1$. Using the basis for $\mathrm{GF}(2^8)$ that is employed in the Rijndael specification, we obtain

$$\begin{aligned}\mathrm{S_{RD}}[x] = {} & \mathsf{05} \cdot x^{254} + \mathsf{09} \cdot x^{253} + \mathsf{F9} \cdot x^{251} + \mathsf{25} \cdot x^{247} \\ & + \mathsf{F4} \cdot x^{239} + \mathsf{01} \cdot x^{223} + \mathsf{B5} \cdot x^{191} + \mathsf{8F} \cdot x^{127} + \mathsf{63}.\end{aligned} \tag{A.4}$$

The coefficients are elements of $\mathrm{GF}(2^8)$. Note that this representation can also be derived by means of the *Lagrange interpolation formula*.

Table A.3. Tabular representation of $\mathrm{Aff}_8(xy)$

		y															
		0	1	2	3	4	5	6	7	8	9	A	B	C	D	E	F
	0	63	7C	5D	42	1F	00	21	3E	9B	84	A5	BA	E7	F8	D9	C6
	1	92	8D	AC	B3	EE	F1	D0	CF	6A	75	54	4B	16	09	28	37
	2	80	9F	BE	A1	FC	E3	C2	DD	78	67	46	59	04	1B	3A	25
	3	71	6E	4F	50	0D	12	33	2C	89	96	B7	A8	F5	EA	CB	D4
	4	A4	BB	9A	85	D8	C7	E6	F9	5C	43	62	7D	20	3F	1E	01
	5	55	4A	6B	74	29	36	17	08	AD	B2	93	8C	D1	CE	EF	F0
	6	47	58	79	66	3B	24	05	1A	BF	A0	81	9E	C3	DC	FD	E2
	7	B6	A9	88	97	CA	D5	F4	EB	4E	51	70	6F	32	2D	0C	13
x	8	EC	F3	D2	CD	90	8F	AE	B1	14	0B	2A	35	68	77	56	49
	9	1D	02	23	3C	61	7E	5F	40	E5	FA	DB	C4	99	86	A7	B8
	A	0F	10	31	2E	73	6C	4D	52	F7	E8	C9	D6	8B	94	B5	AA
	B	FE	E1	C0	DF	82	9D	BC	A3	06	19	38	27	7A	65	44	5B
	C	2B	34	15	0A	57	48	69	76	D3	CC	ED	F2	AF	B0	91	8E
	D	DA	C5	E4	FB	A6	B9	98	87	22	3D	1C	03	5E	41	60	7F
	E	C8	D7	F6	E9	B4	AB	8A	95	30	2F	0E	11	4C	53	72	6D
	F	39	26	07	18	45	5A	7B	64	C1	DE	FF	E0	BD	A2	83	9C

Table A.4. Tabular representation of $\mathrm{Aff}_8^{-1}(xy)$

		y															
		0	1	2	3	4	5	6	7	8	9	A	B	C	D	E	F
	0	05	4F	91	DB	2C	66	B8	F2	57	1D	C3	89	7E	34	EA	A0
	1	A1	EB	35	7F	88	C2	1C	56	F3	B9	67	2D	DA	90	4E	04
	2	4C	06	D8	92	65	2F	F1	BB	1E	54	8A	C0	37	7D	A3	E9
	3	E8	A2	7C	36	C1	8B	55	1F	BA	F0	2E	64	93	D9	07	4D
	4	97	DD	03	49	BE	F4	2A	60	C5	8F	51	1B	EC	A6	78	32
	5	33	79	A7	ED	1A	50	8E	C4	61	2B	F5	BF	48	02	DC	96
	6	DE	94	4A	00	F7	BD	63	29	8C	C6	18	52	A5	EF	31	7B
	7	7A	30	EE	A4	53	19	C7	8D	28	62	BC	F6	01	4B	95	DF
x	8	20	6A	B4	FE	09	43	9D	D7	72	38	E6	AC	5B	11	CF	85
	9	84	CE	10	5A	AD	E7	39	73	D6	9C	42	08	FF	B5	6B	21
	A	69	23	FD	B7	40	0A	D4	9E	3B	71	AF	E5	12	58	86	CC
	B	CD	87	59	13	E4	AE	70	3A	9F	D5	0B	41	B6	FC	22	68
	C	B2	F8	26	6C	9B	D1	0F	45	E0	AA	74	3E	C9	83	5D	17
	D	16	5C	82	C8	3F	75	AB	E1	44	0E	D0	9A	6D	27	F9	B3
	E	FB	B1	6F	25	D2	98	46	0C	A9	E3	3D	77	80	CA	14	5E
	F	5F	15	CB	81	76	3C	E2	A8	0D	47	99	D3	24	6E	B0	FA

Table A.5. Tabular representation of $\mathrm{Inv}_8(xy)$

		y															
		0	1	2	3	4	5	6	7	8	9	A	B	C	D	E	F
	0	00	01	8D	F6	CB	52	7B	D1	E8	4F	29	C0	B0	E1	E5	C7
	1	74	B4	AA	4B	99	2B	60	5F	58	3F	FD	CC	FF	40	EE	B2
	2	3A	6E	5A	F1	55	4D	A8	C9	C1	0A	98	15	30	44	A2	C2
	3	2C	45	92	6C	F3	39	66	42	F2	35	20	6F	77	BB	59	19
	4	1D	FE	37	67	2D	31	F5	69	A7	64	AB	13	54	25	E9	09
	5	ED	5C	05	CA	4C	24	87	BF	18	3E	22	F0	51	EC	61	17
	6	16	5E	AF	D3	49	A6	36	43	F4	47	91	DF	33	93	21	3B
	7	79	B7	97	85	10	B5	BA	3C	B6	70	D0	06	A1	FA	81	82
x	8	83	7E	7F	80	96	73	BE	56	9B	9E	95	D9	F7	02	B9	A4
	9	DE	6A	32	6D	D8	8A	84	72	2A	14	9F	88	F9	DC	89	9A
	A	FB	7C	2E	C3	8F	B8	65	48	26	C8	12	4A	CE	E7	D2	62
	B	0C	E0	1F	EF	11	75	78	71	A5	8E	76	3D	BD	BC	86	57
	C	0B	28	2F	A3	DA	D4	E4	0F	A9	27	53	04	1B	FC	AC	E6
	D	7A	07	AE	63	C5	DB	E2	EA	94	8B	C4	D5	9D	F8	90	6B
	E	B1	0D	D6	EB	C6	0E	CF	AD	08	4E	D7	E3	5D	50	1E	B3
	F	5B	23	38	34	68	46	03	8C	DD	9C	7D	A0	CD	1A	41	1C

A.2 Other Tables

A.2.1 xtime

More explanation about the mapping **xtime** can be found in Sect. 4.1.1. The tabular representation is given in Table A.6.

Table A.6. Tabular representation of xtime(xy)

		y															
		0	1	2	3	4	5	6	7	8	9	A	B	C	D	E	F
	0	00	02	04	06	08	0A	0C	0E	10	12	14	16	18	1A	1C	1E
	1	20	22	24	26	28	2A	2C	2E	30	32	34	36	38	3A	3C	3E
	2	40	42	44	46	48	4A	4C	4E	50	52	54	56	58	5A	5C	5E
	3	60	62	64	66	68	6A	6C	6E	70	72	74	76	78	7A	7C	7E
	4	80	82	84	86	88	8A	8C	8E	90	92	94	96	98	9A	9C	9E
	5	A0	A2	A4	A6	A8	AA	AC	AE	B0	B2	B4	B6	B8	BA	BC	BE
	6	C0	C2	C4	C6	C8	CA	CC	CE	D0	D2	D4	D6	D8	DA	DC	DE
	7	E0	E2	E4	E6	E8	EA	EC	EE	F0	F2	F4	F6	F8	FA	FC	FE
x	8	1B	19	1F	1D	13	11	17	15	0B	09	0F	0D	03	01	07	05
	9	3B	39	3F	3D	33	31	37	35	2B	29	2F	2D	23	21	27	25
	A	5B	59	5F	5D	53	51	57	55	4B	49	4F	4D	43	41	47	45
	B	7B	79	7F	7D	73	71	77	75	6B	69	6F	6D	63	61	67	65
	C	9B	99	9F	9D	93	91	97	95	8B	89	8F	8D	83	81	87	85
	D	BB	B9	BF	BD	B3	B1	B7	B5	AB	A9	AF	AD	A3	A1	A7	A5
	E	DB	D9	DF	DD	D3	D1	D7	D5	CB	C9	CF	CD	C3	C1	C7	C5
	F	FB	F9	FF	FD	F3	F1	F7	F5	EB	E9	EF	ED	E3	E1	E7	E5

A.2.2 Round Constants

The key expansion routine uses round constants. Further explanation can be found in Sect. 3.6. Table A.7 lists the first 30 round constants. Note that RC[0] is never used. In the unlikely case that more values are required, they should be generated according to (3.21).

Table A.7. Round constants for the key generation

i	0	1	2	3	4	5	6	7
RC[i]	00	01	02	04	08	10	20	40
i	8	9	10	11	12	13	14	15
RC[i]	80	1B	36	6C	D8	AB	4D	9A
i	16	17	18	19	20	21	22	23
RC[i]	2F	5E	BC	63	C6	97	35	6A
i	24	25	26	27	28	29	30	31
$RC[i]$	D4	B3	7D	FA	EF	C5	91	39

B. Test Vectors

B.1 KeyExpansion

In this section we give test vectors for the key expansion in the case where both block length and key length are equal to 128. The all-zero key is expanded into the following:

```
 0 00000000000000000000000000000000
 1 62636363626363636263636362636363
 2 9B9898C9F9FBFBAA9B9898C9F9FBFBAA
 3 90973450696CCFFAF2F457330B0FAC99
 4 EE06DA7B876A1581759E42B27E91EE2B
 5 7F2E2B88F8443E098DDA7CBBF34B9290
 6 EC614B851425758C99FF09376AB49BA7
 7 217517873550620BACAF6B3CC61BF09B
 8 0EF903333BA9613897060A04511DFA9F
 9 B1D4D8E28A7DB9DA1D7BB3DE4C664941
10 B4EF5BCB3E92E21123E951CF6F8F188E
```

B.2 Rijndael(128,128)

In this section we give test vectors for all intermediate steps of one encryption. A 128-bit plaintext is encrypted under a 128-bit key. These test vectors are a subset of the extensive set of test vectors generated by Brian Gladman.

```
LEGEND - round r = 0 to 10
input:   cipher input
start:   state at start of round[r]
s_box:   state after s_box substitution
s_row:   state after shift row transformation
m_col:   state after mix column transformation
k_sch:   key schedule value for round[r]
output:  cipher output
```

```
PLAINTEXT:    3243f6a8885a308d313198a2e0370734
KEY:          2b7e151628aed2a6abf7158809cf4f3c
```


J. Daemen, V. Rijmen, *The Design of Rijndael*, Information Security and Cryptography,
https://doi.org/10.1007/978-3-662-60769-5

```
ENCRYPT        16 byte block, 16 byte key
R[00].input    3243f6a8885a308d313198a2e0370734
R[00].k_sch    2b7e151628aed2a6abf7158809cf4f3c
R[01].start    193de3bea0f4e22b9ac68d2ae9f84808
R[01].s_box    d42711aee0bf98f1b8b45de51e415230
R[01].s_row    d4bf5d30e0b452aeb84111f11e2798e5
R[01].m_col    046681e5e0cb199a48f8d37a2806264c
R[01].k_sch    a0fafe1788542cb123a339392a6c7605
R[02].start    a49c7ff2689f352b6b5bea43026a5049
R[02].s_box    49ded28945db96f17f39871a7702533b
R[02].s_row    49db873b453953897f02d2f177de961a
R[02].m_col    584dcaf11b4b5aacdbe7caa81b6bb0e5
R[02].k_sch    f2c295f27a96b9435935807a7359f67f
R[03].start    aa8f5f0361dde3ef82d24ad26832469a
R[03].s_box    ac73cf7befc111df13b5d6b545235ab8
R[03].s_row    acc1d6b8efb55a7b1323cfdf457311b5
R[03].m_col    75ec0993200b633353c0cf7cbb25d0dc
R[03].k_sch    3d80477d4716fe3e1e237e446d7a883b
R[04].start    486c4eee671d9d0d4de3b138d65f58e7
R[04].s_box    52502f2885a45ed7e311c807f6cf6a94
R[04].s_row    52a4c89485116a28e3cf2fd7f6505e07
R[04].m_col    0fd6daa9603138bf6fc0106b5eb31301
R[04].k_sch    ef44a541a8525b7fb671253bdb0bad00
R[05].start    e0927fe8c86363c0d9b1355085b8be01
R[05].s_box    e14fd29be8fbfbba35c89653976cae7c
R[05].s_row    e1fb967ce8c8ae9b356cd2ba974ffb53
R[05].m_col    25d1a9adbd11d168b63a338e4c4cc0b0
R[05].k_sch    d4d1c6f87c839d87caf2b8bc11f915bc
R[06].start    f1006f55c1924cef7cc88b325db5d50c
R[06].s_box    a163a8fc784f29df10e83d234cd503fe
R[06].s_row    a14f3dfe78e803fc10d5a8df4c632923
R[06].m_col    4b868d6d2c4a8980339df4e837d218d8
R[06].k_sch    6d88a37a110b3efddbf98641ca0093fd
R[07].start    260e2e173d41b77de86472a9fdd28b25
R[07].s_box    f7ab31f02783a9ff9b4340d354b53d3f
R[07].s_row    f783403f27433df09bb531ff54aba9d3
R[07].m_col    1415b5bf461615ec274656d7342ad843
R[07].k_sch    4e54f70e5f5fc9f384a64fb24ea6dc4f
R[08].start    5a4142b11949dc1fa3e019657a8c040c
R[08].s_box    be832cc8d43b86c00ae1d44dda64f2fe
R[08].s_row    be3bd4fed4e1f2c80a642cc0da83864d
R[08].m_col    00512fd1b1c889ff54766dcdfa1b99ea
R[08].k_sch    ead27321b58dbad2312bf5607f8d292f
R[09].start    ea835cf00445332d655d98ad8596b0c5
R[09].s_box    87ec4a8cf26ec3d84d4c46959790e7a6
R[09].s_row    876e46a6f24ce78c4d904ad897ecc395
R[09].m_col    473794ed40d4e4a5a3703aa64c9f42bc
R[09].k_sch    ac7766f319fadc2128d12941575c006e
R[10].start    eb40f21e592e38848ba113e71bc342d2
R[10].s_box    e9098972cb31075f3d327d94af2e2cb5
R[10].s_row    e9317db5cb322c723d2e895faf090794
R[10].k_sch    d014f9a8c9ee2589e13f0cc8b6630ca6
R[10].output   3925841d02dc09fbdc118597196a0b32
```

B.3 Other Block Lengths and Key Lengths

The values in this section correspond to the ciphertexts obtained by encrypting the all-zero string with the all-zero key (values on the first lines), and by encrypting the result again with the all-zero key (values on the second lines). The values are given for the five different block lengths and the five different key lengths. The values were generated with the program listed in Appendix C.

```
block length 128  key length 128
66E94BD4EF8A2C3B884CFA59CA342B2E
F795BD4A52E29ED713D313FA20E98DBC

block length 160  key length 128
9E38B8EB1D2025A1665AD4B1F5438BB5CAE1AC3F
939C167E7F916D45670EE21BFC939E1055054A96

block length 192  key length 128
A92732EB488D8BB98ECD8D95DC9C02E052F250AD369B3849
106F34179C3982DDC6750AA01936B7A180E6B0B9D8D690EC

block length 224  key length 128
0623522D88F7B9C63437537157F625DD5697AB628A3B9BE2549895C8
93F93CBDABE23415620E6990B0443D621F6AFBD6EDEFD6990A1965A8

block length 256  key length 128
A693B288DF7DAE5B1757640276439230DB77C4CD7A871E24D6162E54AF434891
5F05857C80B68EA42CCBC759D42C28D5CD490F1D180C7A9397EE585BEA770391

block length 128  key length 160
94B434F8F57B9780F0EFF1A9EC4C112C
35A00EC955DF43417CEAC2AB2B3F3E76

block length 160  key length 160
33B12AB81DB7972E8FDC529DDA46FCB529B31826
97F03EB018C0BB9195BF37C6A0AECE8E4CB8DE5F

block length 192  key length 160
528E2FFF6005427B67BB1ED31ECC09A69EF41531DF5BA5B2
71C7687A4C93EBC35601E3662256E10115BEED56A410D7AC

block length 224  key length 160
58A0C53F3822A32464704D409C2FD0521F3A93E1F6FCFD4C87F1C551
D8E93EF2EB49857049D6F6E0F40B67516D2696F94013C065283F7F01

block length 256  key length 160
938D36E0CB6B7937841DAB7F1668E47B485D3ACD6B3F6D598B0A9F923823331D
7B44491D1B24A93B904D171F074AD69669C2B70B134A4D2D773250A4414D78BE

block length 128  key length 192
AAE06992ACBF52A3E8F4A96EC9300BD7
52F674B7B9030FDAB13D18DC214EB331
```

```
block length 160  key length 192
33060F9D4705DDD2C7675F0099140E5A98729257
012CAB64982156A5710E790F85EC442CE13C520F

block length 192  key length 192
C6348BE20007BAC4A8BD62890C8147A2432E760E9A9F9AB8
EB9DEF13C253F81C1FC2829426ED166A65A105C6A04CA33D

block length 224  key length 192
3856B17BEA77C4611E3397066828AADDA004706A2C8009DF40A811FE
160AD76A97AE2C1E05942FDE3DA2962684A92CCC74B8DC23BDE4F469

block length 256  key length 192
F927363EF5B3B4984A9EB9109844152EC167F08102644E3F9028070433DF9F2A
4E03389C68B2E3F623AD8F7F6BFC88613B86F334F4148029AE25F50DB144B80C

block length 128  key length 224
73F8DFF62A36F3EBF31D6F73A56FF279
3A72F21E10B6473EA9FF14A232E675B4

block length 160  key length 224
E9F5EA0FA39BB6AD7339F28E58E2E7535F261827
06EF9BC82905306D45810E12D0807796A3D338F9

block length 192  key length 224
ECBE9942CD6703E16D358A829D542456D71BD3408EB23C56
FD10458ED034368A34047905165B78A6F0591FFEEBF47CC7

block length 224  key length 224
FE1CF0C8DDAD24E3D751933100E8E89B61CD5D31C96ABFF7209C495C
515D8E2F2B9C5708F112C6DE31CACA47AFB86838B716975A24A09CD4

block length 256  key length 224
BC18BF6D369C955BBB271CBCDD66C368356DBA5B33C0005550D2320B1C617E21
60ABA1D2BE45D8ABFDCF97BCB39F6C17DF29985CF321BAB75E26A26100AC00AF

block length 128  key length 256
DC95C078A2408989AD48A21492842087
08C374848C228233C2B34F332BD2E9D3

block length 160  key length 256
30991844F72973B3B2161F1F11E7F8D9863C5118
EEF8B7CC9DBE0F03A1FE9D82E9A759FD281C67E0

block length 192  key length 256
17004E806FAEF168FC9CD56F98F070982075C70C8132B945
BED33B0AF364DBF15F9C2F3FB24FBDF1D36129C586EEA6B7

block length 224  key length 256
9BF26FAD5680D56B572067EC2FE162F449404C86303F8BE38FAB6E02
658F144A34AF44AAE66CFDDAB955C483DFBCB4EE9A19A6701F158A66
```

```
block length 256  key length 256
C6227E7740B7E53B5CB77865278EAB0726F62366D9AABAD908936123A1FC8AF3
9843E807319C32AD1EA3935EF56A2BA96E4BF19C30E47D88A2B97CBBF2E159E7
```

C. Reference Code

```
/* Rijndael code   August '01
 *
 * author: Vincent Rijmen,

 * This code is based on the official reference code
 * by Paulo Barreto and Vincent Rijmen
 *
 * This code is placed in the public domain.
 * Without any warranty of fitness for any purpose.
 */

#include <stdio.h>

typedef unsigned char word8;
typedef unsigned int word32;

  /* The tables Logtable and Alogtable are used to perform
   * multiplications in GF(256)
   */
word8 Logtable[256] = {
  0,  0, 25,  1, 50,  2, 26,198, 75,199, 27,104, 51,238,223,  3,
100,  4,224, 14, 52,141,129,239, 76,113,  8,200,248,105, 28,193,
125,194, 29,181,249,185, 39,106, 77,228,166,114,154,201,  9,120,
101, 47,138,  5, 33, 15,225, 36, 18,240,130, 69, 53,147,218,142,
150,143,219,189, 54,208,206,148, 19, 92,210,241, 64, 70,131, 56,
102,221,253, 48,191,  6,139, 98,179, 37,226,152, 34,136,145, 16,
126,110, 72,195,163,182, 30, 66, 58,107, 40, 84,250,133, 61,186,
 43,121, 10, 21,155,159, 94,202, 78,212,172,229,243,115,167, 87,
175, 88,168, 80,244,234,214,116, 79,174,233,213,231,230,173,232,
 44,215,117,122,235, 22, 11,245, 89,203, 95,176,156,169, 81,160,
127, 12,246,111, 23,196, 73,236,216, 67, 31, 45,164,118,123,183,
204,187, 62, 90,251, 96,177,134, 59, 82,161,108,170, 85, 41,157,
151,178,135,144, 97,190,220,252,188,149,207,205, 55, 63, 91,209,
 83, 57,132, 60, 65,162,109, 71, 20, 42,158, 93, 86,242,211,171,
 68, 17,146,217, 35, 32, 46,137,180,124,184, 38,119,153,227,165,
103, 74,237,222,197, 49,254, 24, 13, 99,140,128,192,247,112,  7};
```


J. Daemen, V. Rijmen, *The Design of Rijndael*, Information Security and Cryptography,
https://doi.org/10.1007/978-3-662-60769-5

```
word8 Alogtable[256] = {
  1,  3,  5, 15, 17, 51, 85,255, 26, 46,114,150,161,248, 19, 53,
 95,225, 56, 72,216,115,149,164,247,  2,  6, 10, 30, 34,102,170,
229, 52, 92,228, 55, 89,235, 38,106,190,217,112,144,171,230, 49,
 83,245,  4, 12, 20, 60, 68,204, 79,209,104,184,211,110,178,205,
 76,212,103,169,224, 59, 77,215, 98,166,241,  8, 24, 40,120,136,
131,158,185,208,107,189,220,127,129,152,179,206, 73,219,118,154,
181,196, 87,249, 16, 48, 80,240, 11, 29, 39,105,187,214, 97,163,
254, 25, 43,125,135,146,173,236, 47,113,147,174,233, 32, 96,160,
251, 22, 58, 78,210,109,183,194, 93,231, 50, 86,250, 21, 63, 65,
195, 94,226, 61, 71,201, 64,192, 91,237, 44,116,156,191,218,117,
159,186,213,100,172,239, 42,126,130,157,188,223,122,142,137,128,
155,182,193, 88,232, 35,101,175,234, 37,111,177,200, 67,197, 84,
252, 31, 33, 99,165,244,  7,  9, 27, 45,119,153,176,203, 70,202,
 69,207, 74,222,121,139,134,145,168,227, 62, 66,198, 81,243, 14,
 18, 54, 90,238, 41,123,141,140,143,138,133,148,167,242, 13, 23,
 57, 75,221,124,132,151,162,253, 28, 36,108,180,199, 82,246,  1};

word8 S[256] = {
 99,124,119,123,242,107,111,197, 48,  1,103, 43,254,215,171,118,
202,130,201,125,250, 89, 71,240,173,212,162,175,156,164,114,192,
183,253,147, 38, 54, 63,247,204, 52,165,229,241,113,216, 49, 21,
  4,199, 35,195, 24,150,  5,154,  7, 18,128,226,235, 39,178,117,
  9,131, 44, 26, 27,110, 90,160, 82, 59,214,179, 41,227, 47,132,
 83,209,  0,237, 32,252,177, 91,106,203,190, 57, 74, 76, 88,207,
208,239,170,251, 67, 77, 51,133, 69,249,  2,127, 80, 60,159,168,
 81,163, 64,143,146,157, 56,245,188,182,218, 33, 16,255,243,210,
205, 12, 19,236, 95,151, 68, 23,196,167,126, 61,100, 93, 25,115,
 96,129, 79,220, 34, 42,144,136, 70,238,184, 20,222, 94, 11,219,
224, 50, 58, 10, 73,  6, 36, 92,194,211,172, 98,145,149,228,121,
231,200, 55,109,141,213, 78,169,108, 86,244,234,101,122,174,  8,
186,120, 37, 46, 28,166,180,198,232,221,116, 31, 75,189,139,138,
112, 62,181,102, 72,  3,246, 14, 97, 53, 87,185,134,193, 29,158,
225,248,152, 17,105,217,142,148,155, 30,135,233,206, 85, 40,223,
140,161,137, 13,191,230, 66,104, 65,153, 45, 15,176, 84,187, 22};

word8 Si[256] = {
 82,  9,106,213, 48, 54,165, 56,191, 64,163,158,129,243,215,251,
124,227, 57,130,155, 47,255,135, 52,142, 67, 68,196,222,233,203,
 84,123,148, 50,166,194, 35, 61,238, 76,149, 11, 66,250,195, 78,
  8, 46,161,102, 40,217, 36,178,118, 91,162, 73,109,139,209, 37,
114,248,246,100,134,104,152, 22,212,164, 92,204, 93,101,182,146,
108,112, 72, 80,253,237,185,218, 94, 21, 70, 87,167,141,157,132,
144,216,171,  0,140,188,211, 10,247,228, 88,  5,184,179, 69,  6,
208, 44, 30,143,202, 63, 15,  2,193,175,189,  3,  1, 19,138,107,
 58,145, 17, 65, 79,103,220,234,151,242,207,206,240,180,230,115,
150,172,116, 34,231,173, 53,133,226,249, 55,232, 28,117,223,110,
 71,241, 26,113, 29, 41,197,137,111,183, 98, 14,170, 24,190, 27,
252, 86, 62, 75,198,210,121, 32,154,219,192,254,120,205, 90,244,
 31,221,168, 51,136,  7,199, 49,177, 18, 16, 89, 39,128,236, 95,
 96, 81,127,169, 25,181, 74, 13, 45,229,122,159,147,201,156,239,
160,224, 59, 77,174, 42,245,176,200,235,187, 60,131, 83,153, 97,
 23, 43,  4,126,186,119,214, 38,225,105, 20, 99, 85, 33, 12,125};
```

```
word32 RC[30] = {
0x00,0x01,0x02,0x04,0x08,0x10,0x20,0x40,0x80,
0x1B,0x36,0x6C,0xD8,0xAB,0x4D,0x9A,0x2F,0x5E,
0xBC,0x63,0xC6,0x97,0x35,0x6A,0xD4,0xB3,0x7D,
0xFA,0xEF,0xC5};

#define MAXBC           8
#define MAXKC           8
#define MAXROUNDS       14

static word8 shifts[5][4] = {
   0, 1, 2, 3,
   0, 1, 2, 3,
   0, 1, 2, 3,
   0, 1, 2, 4,
   0, 1, 3, 4};

static int numrounds[5][5] = {
    10, 11, 12, 13, 14,
    11, 11, 12, 13, 14,
    12, 12, 12, 13, 14,
    13, 13, 13, 13, 14,
    14, 14, 14, 14, 14};

int BC, KC, ROUNDS;

word8 mul(word8 a, word8 b) {
   /* multiply two elements of GF(256)
    * required for MixColumns and InvMixColumns
    */
    if (a && b) return Alogtable[(Logtable[a] + Logtable[b])%255];
    else return 0;
}

void AddRoundKey(word8 a[4][MAXBC], word8 rk[4][MAXBC]) {
    /* XOR corresponding text input and round key input bytes
     */
    int i, j;

    for(i = 0; i < 4; i++)
        for(j = 0; j < BC; j++) a[i][j] ^= rk[i][j];
}

void SubBytes(word8 a[4][MAXBC], word8 box[256]) {
    /* Replace every byte of the input by the byte at that place
     * in the nonlinear S-box
     */
    int i, j;

    for(i = 0; i < 4; i++)
        for(j = 0; j < BC; j++) a[i][j] = box[a[i][j]] ;
}
```

```
void ShiftRows(word8 a[4][MAXBC], word8 d) {
    /* Row 0 remains unchanged
     * The other three rows are shifted a variable amount
     */
    word8 tmp[MAXBC];
    int i, j;

    if (d == 0) {
        for(i = 1; i < 4; i++) {
            for(j = 0; j < BC; j++)
               tmp[j] = a[i][(j + shifts[BC-4][i]) % BC];
            for(j = 0; j < BC; j++) a[i][j] = tmp[j];
        }
    }
    else {
        for(i = 1; i < 4; i++) {
            for(j = 0; j < BC; j++)
               tmp[j] = a[i][(BC + j - shifts[BC-4][i]) % BC];
            for(j = 0; j < BC; j++) a[i][j] = tmp[j];
        }
    }
}

void MixColumns(word8 a[4][MAXBC]) {
    /* Mix the four bytes of every column in a linear way
     */
    word8 b[4][MAXBC];
    int i, j;

    for(j = 0; j < BC; j++)
        for(i = 0; i < 4; i++)
            b[i][j] = mul(2,a[i][j])
                ^ mul(3,a[(i + 1) % 4][j])
                ^ a[(i + 2) % 4][j]
                ^ a[(i + 3) % 4][j];
    for(i = 0; i < 4; i++)
        for(j = 0; j < BC; j++) a[i][j] = b[i][j];
}

void InvMixColumns(word8 a[4][MAXBC]) {
    /* Mix the four bytes of every column in a linear way
     * This is the opposite operation of Mixcolumns
     */
    word8 b[4][MAXBC];
    int i, j;

    for(j = 0; j < BC; j++)
    for(i = 0; i < 4; i++)
        b[i][j] = mul(0xe,a[i][j])
            ^ mul(0xb,a[(i + 1) % 4][j])
            ^ mul(0xd,a[(i + 2) % 4][j])
            ^ mul(0x9,a[(i + 3) % 4][j]);
```

```
    for(i = 0; i < 4; i++)
        for(j = 0; j < BC; j++) a[i][j] = b[i][j];
}

int KeyExpansion (word8 k[4][MAXKC],
          word8 W[MAXROUNDS+1][4][MAXBC]) {
    /* Calculate the required round keys
     */
    int i, j, t, RCpointer = 1;
    word8 tk[4][MAXKC];


    for(j = 0; j < KC; j++)
        for(i = 0; i < 4; i++)
            tk[i][j] = k[i][j];
    t = 0;
    /* copy values into round key array */
    for(j = 0; (j < KC) && (t < (ROUNDS+1)*BC); j++, t++)
        for(i = 0; i < 4; i++) W[t / BC][i][t % BC] = tk[i][j];

    while (t < (ROUNDS+1)*BC) {
        /* while not enough round key material calculated,
         * calculate new values
         */
        for(i = 0; i < 4; i++)
            tk[i][0] ^= S[tk[(i+1)%4][KC-1]];
        tk[0][0] ^= RC[RCpointer++];

        if (KC <= 6)
            for(j = 1; j < KC; j++)
                for(i = 0; i < 4; i++) tk[i][j] ^= tk[i][j-1];
        else {
            for(j = 1; j < 4; j++)
                for(i = 0; i < 4; i++) tk[i][j] ^= tk[i][j-1];
            for(i = 0; i < 4; i++) tk[i][4] ^= S[tk[i][3]];
            for(j = 5; j < KC; j++)
                for(i = 0; i < 4; i++) tk[i][j] ^= tk[i][j-1];
    }
    /* copy values into round key array */
    for(j = 0; (j < KC) && (t < (ROUNDS+1)*BC); j++, t++)
        for(i = 0; i < 4; i++) W[t / BC][i][t % BC] = tk[i][j];
    }

    return 0;
}

int Encrypt (word8 a[4][MAXBC], word8 rk[MAXROUNDS+1][4][MAXBC])
{
    /* Encryption of one block.
     */
    int r;

    /* begin with a key addition
```

```
     */
    AddRoundKey(a,rk[0]);

        /* ROUNDS-1 ordinary rounds
     */
    for(r = 1; r < ROUNDS; r++) {
        SubBytes(a,S);
        ShiftRows(a,0);
        MixColumns(a);
        AddRoundKey(a,rk[r]);
    }

    /* Last round is special: there is no MixColumns
     */
    SubBytes(a,S);
    ShiftRows(a,0);
    AddRoundKey(a,rk[ROUNDS]);

    return 0;
}


int Decrypt (word8 a[4][MAXBC], word8 rk[MAXROUNDS+1][4][MAXBC])
{
    int r;

    /* To decrypt:
     *   apply the inverse operations of the encrypt routine,
     *   in opposite order
     *
     * - AddRoundKey is equal to its inverse)
     * - the inverse of SubBytes with table S is
     *             SubBytes with the inverse table of S)
     * - the inverse of Shiftrows is Shiftrows over
     *       a suitable distance)
     */

    /* First the special round:
     *   without InvMixColumns
     *   with extra AddRoundKey
     */
    AddRoundKey(a,rk[ROUNDS]);
    SubBytes(a,Si);
    ShiftRows(a,1);

    /* ROUNDS-1 ordinary rounds
     */
    for(r = ROUNDS-1; r > 0; r--) {
        AddRoundKey(a,rk[r]);
        InvMixColumns(a);
        SubBytes(a,Si);
        ShiftRows(a,1);
    }
```

```
    /* End with the extra key addition
     */
    AddRoundKey(a,rk[0]);

    return 0;
}


int main() {

    int i, j;
    word8 a[4][MAXBC], rk[MAXROUNDS+1][4][MAXBC], sk[4][MAXKC];

    for(KC = 4; KC <= 8; KC++)
        for(BC = 4; BC <= 8; BC++) {
            ROUNDS = numrounds[KC-4][BC-4];
            for(j = 0; j < BC; j++)
                for(i = 0; i < 4; i++) a[i][j] = 0;
            for(j = 0; j < KC; j++)
                for(i = 0; i < 4; i++) sk[i][j] = 0;
            KeyExpansion(sk,rk);
            Encrypt(a,rk);
            printf("block length %d  key length %d\n",32*BC,32*KC);
            for(j = 0; j < BC; j++)
                for(i = 0; i < 4; i++) printf("%02X",a[i][j]);
            printf("\n");

            Decrypt(a,rk);
            for(j = 0; j < BC; j++)
                for(i = 0; i < 4; i++) printf("%02X",a[i][j]);
            printf("\n");
            printf("\n");
        }

    return 0;
}
```

Bibliography

1. FIPS 46. Data Encryption Standard, Federal Information Processing Standard (FIPS), publication 46. National Bureau of Standards, 1977.
2. Carlisle M. Adams and Stafford E. Tavares. The structured design of cryptographically good S-boxes. *J. Cryptology*, 3(1):27–41, 1990.
3. *Proc. of the third advanced encryption standard candidate conference*, 2000.
4. Ross A. Anderson, Eli Biham, and Lars R. Knudsen. Serpent: a flexible block cipher with maximum assurance. `http://www.cl.cam.ac.uk/~rja14/Papers/ventura.pdf`.
5. Ross J. Anderson, editor. *Fast Software Encryption, Cambridge Security Workshop, Cambridge, UK, December 9-11, 1993, Proceedings*, volume 809 of *Lecture Notes in Computer Science*. Springer, 1994.
6. Kazumaro Aoki. On maximum non-averaged differential probability. In Stafford E. Tavares and Henk Meijer, editors, *Selected Areas in Cryptography '98, SAC'98, Kingston, Ontario, Canada, August 17-18, 1998, Proceedings*, volume 1556 of *Lecture Notes in Computer Science*, pages 118–130. Springer, 1998.
7. Kazumaro Aoki, Tetsuya Ichikawa, Masayuki Kanda, Mitsuru Matsui, Shiho Moriai, Junko Nakajima, and Toshio Tokita. Camellia: A 128-bit block cipher suitable for multiple platforms - design and analysis. In Douglas R. Stinson and Stafford E. Tavares, editors, *Selected Areas in Cryptography, 7th Annual International Workshop, SAC 2000, Waterloo, Ontario, Canada, August 14-15, 2000, Proceedings*, volume 2012 of *Lecture Notes in Computer Science*, pages 39–56. Springer, 2000.
8. Thomas Baignères, Jacques Stern, and Serge Vaudenay. Linear cryptanalysis of non binary ciphers. In Carlisle M. Adams, Ali Miri, and Michael J. Wiener, editors, *Selected Areas in Cryptography*, volume 4876 of *Lecture Notes in Computer Science*, pages 184–211. Springer, 2007.
9. Paulo S.L.M. Barreto and Vincent Rijmen. The Anubis block cipher. `https://www.cosic.esat.kuleuven.be/nessie/`.
10. Olivier Baudron, Henri Gilbert, Louis Granboulan, Helena Handschuh, Antoine Joux, Phong Nguyen, Fabrice Noilhan, David Pointcheval, Thomas Pornin, Guillaume Poupard, Jacques Stern, and Serge Vaudenay. Report on the AES candidates. `https://web.archive.org/web/20020202072155/http://csrc.nist.gov/encryption/aes/round1/conf2/papers/baudron1.pdf`, 1999.
11. Mihir Bellare, Joe Kilian, and Phillip Rogaway. The security of cipher block chaining. In Desmedt [56], pages 341–358.

J. Daemen, V. Rijmen, *The Design of Rijndael*, Information Security and Cryptography,
https://doi.org/10.1007/978-3-662-60769-5

12. Mihir Bellare, Joe Kilian, and Phillip Rogaway. The security of the cipher block chaining message authentication code. *J. Comput. Syst. Sci.*, 61(3):362–399, 2000.

13. Mihir Bellare and Tadayoshi Kohno. A theoretical treatment of related-key attacks: RKA-PRPs, RKA-PRFs, and applications. In Eli Biham, editor, *Advances in Cryptology - EUROCRYPT 2003, International Conference on the Theory and Applications of Cryptographic Techniques, Warsaw, Poland, May 4-8, 2003, Proceedings*, volume 2656 of *Lecture Notes in Computer Science*, pages 491–506. Springer, 2003.

14. Ishai Ben-Aroya and Eli Biham. Differential cryptanalysis of Lucifer. *J. Cryptology*, 9(1):21–34, 1996.

15. Ryad Benadjila, Olivier Billet, Shay Gueron, and Matthew J. B. Robshaw. The Intel AES instructions set and the SHA-3 candidates. In Matsui [107], pages 162–178.

16. Ryad Bendajila. Use of the AES instruction set. `https://www.cosic.esat.kuleuven.be/ecrypt/AESday/slides/Use_of_the_AES_Instruction_Set.pdf`. ECRYPT II AES Day, 18 October 2012, Bruges, Belgium.

17. Guido Bertoni, Vittorio Zaccaria, Luca Breveglieri, Matteo Monchiero, and Gianluca Palermo. AES power attack based on induced cache miss and countermeasure. In *International Symposium on Information Technology: Coding and Computing (ITCC 2005), Volume 1, 4-6 April 2005, Las Vegas, Nevada, USA*, pages 586–591. IEEE Computer Society, 2005.

18. Eli Biham. New types of cryptoanalytic attacks using related keys (extended abstract). In Helleseth [70], pages 398–409.

19. Eli Biham. A fast new DES implementation in software. In *Fast Software Encryption, 4th International Workshop, FSE '97, Haifa, Israel, January 20-22, 1997, Proceedings* [20], pages 260–272.

20. Eli Biham, editor. *Fast Software Encryption, 4th International Workshop, FSE '97, Haifa, Israel, January 20-22, 1997, Proceedings*, volume 1267 of *Lecture Notes in Computer Science*. Springer, 1997.

21. Eli Biham. A note on comparing the AES candidates. `https://web.archive.org/web/20020202072155/http://csrc.nist.gov/encryption/aes/round1/conf2/papers/biham2.pdf`, 1999.

22. Eli Biham, Alex Biryukov, and Adi Shamir. Cryptanalysis of Skipjack reduced to 31 rounds using impossible differentials. In Jacques Stern, editor, *Advances in Cryptology - EUROCRYPT '99, International Conference on the Theory and Application of Cryptographic Techniques, Prague, Czech Republic, May 2-6, 1999, Proceeding*, volume 1592 of *Lecture Notes in Computer Science*, pages 12–23. Springer, 1999.

23. Eli Biham, Orr Dunkelman, and Nathan Keller. Linear cryptanalysis of reduced round Serpent. In Matsui [106], pages 16–27.

24. Eli Biham, Orr Dunkelman, and Nathan Keller. Related-key impossible differential attacks on 8-round AES-192. In David Pointcheval, editor, *CT-RSA*, volume 3860 of *Lecture Notes in Computer Science*, pages 21–33. Springer, 2006.

25. Eli Biham and Nathan Keller. Cryptanalysis of reduced variants of Rijndael. In AES3 [3].

26. Eli Biham and Adi Shamir. Differential cryptanalysis of DES-like cryptosystems. *J. Cryptology*, 4(1):3–72, 1991.

27. Eli Biham and Adi Shamir. Power analysis of the key scheduling of the AES candidates. `https://web.archive.org/web/20020202072155/http://csrc.nist.gov/encryption/aes/round1/conf2/papers/biham3.pdf`, 1999.

28. Alex Biryukov and Dmitry Khovratovich. Related-key cryptanalysis of the full AES-192 and AES-256. In Matsui [107], pages 1–18.

29. Alex Biryukov, Dmitry Khovratovich, and Ivica Nikolic. Distinguisher and related-key attack on the full AES-256. In Shai Halevi, editor, *CRYPTO*, volume 5677 of *Lecture Notes in Computer Science*, pages 231–249. Springer, 2009.

30. Alex Biryukov and Adi Shamir. Structural cryptanalysis of SASAS. In Birgit Pfitzmann, editor, *EUROCRYPT*, volume 2045 of *Lecture Notes in Computer Science*, pages 394–405. Springer, 2001.

31. Alex Biryukov and David Wagner. Slide attacks. In Knudsen [86], pages 245–259.

32. Andrey Bogdanov, Dmitry Khovratovich, and Christian Rechberger. Biclique cryptanalysis of the full AES. In Dong Hoon Lee and Xiaoyun Wang, editors, *Advances in Cryptology - ASIACRYPT 2011 - 17th International Conference on the Theory and Application of Cryptology and Information Security, Seoul, South Korea, December 4-8, 2011. Proceedings*, volume 7073 of *Lecture Notes in Computer Science*, pages 344–371. Springer, 2011.

33. Charles Bouillaguet, Patrick Derbez, Orr Dunkelman, Pierre-Alain Fouque, Nathan Keller, and Vincent Rijmen. Low-data complexity attacks on AES. *IEEE Transactions on Information Theory*, 58(11):7002–7017, 2012.

34. Christina Boura and Anne Canteaut. Another view of the division property. In Robshaw and Katz [134], pages 654–682.

35. R. Canetti, O. Goldreich, and S. Halevi. The random oracle methodology, revisited. In *Proceedings of the 30th Annual ACM Symposium on the Theory of Computing*, pages 209–218. ACM Press, 1998.

36. David Canright. A very compact S-box for AES. In Josyula R. Rao and Berk Sunar, editors, *CHES*, volume 3659 of *Lecture Notes in Computer Science*, pages 441–455. Springer, 2005.

37. David Canright and Dag Arne Osvik. A more compact AES. In Michael J. Jacobson Jr., Vincent Rijmen, and Reihaneh Safavi-Naini, editors, *Selected Areas in Cryptography*, volume 5867 of *Lecture Notes in Computer Science*, pages 157–169. Springer, 2009.

38. Anne Canteaut. Differential cryptanalysis of Feistel ciphers and differentially δ-uniform mappings. Workshop record of Selected Areas in Cryptography SAC '97, 1997.

39. Suresh Chari, Charanjit Jutla, Josyula R. Rao, and Pankaj Rohatgi. A cautionary note regarding evaluation of AES candidates on smart

cards. https://web.archive.org/web/20020202072155/http://csrc.nist.gov/encryption/aes/round1/conf2/papers/chari.pdf, 1999.

40. Carlos Cid, Sean Murphy, and Matthew J. B. Robshaw. *Algebraic aspects of the advanced encryption standard.* Springer, 2006.

41. Craig S.K. Clapp. Instruction-level parallelism in AES candidates. https://web.archive.org/web/20000816072548/http://csrc.nist.gov/archive/aes/round1/conf2/papers/clapp.pdf, 1999.

42. Toshiba corporation. Specification of Hierocrypt-L1. https://www.cosic.esat.kuleuven.be/nessie/.

43. Joan Daemen, René Govaerts, and Joos Vandewalle. A new approach to block cipher design. In Anderson [5], pages 18–32.

44. Joan Daemen, Lars R. Knudsen, and Vincent Rijmen. The block cipher Square. In Biham [20], pages 149–165.

45. Joan Daemen, Michael Peeters, and Gilles Van Assche. Bitslice ciphers and power analysis attacks. In Schneier [137], pages 134–149.

46. Joan Daemen, Michaël Peeters, Vincent Rijmen, and Gilles Van Assche. Nessie proposal: Noekeon. http://gro.noekeon.org.

47. Joan Daemen and Vincent Rijmen. The Rijndael block cipher. https://csrc.nist.gov/CSRC/media/Projects/Cryptographic-Standards-and-Guidelines/documents/aes-development/Rijndael-ammended.pdf.

48. Joan Daemen and Vincent Rijmen. The block cipher BKSQ. In Jean-Jacques Quisquater and Bruce Schneier, editors, *CARDIS*, volume 1820 of *Lecture Notes in Computer Science*, pages 236–245. Springer, 1998.

49. Joan Daemen and Vincent Rijmen. Resistance against implementation attacks: a comparative study of the AES proposals. https://web.archive.org/web/20000816072451/http://csrc.nist.gov/encryption/aes/round1/conf2/papers/daemen.pdf, 1999.

50. Joan Daemen and Vincent Rijmen. Plateau characteristics. *IET Information Security*, 1(1):11–17, 2007.

51. Joan Daemen and Vincent Rijmen. New criteria for linear maps in AES-like ciphers. *Cryptography and Communications*, 1(1):47–69, 2009.

52. Joan Daemen and Vincent Rijmen. Correlation analysis in GF(2^n). In *Advanced Linear Cryptanalysis of Block and Stream Ciphers*, pages 115–131. 2011.

53. Joan Daemen and Vincent Rijmen. On the related-key attacks against AES. *Proceedings of the Romanian Academy - Series A: Mathematics, Physics, Technical Sciences, Information Science*, 13(4):395–400, 2012.

54. Donald W. Davies. Some regular properties of the 'Data Encryption Standard' algorithm. In David Chaum, Ronald L. Rivest, and Alan T. Sherman, editors, *CRYPTO*, pages 89–96. Plenum Press, New York, 1982.

55. Hüseyin Demirci and Ali Aydin Selçuk. A meet-in-the-middle attack on 8-round AES. In Kaisa Nyberg, editor, *FSE*, volume 5086 of *Lecture Notes in Computer Science*, pages 116–126. Springer, 2008.

56. Yvo Desmedt, editor. *Advances in Cryptology - CRYPTO '94, 14th Annual International Cryptology Conference, Santa Barbara, California, USA, August 21-25, 1994, Proceedings*, volume 839 of *Lecture Notes in Computer Science*. Springer, 1994.

57. Hans Dobbertin, Vincent Rijmen, and Aleksandra Sowa, editors. *Advanced Encryption Standard - AES, 4th International Conference, AES 2004, Bonn, Germany, May 10-12, 2004, Revised Selected and Invited Papers*, volume 3373 of *Lecture Notes in Computer Science*. Springer, 2005.

58. Orr Dunkelman, editor. *Fast Software Encryption, 16th International Workshop, FSE 2009, Leuven, Belgium, February 22-25, 2009, Revised Selected Papers*, volume 5665 of *Lecture Notes in Computer Science*. Springer, 2009.

59. Orr Dunkelman and Nathan Keller. A new criterion for nonlinearity of block ciphers. *IEEE Trans. Information Theory*, 53(11):3944–3957, 2007.

60. H. Feistel, W.A. Notz, and J.L. Smith. Some cryptographic techniques for machine-to-machine data communications. *Proc. IEEE*, 63(11):1545–1554, 1975.

61. Niels Ferguson, John Kelsey, Stefan Lucks, Bruce Schneier, Michael Stay, David Wagner, and Doug Whiting. Improved cryptanalysis of Rijndael. In Schneier [137], pages 213–230.

62. Niels Ferguson, Richard Schroeppel, and Doug Whiting. A simple algebraic representation of Rijndael. In Serge Vaudenay and Amr M. Youssef, editors, *Selected Areas in Cryptography*, volume 2259 of *Lecture Notes in Computer Science*, pages 103–111. Springer, 2001.

63. Pierre-Alain Fouque, Jérémy Jean, and Thomas Peyrin. Structural evaluation of AES and chosen-key distinguisher of 9-round AES-128. In Ran Canetti and Juan A. Garay, editors, *CRYPTO (1)*, volume 8042 of *Lecture Notes in Computer Science*, pages 183–203. Springer, 2013.

64. Henri Gilbert and Marine Minier. A collision attack on 7 rounds of Rijndael. In AES3 [3], pages 230–241.

65. Brian Gladman. Implementation experience with the AES candidate algorithms. `https://web.archive.org/web/20000816072444/http://csrc.nist.gov/encryption/aes/round1/conf2/papers/gladman.pdf`, 1999.

66. S.W. Golomb. *Shift Register Sequences*. Holden–Day Inc., 1967.

67. Lorenzo Grassi, Christian Rechberger, and Sondre Rønjom. Subspace trail cryptanalysis and its applications to AES. *IACR Trans. Symmetric Cryptol.*, 2016(2):192–225, 2016.

68. Shay Gueron. Intel's new AES instructions for enhanced performance and security. In Dunkelman [58], pages 51–66.

69. Gaël Hachez, François Koeune, and Jean-Jacques Quisquater. cAESar results: Implementation of four AES candidates on two smart cards. `https://web.archive.org/web/20040421103357/http://csrc.nist.gov/CryptoToolkit/aes/round1/conf2/papers/hachez.pdf`, 1999.

70. Tor Helleseth, editor. *Advances in Cryptology - EUROCRYPT '93, Workshop on the Theory and Application of of Cryptographic Techniques, Lofthus, Nor-*

way, May 23-27, 1993, Proceedings, volume 765 of *Lecture Notes in Computer Science*. Springer, 1994.

71. M. Hellman, R. Merkle, R. Schroeppel, L. Washington, W. Diffie, S. Pohlig, and P. Schweitzer. Results of an initial attempt to cryptanalyze the NBS Data Encryption Standard. Information Systems Lab., Dept. of Electrical Eng., Stanford Univ., 1976.

72. Goce Jakimoski and Yvo Desmedt. Related-key differential cryptanalysis of 192-bit key AES variants. In Mitsuru Matsui and Robert J. Zuccherato, editors, *Selected Areas in Cryptography*, volume 3006 of *Lecture Notes in Computer Science*, pages 208–221. Springer, 2003.

73. Thomas Jakobsen and Lars R. Knudsen. The interpolation attack on block ciphers. In Biham [20], pages 28–40.

74. Thomas Jakobson. Cryptanalysis of block ciphers with probabilistic non-linear relations of low degree. In Hugo Krawczyk, editor, *CRYPTO*, volume 1462 of *Lecture Notes in Computer Science*, pages 212–222. Springer, 1998.

75. J.B. Kam and G.I. Davida. Structured design of substitution-permutation encryption networks. *IEEE Trans. on Computers*, C(28):747–753, 1979.

76. Emilia Käsper and Peter Schwabe. Faster and timing-attack resistant AES-GCM. In Christophe Clavier and Kris Gaj, editors, *Cryptographic Hardware and Embedded Systems - CHES 2009, 11th International Workshop, Lausanne, Switzerland, September 6-9, 2009, Proceedings*, volume 5747 of *Lecture Notes in Computer Science*, pages 1–17. Springer, 2009.

77. Jonathan Katz and Yehuda Lindell. *Introduction to Modern Cryptography, Second Edition.* CRC Press, 2014.

78. Geoffrey Keating. Performance analysis of AES candidates on the 6805 CPU core. `https://web.archive.org/web/20020202072155/http://csrc.nist.gov/encryption/aes/round1/conf2/papers/keating.pdf`, 1999.

79. Liam Keliher. Refined analysis of bounds related to linear and differential cryptanalysis for the AES. In Dobbertin et al. [57], pages 42–57.

80. Liam Keliher and Jiayuan Sui. Exact maximum expected differential and linear probability for two-round Advanced Encryption Standard. *IET Information Security*, 1(2):53–57, 2007.

81. John Kelsey, Bruce Schneier, and David Wagner. Key-schedule cryptoanalysis of IDEA, G-DES, GOST, SAFER, and Triple-DES. In Koblitz [89], pages 237–251.

82. Jongsung Kim, Seokhie Hong, and Bart Preneel. Related-key rectangle attacks on reduced AES-192 and AES-256. In Alex Biryukov, editor, *FSE*, volume 4593 of *Lecture Notes in Computer Science*, pages 225–241. Springer, 2007.

83. Lars Knudsen. Deal - a 128-bit block cipher. In *NIST AES Proposal*, 1998.

84. Lars R. Knudsen. Truncated and higher order differentials. In Preneel [129], pages 196–211.

85. Lars R. Knudsen. A key-schedule weakness in SAFER K-64. In Don Coppersmith, editor, *CRYPTO*, volume 963 of *Lecture Notes in Computer Science*, pages 274–286. Springer, 1995.

86. Lars R. Knudsen, editor. *Fast Software Encryption, 6th International Workshop, FSE '99, Rome, Italy, March 24-26, 1999, Proceedings*, volume 1636 of *Lecture Notes in Computer Science*. Springer, 1999.

87. Lars R. Knudsen and Vincent Rijmen. On the decorrelated fast cipher (DFC) and its theory. In Knudsen [86], pages 81–94.

88. Lars R. Knudsen and David A. Wagner. Integral cryptanalysis. In Joan Daemen and Vincent Rijmen, editors, *Fast Software Encryption, 9th International Workshop, FSE 2002, Leuven, Belgium, February 4-6, 2002, Revised Papers*, volume 2365 of *Lecture Notes in Computer Science*, pages 112–127. Springer, 2002.

89. Neal Koblitz, editor. *Advances in Cryptology - CRYPTO '96, 16th Annual International Cryptology Conference, Santa Barbara, California, USA, August 18-22, 1996, Proceedings*, volume 1109 of *Lecture Notes in Computer Science*. Springer, 1996.

90. Paul C. Kocher. Timing attacks on implementations of Diffie-Hellman, RSA, DSS, and other systems. In Koblitz [89], pages 104–113.

91. Paul C. Kocher, Joshua Jaffe, and Benjamin Jun. Differential power analysis. In Michael J. Wiener, editor, *CRYPTO*, volume 1666 of *Lecture Notes in Computer Science*, pages 388–397. Springer, 1999.

92. Gilles Lachaud and Jacques Wolfmann. The weights of the orthogonals of the extended quadratic binary Goppa codes. *IEEE Transactions on Information Theory*, 36(3):686–692, 1990.

93. Xuejia Lai, James L. Massey, and Sean Murphy. Markov ciphers and differential cryptanalysis. In Donald W. Davies, editor, *EUROCRYPT*, volume 547 of *Lecture Notes in Computer Science*, pages 17–38. Springer, 1991.

94. Mario Lamberger, Florian Mendel, Martin Schläffer, Christian Rechberger, and Vincent Rijmen. The rebound attack and subspace distinguishers: Application to Whirlpool. *J. Cryptology*, 28(2):257–296, 2015.

95. Rudolf Lidl and Harald Niederreiter. *Introduction to finite fields and their applications*. Cambridge University Press, 1986 (Reprinted 1988).

96. Chae Hoon Lim. Crypton: a new 128-bit block cipher. `http://citeseerx.ist.psu.edu/viewdoc/summary?doi=10.1.1.52.5771`.

97. Michael Luby and Charles Rackoff. How to construct pseudorandom permutations from pseudorandom functions. *SIAM J. Comput.*, 17(2):373–386, 1988.

98. Stefan Lucks. Attacking 7 rounds of Rijndael under 192-bit and 256-bit keys. In AES3 [3], pages 215–229.

99. Stefan Lucks. The saturation attack - a bait for Twofish. In Matsui [106], pages 1–15.

100. F.J. MacWilliams and N.J.A. Sloane. *The Theory of Error-Correcting Codes*. North-Holland Publishing Company, 1978.

101. Hamid Mala, Mohammad Dakhilalian, Vincent Rijmen, and Mahmoud Modarres-Hashemi. Improved impossible differential cryptanalysis of 7-round AES-128. In Guang Gong and Kishan Chand Gupta, editors, *INDOCRYPT*, volume 6498 of *Lecture Notes in Computer Science*, pages 282–291. Springer, 2010.

102. Keith Martin. *Everyday Cryptography: Fundamental Principles & Applications.* Oxford University Press, 2nd edition, 2017.

103. James L. Massey. SAFER K-64: A byte-oriented block-ciphering algorithm. In Anderson [5], pages 1–17.

104. Mitsuru Matsui. Linear cryptoanalysis method for DES cipher. In Helleseth [70], pages 386–397.

105. Mitsuru Matsui. The first experimental cryptanalysis of the Data Encryption Standard. In Desmedt [56], pages 1–11.

106. Mitsuru Matsui, editor. *Fast Software Encryption, 8th International Workshop, FSE 2001 Yokohama, Japan, April 2-4, 2001, Revised Papers*, volume 2355 of *Lecture Notes in Computer Science.* Springer, 2002.

107. Mitsuru Matsui, editor. *Advances in Cryptology - ASIACRYPT 2009, 15th International Conference on the Theory and Application of Cryptology and Information Security, Tokyo, Japan, December 6-10, 2009. Proceedings*, volume 5912 of *Lecture Notes in Computer Science.* Springer, 2009.

108. R.J. McEliece. *Finite Fields for Computer Scientists and Engineers.* Kluwer Academic Publishers, 1987.

109. Florian Mendel, Christian Rechberger, Martin Schläffer, and Søren S. Thomsen. The rebound attack: Cryptanalysis of reduced Whirlpool and Grøstl. In Dunkelman [58], pages 260–276.

110. Alfred J. Menezes, Paul C. van Oorschot, and Scott A. Vanstone. *Handbook of Applied Cryptography.* CRC Press, 1996.

111. Nele Mentens, Lejla Batina, Bart Preneel, and Ingrid Verbauwhede. A systematic evaluation of compact hardware implementations for the Rijndael S-box. In Alfred Menezes, editor, *Topics in Cryptology - CT-RSA 2005, The Cryptographers' Track at the RSA Conference 2005, San Francisco, CA, USA, February 14-18, 2005, Proceedings*, volume 3376 of *Lecture Notes in Computer Science*, pages 323–333. Springer, 2005.

112. Marine Minier. A three rounds property of the AES. In Dobbertin et al. [57], pages 16–26.

113. Marine Minier, Raphael C.-W. Phan, and Benjamin Pousse. On integral distinguishers of Rijndael family of ciphers. *Cryptologia*, 36(2):104–118, 2012.

114. Sean Murphy. The effectiveness of the linear hull effect. *J. Mathematical Cryptology*, 6(2):137–147, 2012.

115. Sean Murphy and Matthew J. B. Robshaw. Essential algebraic structure within the AES. In Moti Yung, editor, *CRYPTO*, volume 2442 of *Lecture Notes in Computer Science*, pages 1–16. Springer, 2002.

116. Jorge Nakahara, Jr., Daniel Santana de Freitas, and Raphael Chung-Wei Phan. New multiset attacks on Rijndael with large blocks. In Ed Dawson and Serge Vaudenay, editors, *Mycrypt*, volume 3715 of *Lecture Notes in Computer Science*, pages 277–295. Springer, 2005.

117. James Nechvatal, Elaine Barker, Lawrence Bassham, William Burr, Morris Dworkin, James Foti, and Edward Roback. Report on the Development of the Advanced Encryption Standard (AES). *Journal of Research of the National*

Institute of Standards and Technology, 106(3), 2001. `https://nvlpubs.nist.gov/nistpubs/jres/106/3/j63nec.pdf`.

118. James Nechvatal, Elaine Barker, Donna Dodson, Morris Dworkin, James Foti, and Edward Roback. Status report on the first round of the development of the Advanced Encryption Standard. *Journal of Research of the National Institute of Standards and Technology*, 104(5), 1999. `https://nvlpubs.nist.gov/nistpubs/jres/104/5/j45nec.pdf`.

119. Svetla Nikova, Vincent Rijmen, and Martin Schläffer. Using normal bases for compact hardware implementations of the AES S-box. In Rafail Ostrovsky, Roberto De Prisco, and Ivan Visconti, editors, *SCN*, volume 5229 of *Lecture Notes in Computer Science*, pages 236–245. Springer, 2008.

120. Kaisa Nyberg. Differentially uniform mappings for cryptography. In Helleseth [70], pages 55–64.

121. Kaisa Nyberg. Linear approximation of block ciphers. In Alfredo De Santis, editor, *EUROCRYPT*, volume 950 of *Lecture Notes in Computer Science*, pages 439–444. Springer, 1994.

122. Kaisa Nyberg and Lars R. Knudsen. Provable security against a differential attack. *J. Cryptology*, 8(1):27–37, 1995.

123. Luke O'Connor. On the distribution of characteristics in bijective mappings. *J. Cryptology*, 8(2):67–86, 1995.

124. Christof Paar and Martin Rosner. Comparison of arithmetic architectures for Reed-Solomon decoders in reconfigurable hardware. Fifth annual IEEE symposium on field-programmable custom computing machines (FCCM '97).

125. Dan Page. Theoretical use of cache memory as a cryptanalytic side-channel. *IACR Cryptology ePrint Archive*, 2002:169, 2002.

126. Sangwoo Park, Soo Hak Sung, Seongtaek Chee, E-Joong Yoon, and Jongin Lim. On the security of Rijndael-like structures against differential and linear cryptanalysis. In Yuliang Zheng, editor, *ASIACRYPT*, volume 2501 of *Lecture Notes in Computer Science*, pages 176–191. Springer, 2002.

127. Sangwoo Park, Soo Hak Sung, Sangjin Lee, and Jongin Lim. Improving the upper bound on the maximum differential and the maximum linear hull probability for SPN structures and AES. In Thomas Johansson, editor, *FSE*, volume 2887 of *Lecture Notes in Computer Science*, pages 247–260. Springer, 2003.

128. Bart Preneel. Analysis and design of cryptographic hash functions. Doctoral Dissertation, KU Leuven, 1993.

129. Bart Preneel, editor. *Fast Software Encryption: Second International Workshop. Leuven, Belgium, 14-16 December 1994, Proceedings*, volume 1008 of *Lecture Notes in Computer Science*. Springer, 1995.

130. Vincent Rijmen. Cryptanalysis and design of iterated block ciphers. Doctoral Dissertation, KU Leuven, 1997.

131. Vincent Rijmen, Paulo S. L. M. Barreto, and Décio L. Gazzoni Filho. Rotation symmetry in algebraically generated cryptographic substitution tables. *Inf. Process. Lett.*, 106(6):246–250, 2008.

132. Vincent Rijmen, Joan Daemen, Bart Preneel, Antoon Bosselaers, and Erik De Win. The cipher SHARK. In Dieter Gollmann, editor, *FSE*, volume 1039 of *Lecture Notes in Computer Science*, pages 99–111. Springer, 1996.

133. Ronald L. Rivest, Matthew J.B. Robshaw, Ray Sidney, and Yiqun L. Yin. The RC6 block cipher. `http://people.csail.mit.edu/rivest/pubs/RRSY98.pdf`.

134. Matthew Robshaw and Jonathan Katz, editors. *Advances in Cryptology - CRYPTO 2016 - 36th Annual International Cryptology Conference, Santa Barbara, CA, USA, August 14-18, 2016, Proceedings, Part I*, volume 9814 of *Lecture Notes in Computer Science*. Springer, 2016.

135. Sondre Rønjom, Navid Ghaedi Bardeh, and Tor Helleseth. Yoyo tricks with AES. In Tsuyoshi Takagi and Thomas Peyrin, editors, *Advances in Cryptology - ASIACRYPT 2017 - 23rd International Conference on the Theory and Applications of Cryptology and Information Security, Hong Kong, China, December 3-7, 2017, Proceedings, Part I*, volume 10624 of *Lecture Notes in Computer Science*, pages 217–243. Springer, 2017.

136. Akashi Satoh, Sumio Morioka, Kohji Takano, and Seiji Munetoh. A compact Rijndael hardware architecture with S-box optimization. In Colin Boyd, editor, *ASIACRYPT*, volume 2248 of *Lecture Notes in Computer Science*, pages 239–254. Springer, 2001.

137. Bruce Schneier, editor. *Fast Software Encryption, 7th International Workshop, FSE 2000, New York, NY, USA, April 10-12, 2000, Proceedings*, volume 1978 of *Lecture Notes in Computer Science*. Springer, 2001.

138. Bruce Schneier, John Kelsey, Dough Whiting, David Wagner, Chris Hall, and Niels Ferguson. Performance comparison of the AES submissions. `https://web.archive.org/web/20010603004848/http://csrc.nist.gov/archive/aes/round1/conf2/papers/schneier1.pdf`, 1999.

139. Claude E. Shannon. A mathematical theory of communication. *Bell Syst. Tech. Journal*, 27(3):379–423, 623–656, 1948.

140. Claude E. Shannon. Communication theory of secrecy systems. *Bell Syst. Tech. Journal*, 28:656–715, 1949.

141. Bing Sun, Meicheng Liu, Jian Guo, Longjiang Qu, and Vincent Rijmen. New insights on AES-like SPN ciphers. In Robshaw and Katz [134], pages 605–624.

142. Jonathan Swift. Gulliver's travels.

143. Yosuke Todo. Structural evaluation by generalized integral property. In Elisabeth Oswald and Marc Fischlin, editors, *Advances in Cryptology - EUROCRYPT 2015 - 34th Annual International Conference on the Theory and Applications of Cryptographic Techniques, Sofia, Bulgaria, April 26-30, 2015, Proceedings, Part I*, volume 9056 of *Lecture Notes in Computer Science*, pages 287–314. Springer, 2015.

144. Serge Vaudenay. On the need for multipermutations: Cryptanalysis of MD4 and SAFER. In Preneel [129], pages 286–297.

145. David A. Wagner. The boomerang attack. In Knudsen [86], pages 156–170.

146. Qingju Wang, Dawu Gu, Vincent Rijmen, Ya Liu, Jiazhe Chen, and Andrey Bogdanov. Improved impossible differential attacks on large-block Rijndael. In

Taekyoung Kwon, Mun-Kyu Lee, and Daesung Kwon, editors, *ICISC*, volume 7839 of *Lecture Notes in Computer Science*, pages 126–140. Springer, 2012.

147. Xiaoyun Wang and Hongbo Yu. How to break MD5 and other hash functions. In Ronald Cramer, editor, *Advances in Cryptology - EUROCRYPT 2005, 24th Annual International Conference on the Theory and Applications of Cryptographic Techniques, Aarhus, Denmark, May 22-26, 2005, Proceedings*, volume 3494 of *Lecture Notes in Computer Science*, pages 19–35. Springer, 2005.

148. Eric W. Weisstein. Hypergeometric distribution. Visited October 2018.

149. Wentao Zhang, Wenling Wu, Lei Zhang, and Dengguo Feng. Improved related-key impossible differential attacks on reduced-round AES-192. In Eli Biham and Amr M. Youssef, editors, *Selected Areas in Cryptography*, volume 4356 of *Lecture Notes in Computer Science*, pages 15–27. Springer, 2006.

150. Yuliang Zheng and Xian-Mo Zhang. Plateaued functions. In Vijay Varadharajan and Yi Mu, editors, *Information and Communication Security, Second International Conference, ICICS'99, Sydney, Australia, November 9-11, 1999, Proceedings*, volume 1726 of *Lecture Notes in Computer Science*, pages 284–300. Springer, 1999.

Index

J. Daemen, V. Rijmen, *The Design of Rijndael*, Information Security and Cryptography,
https://doi.org/10.1007/978-3-662-60769-5

Printed in the United States
by Baker & Taylor Publisher Services